Illuminated Paris

The University of Chicago Press CHICAGO & LONDON

Publication is made possible in part by a gift from Liz Warnock to the Department of Art History at Northwestern University

Hollis
Clayson

ESSAYS ON ART
AND LIGHTING IN
THE BELLE ÉPOQUE

ILLUMINATED
PARIS

The University of Chicago Press, Chicago 60637
The University of Chicago Press, Ltd., London

Published 2019
Printed in China

28 27 26 25 24   2 3 4 5

ISBN-13: 978-0-226-59386-9 (cloth)
ISBN-13: 978-0-226-59405-7 (e-book)
DOI: https://doi.org/10.7208/chicago/9780226594057.001.0001

Library of Congress Cataloging-in-Publication Data

Names: Clayson, Hollis, 1946– author.
Title: Illuminated Paris : essays on art and lighting in the belle époque / Hollis Clayson.
Description: Chicago ; London : The University of Chicago Press, 2019. | Includes bibliographical references and index.
Identifiers: LCCN 2018029034 | ISBN 9780226593869 (cloth : alk. paper) | ISBN 9780226594057 (ebook)
Subjects: LCSH: Paris (France)—in art. | Lighting, Architectural and decorative, in art. | Art, Modern—19th century.
Classification: LCC N8214.5.F8 C539 2019 | DDC 709.04—dc23
LC record available at https://lccn.loc.gov/2018029034

⊗ This paper meets the requirements of ANSI/NISO Z39.48-1992 (Permanence of Paper).

FOR MAXWELL DEAN COGBILL *(born 2017)*

Contents

Acknowledgments

Books that take a long time incur many debts, but a recent witty review made me rethink extended acknowledgments. The reviewer dispatched a baggy set of thank-yous with a waggish putdown: if only half of those mentioned bought the book, he observed, it would be a best seller. I was chastened by this witticism, and, as a result, I will be brief. Please know, however, that the institutions and individuals I name have my deepest gratitude.

The Clark Art Institute and the Getty Research Institute awarded fellowships to the project when it centered exclusively on Americans in Paris, emphasizing Mary Cassatt. As I began to focus on artificial illumination, fellowships from the Clark Art Institute (again), Reid Hall, Columbia University (Paris), and the Huntington Library were indispensable. During the course of work on the book, three invited professorships put me in dialogue with generous and brilliant interlocutors: the Sterling Clark Visiting Professorship, Williams Graduate Program in Art History; the Samuel H. Kress Professorship, Center for Advanced Study in the Visual Arts, National Gallery of Art; and the Kirk Varnedoe Visiting Professorship, Institute of Fine Arts, New York University. Two exhibitions, both entitled *Electric Paris*, refined my approach to art and illumination in critical ways. At the Clark Art Institute (2013), the show I co-curated with Sarah Lees was brought into being by Jay Clarke with support from Elizabeth Liebman, and at the Bruce Museum, Greenwich, Connecticut (2016), Susan Ball and Margarita Karasoulas spearheaded the exhibition, for which I served as advisor. During intensive research campaigns, the librarians and archivists of the Bibliothèque Nationale de France (Paris), the Bibliothèque du Musée des Arts Décoratifs (Paris), the Huntington Library (San Marino, California), the Getty Research Institute (Los Angeles), the Clark Art Institute Library (Williamstown, Massachusetts), and the Charles Deering McCormick Library of Special Collections, Northwestern University Libraries (Evanston, Illinois) were flawlessly helpful.

Illuminated Paris benefited from the research assistance, transformative suggestions, and kind invitations of a far-flung constellation of individuals. In alphabetical order, I thank Caroline Arscott, Patricia Berman, Sarah Betzer, Alexander Bigman, John Brewer, Bill Brown, François Brunet, Matti Bunzl, Emmelyn Butterfield-Rosen, Catherine Clark, Jean-Louis Cohen, Elizabeth Cropper, Thomas Crow, Justine De Young, André Dombrowski, Arne Eggum, Hannah Feldman, Lily Foster, Caroline Fowler, Ernie Freeberg, David Getsy, Marc Gotlieb, Linda Green, Joseph Hammond, Jodi Hauptman, Mark Haxthausen, Lee Hendrix, Steve Hindle, Michael Ann Holly,

Katie Hornstein, Sandy Isenstadt, Heather Belnap Jensen, Jonathan D. Katz, Ara Kebapçioğlu, Sarah Kennel, Scott Krafft, Jason La Fountain, Keith Leitsche, Michael Leja, Rob Linrothe, Peter Lukehart, Régis Michel, Kent Minturn, Mary Morton, John Murphy, Kevin D. Murphy, Dietrich Neumann, Andrew Nogal, Therese O'Malley, Peter Parshall, Todd Porterfield, Dominique Poulot, Hector Reyes, Françoise Reynaud, Jennifer Roberts, Mary Roberts, Patricia Rubin, Tina Rivers Ryan, Vanessa Schwartz, Mark Simpson, Nancy Spector, Harriet Stratis, Veerle Thielemans, Krista Thompson, Hélène Valance, David Van Zanten, Susan Wager, and Andrés Zervigón. At the University of Chicago Press, I worked with the best: nonpareil editor Susan Bielstein, her right hand James Whitman Toftness, and two anonymous readers who made all the difference.

INTRODUCTION

Paris, City of Éclairage

Granted that the art of lighting [*éclairage*] cannot be the monopoly of any country or capital, it is certain that it owes its development principally to Paris.

HENRI MARÉCHAL, 1894[1]

Interest in the aesthetics of light shaped the work of countless modern artists on both sides of the Atlantic. The bond between innovative art and light was especially pronounced in France. The epigraph to this introduction sets a new priority for the study of the illuminations of the Paris region that roused artists, by informing us that lighting in the French capital was consequential at the time when artists were focusing on light. Henceforth, *éclairage* (lighting) and *lumière* (light) should be uncoupled. Recognition of the difference is logical but also imperative, because *éclairage* was a principal characteristic of nineteenth-century Paris—one of the most modern large cities in the world, the international headquarters of contemporary art, and *La Ville Lumière* (The City of Light). Improvements in lighting helped to define the cultural and technological landscape of the French capital city during the entire course of the 1800s.[2] For some artists enthralled by light, aesthetic curiosity about night and its artificial illuminations coexisted with the better-known love affair with the nuances of daylight. This book homes in on a sequence of Paris-based art practices that were guided by modern illumination. The resultant art works interrogated the visibility/invisibility dialectic, a central preoccupation of the era, from a wide range of approaches in a variety of aesthetic forms.

The argument that cuts across all the chapters is not that new lighting simply provided new motifs (new subject matter) for the fact-based visual arts, but rather that the lights plus the heated discussions they provoked—what I call illumination discourse—helped to shape many modernity-oriented art practices centered on Paris. We will discover that this convergence of technology and discourse defined a philosophical and visual matrix of paramount importance to "the New Painting," a term used here to embrace painting and the graphic arts.

A primary aim of the book is to come to grips with the idiosyncrasies of diverse representations of Parisian spaces, both outdoor and indoor, aglow with *éclairage,* in order to attend closely to patterns of artistic interest and disinterest in the lights, as well as relevant routines of topographic and social inclusion and exclusion. The chapters that follow track the aesthetic results of a specific episode of passion for *éclairage* that took hold of the city and its artists during the final quarter of the nineteenth century and the opening years of the twentieth. They investigate diverse ways that visual representations of night in the City of Light both disenchanted and reenchanted the French capital city during the era of gaslight and its coexistence and competition with new electric illuminations. This study of art from the later

1800s thus departs from custom by getting out of the sunshine to map responses to the illuminated darkness.

Éclairage was a central component of the Parisian spectacle and indispensable to the city's capacities for display that led Walter Benjamin, the influential twentieth-century German critic, to call it the "Capital of the Nineteenth Century."[3] The historians Martin Bressani and Marc Grignon add a crucial amendment: "Lighting is one of the key components of the Parisian enchantment that Benjamin examines, the most basic but also the most efficient way to attract attention. . . . The alternation between light and shadow is the basis of the visual spectacle of the nineteenth century."[4] Or rather, to paraphrase night historian Simone Delattre, the Parisian nights of the nineteenth century became objects of discourse and sites of experience when technical means of vanquishing the darkness were gaining in sophistication.[5] Tracking responses to the dialogue between natural darkness and new forms and quantities of artificial light is the spine of this book.

All of the foregoing points to the importance of closely scrutinizing the historical status of the French capital's nickname, *La Ville Lumière* (The City of Light). Before the city became known for pioneering *éclairage*—first gas and then electric—it was linked metaphorically to *lumière*. In symbolic recognition of its Enlightenment, it was first called *La Ville Lumière* in the eighteenth century. The city's increasingly abundant industrialized light, starting in the mid-1800s, caused the figurative honorific of the 1700s to morph into a descriptive epithet. All that light formed a cornerstone of its claim to metropolitan modernity and constituted one of the salient bases of the city's ascendant reputation as the Capital of the Nineteenth Century, which was founded, as we have already noted, in large measure on its capacity for display.[6] As everyone knows, the label City of Light stuck. The chapters of this book examine the role played by selected examples of visual culture in upholding the city's reputation for luminosity even after its claim to leadership in lighting innovation, not to mention absolute brightness, expired late in the nineteenth century.

Paris had been illuminated by gas, the successor to lighting by oil lamps, since the July Monarchy (1830–48). Actually, the first gaslight was lit in Paris in the place du Carrousel at the very end of the Restoration, in 1829.[7] Paris, unlike London, was slow to expand the number of gaslights in the city, but they did proliferate during the July Monarchy, especially in the 1840s, to the degree that Paris had 13,733 *réverbères* (streetlamps) by 1852.[8] The Second Empire (1851–70) made all the difference when it ushered in arresting and abundant gaslight, whose practical and symbolic benefits were much remarked and generally well liked.

It was the blossoming of lights in the 1840s and 1850s that pushed the French capital's old eighteenth-century Enlightenment sobriquet, *La Ville Lumière*, into the realm of the descriptive. To be clearer: Paris was regarded as an exceptionally brightly and beautifully lit city years before the key players in the city's mid-century modernization—Emperor Napoléon III, Prefect of the Seine Georges-Eugène Haussmann, engineer Jean-Charles Adolphe Alphand, and architect Gabriel Davioud—

went to work, but the Second Empire endeavor enhanced that reputation dramatically. The expanded system of *éclairage* put in place during the Second Empire was intended to guarantee the security of the city's inhabitants, but it also—especially as increasingly conjoined with the burgeoning lighting of private commerce—enhanced the status of Paris as a headquarters of display, luxury, and enjoyment, and gave birth to the notion of night life as a sociological and discursive category.[9] By 1870, when Haussmann was relieved of his duties, Paris had 20,766 *réverbères*.[10]

The streetlamps were painstakingly and elegantly designed, and quite varied. During the Second Empire construction of street furniture, there were seventy-eight variants of lamp distributed across the city, a rich declension (*déclinaison*) of seven basic models.[11] The systematic use of cast iron and the consistent brown paint color (sometimes heightened with touches of gold) facilitated the integration of the variants into a coherent ensemble, according to François Loyer, the dean of specialists on the urban fabric of Paris.[12] A few lamps in prestigious locations—in the courtyard of the Louvre, for example—were cast bronze. For the most part, Parisians embraced blazing gaslight as a welcome new metropolitan signature.

•

Does that mean . . . that electric light will make gas lighting disappear? That thought must be entirely discarded.

LOUIS FIGUIER, 1882[13]

A markedly heightened awareness of and burgeoning preoccupation with artificial lighting swept Paris only later, however, when electric light began to flood the public eye. The French capital was one of the first cities to experiment with electric street lighting. Its prominent Right Bank space, the place de la Concorde—lit with huge, blinding arc lights (early electric lamps in which sparks ignited between parallel carbon rods) in 1840 and again, more spectacularly, in 1844—was the first urban space anywhere illuminated by electricity.[14] Glaring white lights abruptly blazed, if only briefly in the 1840s, in a city illumined only by wavering orange flames. Thus, the opening decades of the electric-light era coincided with the glory days of gaslight.

Public interest in electric light reached fever pitch thirty years later. The period 1878 to 1881 was a short but significant era in the history of illumination in the City of Light, when the French capital's lights intensified in both brightness and topicality. In 1878 the Exposition Universelle (World's Fair) was dazzlingly lit, the international press excitedly bruited the imminence of the perfection of the American Thomas Edison's incandescent electric lightbulb, and eye-catching electric-arc street lights illumined high-profile Parisian spaces in a systematic way for the first time. The intensity of Parisian curiosity about lighting in these years was also shown by the rise of specialist periodicals like *La Lumière électrique* (founded April 15, 1879). The year 1881, when the Exposition Internationale de l'Électricité (International Exhibition of Electricity) was held in the Palais de l'Industrie (the Palace of Industry, built in 1855), was an *annus mirabilis* for *La Ville Lumière*. The exposition was the first such event of its kind in the world; it ran from August 11 to November 30, and drew 900,000 visitors. The exposition, discussed in detail in chapter 3, secured the symbolic centrality of Paris in the emergent field of electric illumination, and placed "light" on ever more Parisian tongues.

The 1881 fair also inaugurated the city's brief run as *La Capitale Électrique* (The Electric Capital) of the world. That designation was enhanced by the Jablochkoff candles (electric-arc lights named for their Russian inventor, Pavel Jablochkoff [Yablochkov], an expatriate electrical engineer) that bloomed as experimental streetlights along prominent thoroughfares between 1878 and 1882, introducing startling patches of intense white light into the gaslit city. While the rise and fall of electric-arc streetlamps played a critical role in fostering innovative art in Illuminated Paris, as discussed in chapters 2 and 4, the new electric *éclairage*, whatever its technology, shared the night with older lights (gas and some oil). When first tested in the place de la Concorde in 1840, outdoor arc lighting heaved a visual jolt of radiant white light against a tapestry of darkness and yellow-orange gaslights, a singular nocturnal visual shock repeated and expanded in 1878. The arc light resembled a sun, inasmuch as the light it cast had a spectrum similar to that of daylight.[15] Gaslit districts, though routinely described as ablaze with light, appeared red-orange, twilit, and sooty by comparison.

The coexistence and clash of different kinds of nocturnal illumination (frequently short-lived and inchoate) became basic components of the city's urban identity. The new mixture of lights made the French capital of the late nineteenth century a textbook example of Jonathan Crary's urban sensorium that constantly changes in conditions of capitalist modernity.[16] A certain logic of competition framed the changes that defined the volatile lighting environment of Paris in this era. The antagonism between the forces promoting and devising experimental electric lights and the more powerful supporters of the city's pervasive gas illumination molded and inhibited electric lighting innovation throughout the balance of the 1800s and beyond.[17]

As a consequence, the diverse flirtations with electric lighting of the later century—whether inaugurated by the city council, the state, or individual entrepreneurs—unfolded against the backdrop of an expansive nocturnal fabric of gas streetlights, including the thousands of lamps erected during the Second Empire. Marc Armengaud, a founding partner of AWP (a French architecture and design firm), argues that it was in the period from 1830 to 1860, the heyday of gas illumination, that *Paris la nuit* (Paris by night) first arose as an expression that denoted a time for excitement and pleasure that coincided with the hours after sundown. Gas streetlamps continued to function straight through to the liberation of Paris in 1945. Notwithstanding the erratic but ultimately widespread electrification of the city's lights between the late 1800s and the first decades of the twentieth century, the very last gas lamps did not go dark until 1962.

The baptism of Paris as the Electric Capital in 1881 ruptured the sovereignty of gas. Yet the new alias for the City of Light, Electric Paris, proved precarious and fleeting. Marc Gaillard, a specialist in the material fabric of Paris, destroys any hope that the City of Light deserved to retain its novel epithet, the Electric Capital: "In reality, despite the progress made by electricity, the final decades of the nineteenth century and the Belle Époque marked the apogee of gas lighting."[18] To wit, in 1894 the city had only 461 electric streetlamps versus 53,000 gaslights.[19] The *fin de siècle* was indisputably an era of gaslight.

Spectacular electric light nonetheless continued to draw attention and to add to the city's increasing luminosity. Commentators on the city's modernity were dazzled, for example, by the unparalleled array of electric lights on show in the 1881 International Exhibition of Electricity, mentioned above, and by the abundance of spectacular electric *éclairage* at the world's fairs (in 1878, 1889, and again in 1900). Any increase nurtured illumination discourse and created optical conditions that made it possible for even government and industry insiders to believe that electric light really was in the process of eclipsing gaslight as the *fin de siècle* approached. No less an expert on urban transformation than Haussmann, the overseer of the recent modernization of the French capital, made just such a faulty assessment in his 1890s memoir: "At the very moment that I write these lines, a radical revolution appears at the point of transforming the lighting of the Public Thoroughfares of Paris: the substitution of electric light for that of gas."[20] The later nineteenth-century French capital was *La Ville Lumière* to be sure, but, to repeat, in the register of the technical underpinnings of its outdoor appearance, its more accurate moniker would have been Gaseous Paris.[21]

The electrification of city street lighting was governed by somewhat centralized arrangements, but not consistently. It was eventually enabled but not actively fostered by the city council (*le Conseil municipal*). Having renounced undertaking the electrification of the capital city itself, the council decided (in late 1888 and early 1889) to grant "permissions" to private concerns for eighteen years by carving up the city into *secteurs*, a bad decision by all accounts.[22] The process was fraught, protracted, and hobbled by government bureaucracy and indecision. It was indeed absurd, and, as a result of the city council's dithering oversight, Paris, the erstwhile *Capitale Électrique*, became one of the slowest big cities to electrify its street lighting.

There were, nonetheless, strong ideological forces that loudly insisted upon the untenable old-fashionedness of gas. Marc Armengaud dates the strong change of opinion about gas to the 1880s: "The criteria for the appreciation of luminosity were unstable. Thus, gas, a revolution inseparable from the Baudelairean nocturnal climate of Paris ("the smoky gleams"), became progressively unacceptable starting in the 1880s: irregular, nauseating, dangerous, this technology of lighting was perceived as being as archaic as oil lighting."[23] Electric light was substituted for gaslight in many spaces of entertainment in Paris, owing to the fear of gas created by an 1887 fire. To quote Ernest Freeberg, twentieth-century historian of lighting:

> The movement [towards the adoption of electric light] was slow during the five years that followed the [1881] exposition; one can only cite, as genuinely interesting, the installations carried out by the Edison Society in various theatres, the principal of which is the Grand Opera. . . . The terrible catastrophe at the Opéra-Comique, this disastrous fire [due to gas lighting, May 27, 1887] provided the primary impulse. Following that event a general cry went up: "we must put electric lighting in all theaters" and naturally people went wild with the idea and called for electric light everywhere for everyone. The realization of this wish for the theaters was not long in coming.[24]

It does appear that on the heels of the theater fire disaster, the Paris city council took decisive action giving all theaters, cafés, and concert halls three months to replace their gaslights with electric lights.[25] The benighted city council and the omnipotent gas company remained the critical political and economic obstacles to the comprehensive removal of the whiff of gas from the atmosphere of Paris. The result? The lighting remained a mosaic of gas and electric light in the streets, and a montage of candles, oil lamps, gaslights, and some electric lights in the city's private dwellings—while an accelerating awareness of technologized illumination fostered a robust public conversation about lighting in the French capital.

Illumination discourse was an altogether novel nexus of knowledge and belief that helped to mold the character of the City of Light at a moment of unparalleled volatility in its light environment. A coming-together of intense and occasionally conflicting but tightly thematized ingredients, illumination discourse encompassed fervor and understanding alongside fear and anxiety in the social, technological, physiological, psychological, and aesthetic realms. Alongside the partisans of the new lights, the electro-enthusiasts (some of them visual artists), arose a core of naysayers (including artists), avatars of fear and anxiety; this helps to explain the multiplicity of standpoints about the new lighting that coalesced in the varied artworks discussed in this book.[26] Among the enthusiasts for the new streetlights of the 1910s was Sonia Terk Delaunay whose *Electric Prisms* paintings of 1913–14 are dazzling examples of abstract painting directly motivated by the visual properties of electric luminosities. At the same time, the light cult sparked worries about a dark side to the new lights. Concerns ranged from fear of blindness and panic about surveillance to melancholy about the dispersal of the magic of the dark. Both sides of the culture-wide fascination with artificial light and its infrastructure helped to foster interest in the visual properties of light among artists in Paris, a multifaceted interest tracked in the chapters that follow.

•

That's how it is with electric light: it dazzles but does not clarify.

JULES LUQUIENS, 1893[27]

Visual artists were not alone in taking note of and keeping up with the new lights. When novelist Émile Zola, for example, wrote his department-store novel, *Au Bonheur des dames*, he updated the lighting system to electric lamps for publication in 1883, when it had been lit by gas in the earlier 1876 phase of writing.[28] The expressions of various other writers make it clear that no matter what lighting technology was emphasized, the Parisian "eye" was accustomed to blazing illumination.

Novelist Henry James, for example, remarked upon the astonishing brightness of the gaslight in Paris. In one of his 1876 letters to the *New York Tribune*, he wrote that "of a summer evening you pay a penalty for living in the best lighted capital in the world. The inordinate amount of gas in all the thoroughfares heats and thickens the atmosphere, and makes you feel of a July night as if you were in a vast concert hall."[29] The observation is uncanny in its anticipation of Walter Benjamin's proposition regarding the boulevard turning into an interior in modernity. Wolfgang Schivelbusch's gloss is helpful: "As the boulevard at night developed in the nineteenth century, it did in fact look like an interior out of doors. . . . There is a simple psychological explanation for the fact that the street looks like an 'Interieur,' to borrow Walter Benjamin's expression. Any artificially lit area out of doors is experienced as an interior because it is marked off from the surrounding darkness as if by walls, which run along the edges of the lit up area."[30]

Richard Harding Davis, the Pennsylvania journalist, was also smitten by Parisian gaslight. He became managing editor of *Harper's Weekly* in 1890 (at the tender age of twenty-six) allowing him to travel and write. In his *About Paris*, published in 1895, he observed: "The Parisian may economize in household matters . . . but in public he is most generous; and he is in nothing so generous as in his reckless use of gas. He raises ten lamp-posts to every one that is put up in London or New York, and he does not plant them only to light some thing or some person, but because they are pleasing to look at in themselves."[31]

In his multivolume study of the city's mechanisms, Maxime du Camp, writer and photographer, made a shrewd appraisal in 1875 of the visual traits of intractable electric lighting, presumably arc light:

> Every light, in order to be expediently used for general public functions, must be able to disperse, to subdivide or fractionate infinitely; without that, it remains a restricted fire, dazzling, but improper for satisfying the needs of a large city. Thus it

> is with electric light: it dazzles and does not illuminate; it may be used in many circumstances, but we have not yet arrived at the point of making it a regular instrument of illumination.[32]

The Italian travel book writer, Edmondo de Amicis, responded similarly to the electric-arc lights he saw during that year's Exposition Universelle. In his 1878 guide to Paris, he observed:

> In the heart of the city it seems as if day were beginning again. It is not an illumination, but a fire. The Boulevards are blazing. Half closing the eyes it seems as if one saw on the right and left two rows of flaming furnaces. . . . All this broken light, refracted, variegated, and mobile, falling in showers, gathered in torrents, and scattered in stars and diamonds, produces the first time an impression of which no idea can possibly be given.[33]

It was certainly the brilliancy of arc lights planted here and there in Paris in 1878 that inspired Robert Louis Stevenson's matchless screed against Parisian electricity, "A Plea for Gas Lamps":

> The word electricity now sounds the note of danger. In Paris, at the mouth of the Passage des Princes, in the place before the Opera portico, and in the Rue Drouot at the Figaro office, a new sort of urban star now shines out nightly, horrible, unearthly, obnoxious to the human eye; a lamp for a nightmare! Such a light as this should shine only on murders and public crime, or along the corridors of lunatic asylums, a horror to heighten horror. . . . That ugly blinding glare may not improperly advertise the home of slanderous Figaro . . . but where soft joys prevail, where people are convoked to pleasure and the philosopher looks on smiling and silent, where love and laughter and deifying wine abound, there, at least, let the old mild lustre shine upon the ways of man.[34]

Managing the dazzle, the often blinding brilliance, was a constant concern during the early years of the history of electric light out of doors. Indeed, illumination and dazzlement were construed as antagonistic components of the new light.[35] Rather than permitting one to see better, it prevented seeing altogether. One of the striking features of Edison's 1881 display, which presented networked incandescent electric lights, was that, according to his devotees, his lights domesticated the blazing electric beast. According to Henri de Parville, a prominent science journalist and an astute observer of the exhibition:

> We are accustomed to representing electric light to ourselves in the form of dazzling, scintillating fires, hard on the eye, burning, constantly changing in intensity, in variable and wan tones. Here, on the contrary, one has a light that has been in

some sense civilized, accommodated to our customs, put at our disposal (or level); each lamp illuminates like gas, but like a gas that had to be invented, a gas providing light of a perfect fixity, gay and brilliant without bothering the retina.[36]

Illuminated Paris and the History of Art

As I suggested at the outset, at the heart of the history of modern art is the secure knowledge—whether gained intuitively or through research—that the investigation of light shaped numerous practices in the art worlds of Europe and the US in the later 1800s. This article of faith has led historians of nineteenth-century art to focus repeatedly upon the methods used by artists to capture and analyze natural light, a focus that has molded our understanding of modern art's essential touchstones and norms. The most conspicuous examples of the close bond between modern art and light in France are found in landscape art including photographs of the countryside and the city,[37] and the innovative paintings eventually called Impressionist that probingly investigated natural light, especially the flux of sunlight and clouds in the Paris region in the later nineteenth century. Such paintings are at the heart of the art history of the period because they matter, but also for the reason that scholars have taken their cue from the many modern artists who were heliophile landscapists and genre painters, and stressed the critical role of *plein air* movements in the aesthetics of the era to the exclusion of nighttime and indoor works.[38]

My discussion of *éclairage* in the domain of the history of later nineteenth-century art is intended to counterbalance art history's century-and-a-half-long romance with natural light. Although my intention is not to re-valence sunshine-based art practices as Luddite, it is possible that certain *plein air* practices deserve to be thought of as instances of romantic anticapitalism in view of the lights not chosen. But the main drive of this discussion is toward the straightforward acknowledgment that in some quarters aesthetic inquisitiveness about night and its artificial lights paralleled, complemented, and inflected the better-known love affair with the nuances of daylight in the pictorial arts of the later nineteenth century. This recognition will effectively round out the picture we have of the era's visual culture, making daytime *plein air* painting a much more deliberate choice. Indeed, my inquiry seeks to denaturalize *plein air* painting.

The art historian Bruno Foucart taught us to be cautious when attempting to align lighting with its evocation in modernist painting. In an important yet little-known 1985 essay, Foucart was the first to throw down the gauntlet of empirical perspicuity with regard to a technical understanding of lights and lighting in an art-history context: "The archaeology of representations of electricity remains still to be done.... Consider that we are almost incapable of saying in certain of the greatest Impressionist tableaux if the light is gas or already electric!"[39] His challenge contained specific worries about the lighting in an 1873 canvas by Édouard Manet, *Masked Ball at the Opera* (fig. 0.1):

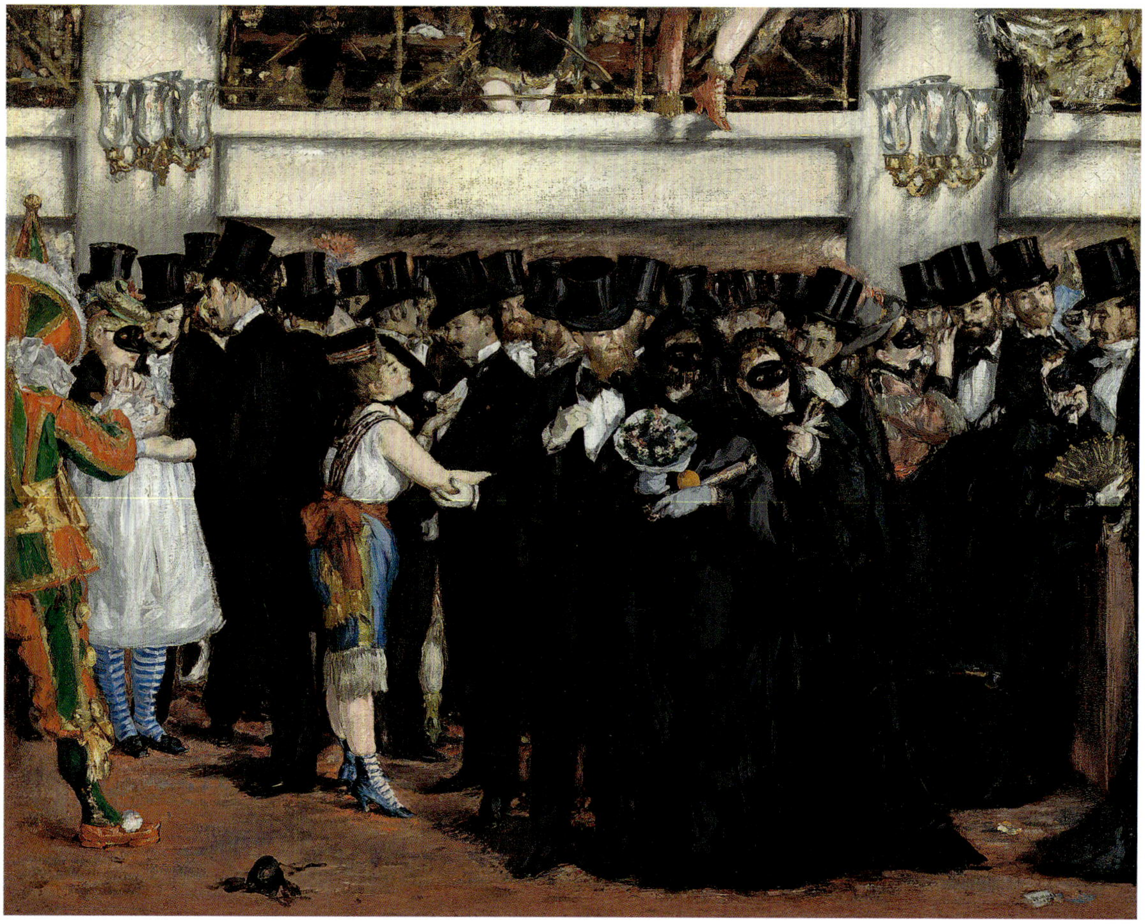

0.1 Edouard Manet, *Masked Ball at the Opera*, 1873. Oil on canvas. Image: 59.1 × 72.5 cm (23¼ × 28⁹⁄₁₆ in.); framed: 80 × 94 cm (31½ × 37 in.). Courtesy National Gallery of Art, Washington, DC.

> The art historian must fundamentally call upon his historian faculties and gather the information that would permit him to know if such and such place painted by the artist was lit by gas or electricity at the time. For example, in *The Masked Ball at the Opera* by Manet, we are more or less certain that the illumination of the wall sconces attached to the columns holding up the balcony was gas. But very quickly electricity must have followed gas in this opera house on the rue Lepelletier as it did in the Opéra of Charles Garnier. Degas's work poses comparable questions.[40]

The task that Foucart dared art historians to fulfill more than thirty years ago remains unfulfilled, as far as I know. His strong and rousing asseveration was, however, problematic. It provides an unintended example of the kind of wrongheadedness that can threaten to shove a project like mine into the empirical or interpretative ditch—or both. Manet's complex approach to depiction could easily raise eyebrows, as he courted connotation and openness, not to say indeterminacy, where other artists sought denotation and closure. The irony is that Manet precisely limned the yellow and orange of the wavering gas flames within the glass globes of the wall sconces at the masked ball. Foucart's call to arms thus exemplifies a paradox. While urging the art historian to grasp hold of the history of indoor commercial *éclairage*, his own grasp fell short. The lesson is clear: knowing whether such-and-such lighting was installed in this or that space is foundational, but not equivalent to understanding its painted representation.[41]

•

The heart of *Illuminated Paris* is six detailed studies (chapters 1 through 6), which are arranged in chronological order. Each is a focused essay concentrating upon artworks made between 1860 and 1890. The conclusion endeavors to clarify the book's compass, stakes, and shortcomings. Chapter 1, "*Cherchez la lampe*," examines the role of the unlit gas streetlight in a series of Second Empire photographs by Charles Marville and in Gustave Caillebotte's signature painting, *Temps de pluie* (1877). The resonance between the treatment of the lamps in Marville's photographs and the prominence of the daytime gas lamp at the center of Caillebotte's painting serves to remind us that all of the canonical vanguard images of the 1870s and earlier 1880s of the Paris boulevard or square (*place*) are daytime pictures.

The nocturne takes center stage in chapter 2, "Losing the Moon." It conducts a painstaking examination of a pair of 1879 twilights set in the Jardin du Luxembourg by cosmopolitan artist John Singer Sargent (1856–1925) in his very early career, and compares them to his contemporaneous paintings of brashly lit orchestra rehearsals in the Cirque d'Hiver. The main focus of the chapter is Sargent's singularly modern form of urban nocturne, one that took its cue from the influential inventor of the painted nocturne, another expatriate American, James Abbot McNeill Whistler (1834–1903), by not cleaving exclusively to the luminosities of the moon, the stars, and the planets. Sargent's subtle canvases combine celestial illumination with artificial lights to stage a complex duel between moonlight and industrial light.

Chapter 3, "Bright Lights, Brilliant Wit," combines an analysis of several high-profile, electrically lit spaces—the place de l'Opéra of the late 1870s, the electric Salon exhibitions of 1880 and 1881, and the International Exhibition of Electricity of 1881—with an investigation of the response to them by several of the era's greatest artists working in the mode of graphic satire: Cham (Amédée Charles Henri de Noé; 1818–79), Draner (Jules Jean Georges Renard; 1833–1926), and Albert Robida (1848–1926). The application of the protocols of caricature to the sight of Parisians bathed in electric light in spectacular settings triggered imaginative and hilarious perspectives on the social and optical effects of the new illuminations, while at the same time securing certain tropes of Paris boulevard humor forged in earlier decades. The chapter also includes a concise history of the practice of the artificial lighting of art spaces. It ends with a discussion of the impact of the 1881 electricity exhibition upon two divergent spheres: the improvement of electric lighting in domestic interiors coincided with a disinclination to embrace the electric illumination of artwork in public exhibition spaces.

Chapter 4, "Night Lights on Paper," examines black-and-white intaglio prints made by Mary Cassatt (1844–1926) and Edgar Degas (1834–1917) between about 1878 and 1882. Their tandem absorption in printmaking in these years emerged directly from their involvement in a planned journal, *Le Jour et la nuit*. I argue that the singular inventiveness of their work in monochromatic media instanced the impact upon their aesthetics of illumination discourse, as well as their familiarity with period lighting technologies, to the point of suggesting that their immersion in monochromatic intaglio printmaking may have been stimulated in hitherto unremarked ways by the new lighting environment.

Chapter 5, "Outsider Nocturnes," considers night pictures by American painters of modern life who worked in Paris for various periods of time in the 1880s, featuring paintings by Charles Courtney Curran (1861–1942) and Childe Hassam (1859–1935), especially the latter. The transatlantic issue first broached in chapter 3—the competition between France and the United States in these years over supremacy in the field of electricity—arises again, but here I argue that the Franco-American rivalry left its traces in the formation of a distinctively American artistic imagination of the City of Light at night. The often sharp clash between the utopic nocturnal French capital coined by American painters and the dystopic *ville lumière* fashioned by Parisian modernists is a central focus. The chapter also maps contrasts between pictures of the French capital by night and those of New York City produced by some of the same American artists when they returned home. Visualizations of "disenchanted night" (Schivelbusch's term) in two of the great cities of the transatlantic world, both headquarters of electricity and of modernist urban art, are thus placed in dialogue.

Chapter 6, "Man at the Window," investigates the artificially lit world of Paris limned by another artist visitor to the French capital, the Norwegian Edvard Munch (1863–1944), who devised a set of nighttime interiors structured as views of his rented riverside flat in the suburb of Saint-Cloud starting in 1890, pointing both toward and beyond the window. I argue that his singularly moody and blue window pictures, all set in threshold space at night, exemplify a pre-Symbolist mode, construct an alignment between artificial light and sexual energy, and instate longing that coexists with anxiety about the penetration of the home by outside forces.

· 1 ·

Cherchez la lampe

CHARLES MARVILLE, GUSTAVE CAILLEBOTTE, AND THE GAS LAMPPOST

With the arrival of gaslight, the Parisian night is profoundly transformed. . . . The new illumination is able to create, for the first time on an urban scale, an all-embracing visual and sensory spectacle which permanently affects city dwellers' perception of their daily ritual.

MARTIN BRESSANI, 2015

Charles Marville's Streetlamps

The series of lamppost photos taken by Charles Marville (1813–1879), a celebrated photographer of the city, was the most tightly defined in his entire œuvre.[1] He took about ninety different photographs of the range of gas lamps installed in the city of Paris, focusing upon those erected during the Second Empire.[2] It appears he inaugurated the work when he assumed the title, *Photographe de la ville de Paris*, in 1862. His photographs were thus taken between about 1861 and the early 1870s for the most part, though some were made all the way into the later 1870s. This means that while most of the streetlamp photos were taken during the intense phase of public works enhancements of the second half of the twenty-year-long Second Empire, Marville's photographic account of the city's street furniture, *le mobilier urbain* (hereafter MU), did not end with the fall of the empire.[3]

According to the Service des Promenades et Plantations, which was charged with installing the new street furniture (including the streetlamps), the MU was defined as "the ensemble of objects or devices or apparatuses, public or private, installed in public space and tied to a function or a service offered to the collectivity."[4] The architect Gabriel Davioud, the chief architect for the service and indeed the chief architect of the city of Paris during the Second Empire and again in the 1880s, was a key figure in this realm.[5] According to Marie de Thézy, he was the "véritable père du MU" (the true father of street furniture).[6] Jean-Charles Adolphe Alphand, engineer and chief of the Service des Promenades et Plantations from 1854, and from 1867, director of Voierie et Plantations (roads and plantings), played a major role in the specific realm of *éclairage*. While Marville's photos of newer gas lampposts—the *réverbères* and *candélabres* of the Second Empire (the former single, the latter multiple)—dominate his streetlamp corpus, his photographs record older streetlamp forms and technologies as well, including *torchères* that appear in his *Le Vieux Paris* series of the 1860s (an album of 425 *vues*.)[7]

Marville's photographs give the streetlights a human scale (figs. 1.1, 1.2, 1.3). The combination of the artist's angle of address (motif and photographer, eye to eye) and the framing tactic that he used in shooting the individual lampposts gave them an arresting presence. He often framed the individual machines dexterously: each lamp is positioned symmetrically against a rectangle of pale backdrop (in both figs. 1.1 and 1.2), and is just about bordered and embraced by identical columns in

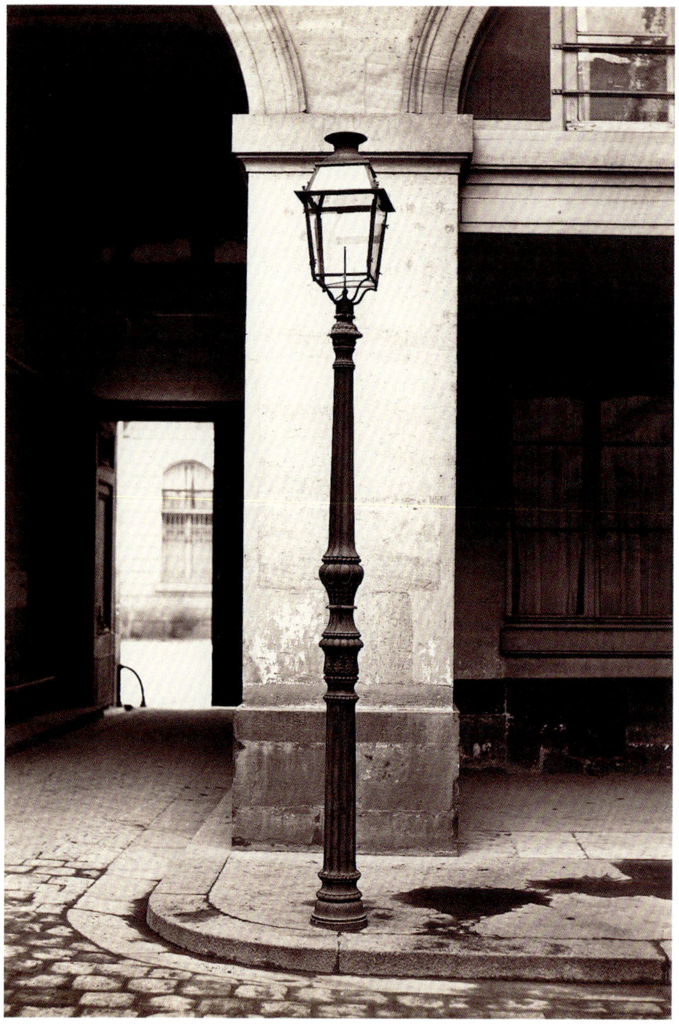

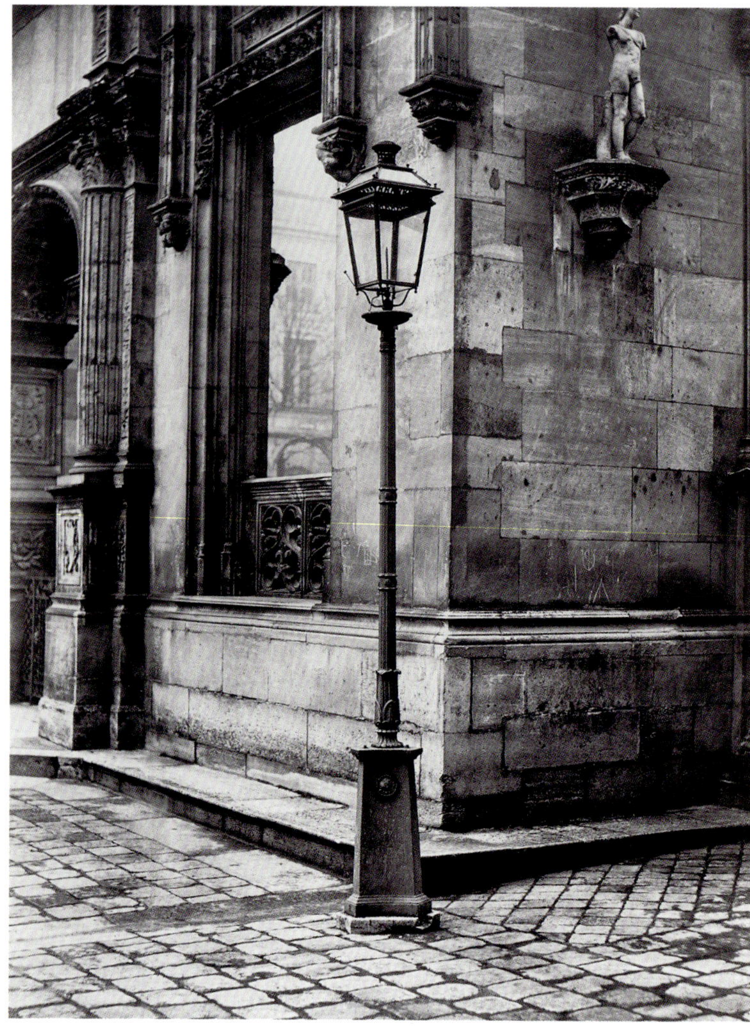

1.1 Charles Marville, *Hôtel de la Marine*, 1864–70. Albumen print. Image: 36.2 × 23.5 cm (14¼ × 9¼ in.); mount: 60.3 × 45.1 cm (23¾ × 17¾ in.). Courtesy National Gallery of Art, Washington, DC.

1.2 Charles Marville, *École des Beaux Arts, rue Bonaparte, Lamppost, Paris* (6th arrondissement), c. 1860. Paris, Musée Carnavalet. Image © Charles Marville/Musée Carnavalet/Roger-Viollet/The Image Works.

figure 1.3. Lamp photographs like these deserve to be regarded as portraits, in the full sense of the word. The lamps are enough like us to engage our curiosity and sympathy, yet sufficiently distinct to fascinate as singular individuals that could activate our capacity for empathy. Not only are they unexpectedly sensate and alert, for iron-and-glass appliances, but also their isolation from their "peers" in Marville's series serves to individualize each lamp, making each post appear well poised for an interaction with, and equivalent in presence to, any passerby.[8] Each light machine has a head, after all.

The visual language of many of Marville's lamp photos is noteworthy for its consistency. The photographic style used for the gas lamp pictures was honed by the aesthetic of his earlier photos of very different motifs, including, for example, the Treasury of Reims Cathedral (1854), a tree in the Bois de Boulogne (1862–72) (fig. 1.4), the Fontaine des Innocents (1858), and the spire of Notre Dame Cathedral (1859–60) (fig. 1.5).[9] As in the lamp photos, the consistent organizational traits are the isolation

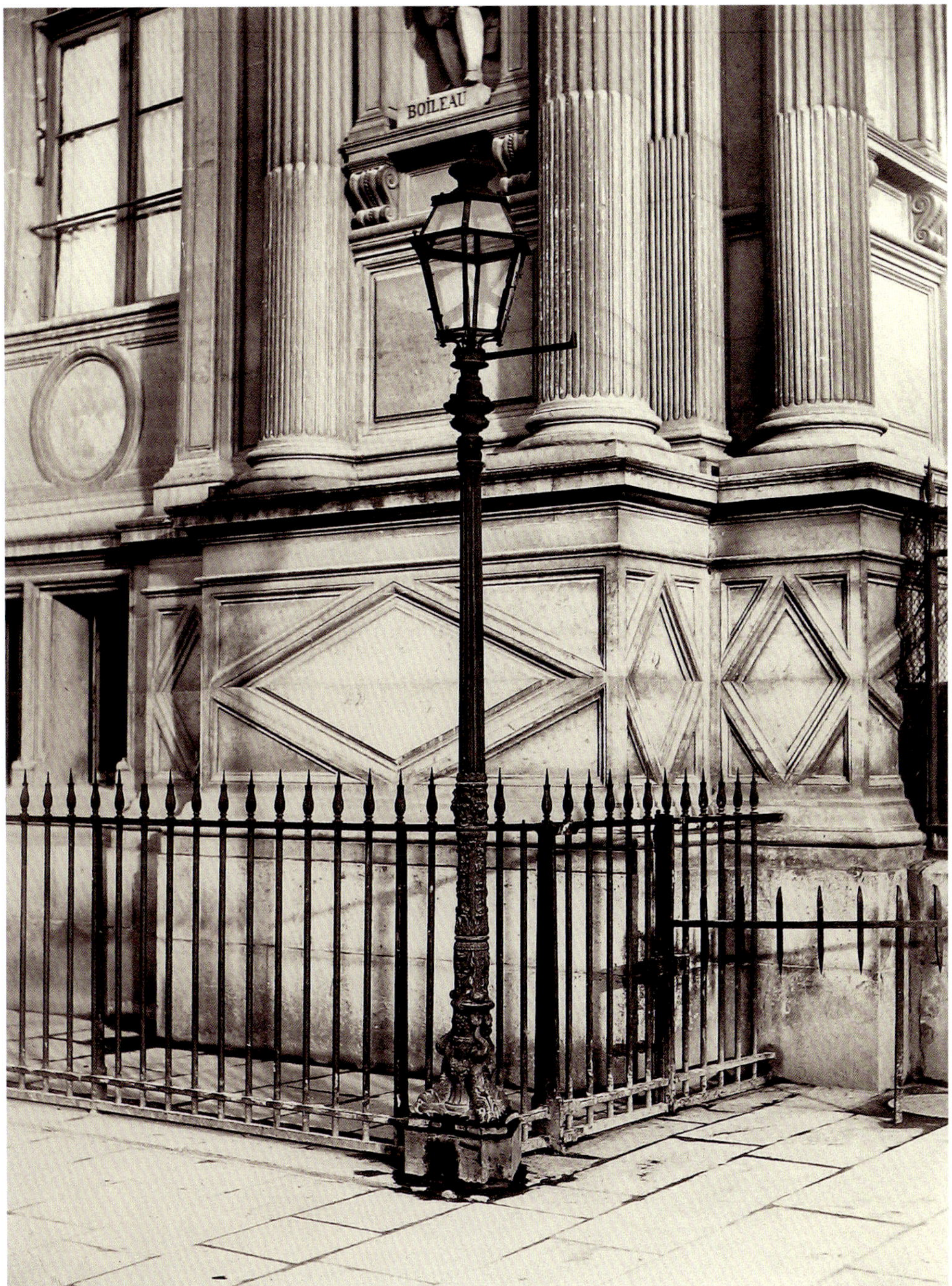

1.3 Charles Marville, *Réverbère, Hôtel de Ville* (4th arrondissement), 1858–71.

and balance of vertical protagonist(s), resulting in frontality, centrality, symmetry, and balance.

Marville's approach to object photography matched Baron Haussmann's lamp aesthetic, marking one of the great convergences in nineteenth-century Parisian visual culture. According to François Loyer, "The Second Empire . . . made the streetlamp a major theme."[10] And thus the set of lamp photographs instantiates a correspondence between a great photographer and one of the celebrity objects of the era. Haussmann had strong views about the look, size, and placement of the new streetlights. He didn't like socles or pedestals. He wanted even to remove them from Jacques-Ignace Hittorff's famous earlier lamps, such as the colossal columnar fixtures in the place de la Concorde. The prefect preferred that the lamps be down on the ground. Just like a person.

Haussmann explained his preferences clearly in his memoirs of around 1890, emphasizing the quality of light obtained by maintaining a modest scale for the lamps:[11]

> A gaslamp that is placed too high up will project its light farther, but will not give adequate light to the immediate area around it. Obviously, that was not our goal. The higher a lamp, the greater the unlighted area at its base. By reducing the height of streetlamps and the distance between them, and decreasing the intensity of the flame in each lamp so as not to use more gas, we were able to light the city's streets better. Extremely bright lights are useless; they blind people more than they light their way.

Haussmann's technical reasons for preferring shorter lamps were inseparable from an evident preference for their chiaroscuro effects on the illuminated façades. François Loyer concurs: "At night this forest of cast-iron trees created a low layer of light. Lampposts were relatively short not only for technical reasons (turning the lamps on) but also for aesthetic ones. The light they gave off just reached the top of the ground floor; the stories above were no more than shadows."[12] But since capturing a particular quality of illumination was not part of Marville's remit, he responded to their human size as well as their design and ornamental details. The down-on-the-street-ness and the nose-to-nose-with-the-passerby-ness of Marville's individual lamp portraits are their most striking features.[13] For Haussmann, the *réverbères* needed roughly human stature to guarantee the effectiveness of the illumination produced by a group, line, or cluster of lamps thus sized, but Marville divided the forest into individual trees, and upheld the human scale and machinic singularity of each lamp scrupulously, even tenderly, by virtue of its solitude and painstaking composition. The time of day contributes to the humanity of the lamps.

The topic of the quality of the light shed by the gas lamps of the Second Empire drives home the great and irreconcilable paradox of Marville's streetlamp photos. The lamps are all recorded in daylight. They are not lit. They are off duty. Pursuing

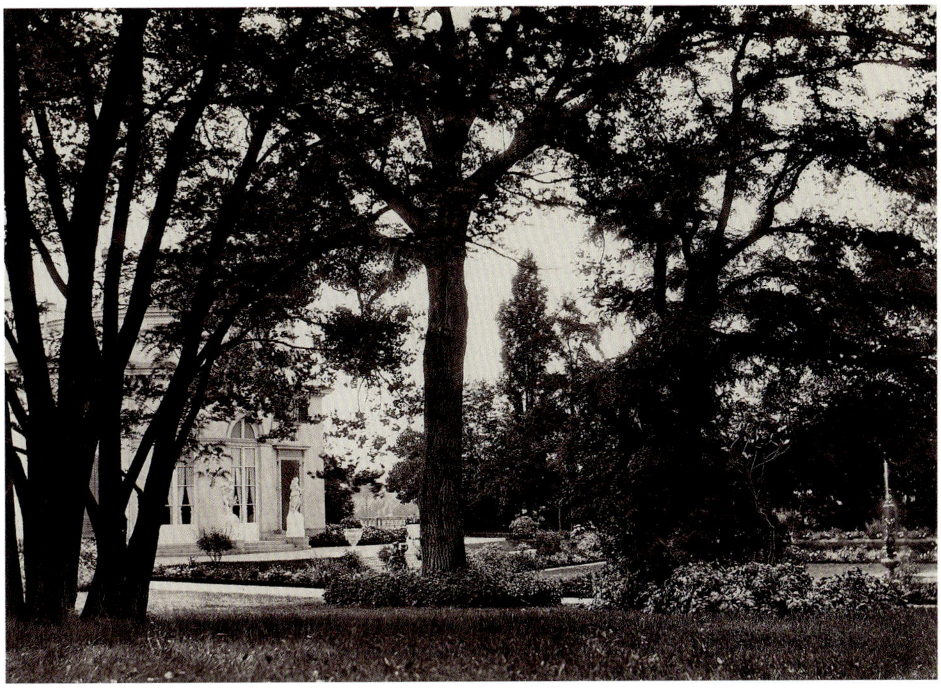

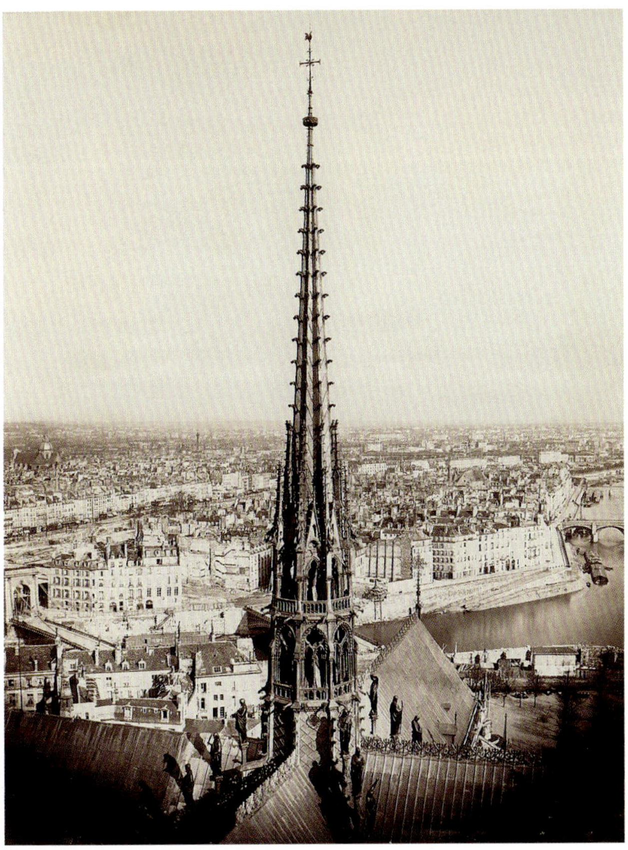

1.4 Charles Marville, *Parc de Bagatelle, Bois de Boulogne* (16th arrondissement), 1862–72. Image © Charles Marville/Musée Carnavalet/Roger-Viollet/The Image Works.

1.5 Charles Marville, *Flèche de Notre Dame, Viollet-le-Duc Ar[chitecte]* (*Spire of Notre Dame, Viollet-le-Duc, Ar[chitect]*), 1859–60. Albumen print from collodion negative, 49.5 × 36.5 cm (19½ × 14⅜ in.). AIA/AAF Collection, Prints and Photographs Division, Library of Congress, Washington, DC.

our characterization of the photographed lamps as sensate beings, their consistent daylight identity places them in a theatrical context complete with its defining temporalities. They resemble actors before they step on the stage, or after they have stepped off. They possess a kind of nonprofessional condition, placing them in performative limbo insofar as each has a day-off or time-out persona. They patiently (impassively) and alertly await the fall of night and their illumination. For all their indispensability to the appearance of Paris, they have no agency. The lamplighter (or later the network) acts; the lamps react. Or they might be seen as frozen in the aftermath of performance, as forlorn has-beens documented when the time of recital—the night—has been erased by the return of day. They are becalmed and functionless urban warriors in iron and glass, about the same size as, and sharing the pavement with, myriad mobile passing strollers. They are outright disenchanted by Marville's plainspoken daylight exposures. Unlit, they played no role in the signature enchantments of the gas era. Marville's photos of streetlamps—for the most part, *réverbères* installed in the central and better districts of the city—show us modernization but not modernity.

The temporality of Marville's streetlamp photos jars into view as strikingly and strangely premodern. Although they are the contraptions that realize the great optical fact of nineteenth-century urban modernity—the alternation between light and dark that defines the night urban experience—in Marville's photographs, they do not.[14] The lamps are off; frozen in prelude or aftermath. The photographer could not have done otherwise; exposures could not be taken at night in the 1860s and 1870s.[15]

Consideration of Marville's photographs of a setting containing older gas lighting fixtures, documented in the *Ancien Paris* project, nuances the conclusions reached above. Marville photographed the passage de l'Opéra, for example, clearly delineating its old-fashioned hanging and wall-mounted gaslights, in 1868 (figs. 1.6 and 1.7). That arcade has not survived, but it endured long enough to house many of the key meetings and conversations of the Surrealists in the 1920s. This and the other arcades that Marville photographed in daytime acquired the capacity to enchant, even bewitch at night when the gas jets were lit. As Wolfgang Schivelbusch observes, "People seem to have been fascinated by the interplay between the brilliance of the light and the wares on display, and the lively crowd."[16] The appearance and function of the *passages* of Paris are at the heart of the most influential account of the modernity of the nineteenth-century French capital, that written by Walter Benjamin during the 1930s.[17] Indeed, the arcade is one of only two key spatial topoi of the nineteenth century that he pinpointed; the other is the domestic interior.

The arcade is a building type that proliferated in early nineteenth-century Paris before and as a precursor to the *grand magasin*. Typically sheltered beneath an iron-and-glass roof, the arcade was a block-long pedestrian passage nestled between two masonry structures, and lined on either side with small shops, tearooms, amusements, and other commercial attractions. At one time, some two to three hundred arcades punctuated the Paris cityscape. About thirty remain, most of them clustered

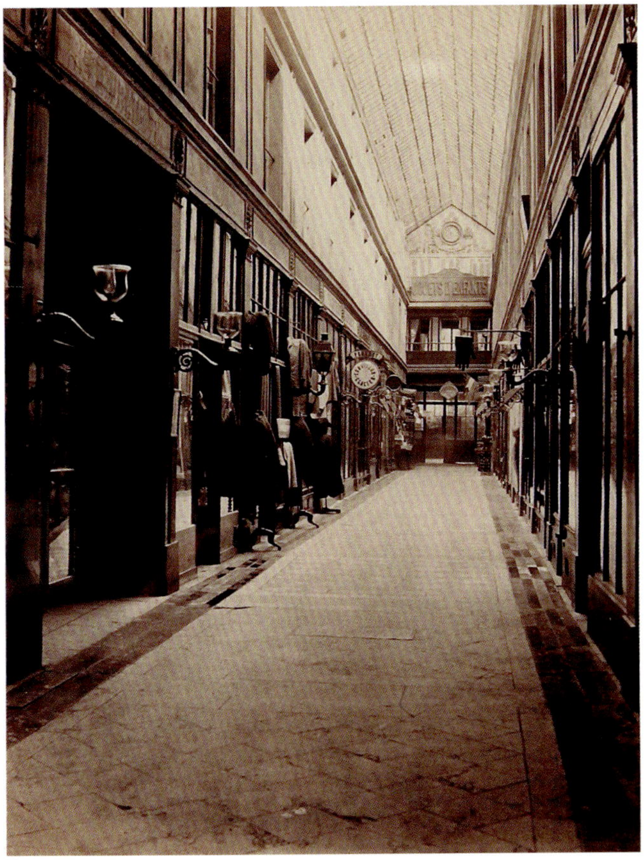

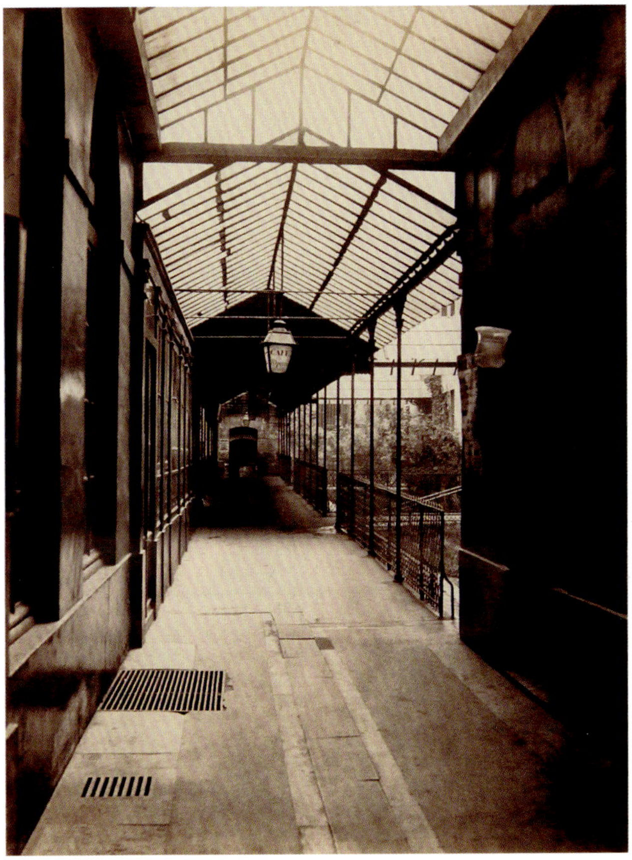

on the Right Bank in the first and second arrondissements. The arcade represented a pivotal moment in modern social and economic history. With it, capitalist society began its transition from a culture of production to one of consumption. Beneath the arcade's greenhouse roof, a technical apparatus of the industrial society furnished objects of consumer desire. In Charles Rice's words: "Through a commercial logic the arcades come into being as phantasmagorias, worlds-unto-themselves which organized the visual experience of commercial splendor for a passing parade of citizens who experienced themselves as part of this splendid object-world, partners in the social relations of consumption."[18] The glittering, transparent arcades chock full of goods on sale encouraged Benjamin to conceive of the nineteenth century as a dream—a nightmare—from which we twenty-first-century moderns have yet to wake. We are still mesmerized by, and at the mercy of, capitalist commerce, and the gaslit arcade was its crucible.

Indeed, the places and institutions most associated with luxury were the first to benefit from gaslight: arcades, *grands boulevards* (the old eighteenth-century network given a new lease on life in the nineteenth century), shops, cafés, restaurants, and theaters offered the greatest concentration of night illumination, and noctur-

1.6 Charles Marville, *Passage de l'Opéra (Galerie de l'Horloge)* (9th arrondissement), c. 1868. Albumen silver print from glass negative. Musée Carnavalet.

1.7 Charles Marville, *Passage de l'Opéra (de la rue Lepeletier)* (9th arrondissement), c. 1868. Albumen silver print from glass negative. Musée Carnavalet.

nal pedestrians of the era (the pioneering respectable noctambulists) recalled the impression of living in conditions of a perpetual fête, as if they had been transported to an illusory city.

Lighting (*éclairage*) was obviously the primary and most efficient way to attract attention to the spaces, and to the goods on sale. Considering that the arcade and its flickering gas jets were an integral totality, Marville's daytime images of a *passage* are unsettling. Only a concentrated effort at projection can endow them with the ability to allure. The arcades, like the *réverbères* in Marville's photographs, are stripped of their ability to captivate. The gas lamps in the *passage*, cut loose from the job of lighting, are unable to generate the sensate conditions of raptness. To put it anachronistically but also admiringly: Marville's representations disenchant, but also critically demystify ground zero of the city's phantasmagorias in the gaslight era.

•

If you want Baudelairean images of Paris . . . it is not in Marville's photos that you should look, but rather in Manet.

ERIC HAZAN, 2010[19]

The gulf between Charles Marville's arcade photographs and Walter Benjamin's phantasmagorias should not block the discovery of resonances between some of his daylight Paris photos and other leading contemporaneous ideologies of urban modernity, both verbal and visual. There are striking parallels, for example, between certain of Marville's pictures of Paris and the attributes and modernities of the city in Charles Baudelaire's *Paris Spleen*, in whose poems the metropolis is often characterized by the juxtaposition of squalor and privilege, a place where pleasure and misery were often disposed cheek by jowl.[20]

Marville photographed the brand-new rue Soufflot in 1876–77 (fig. 1.8). (Its construction is discussed in chapter 2.) The photo records the *quartier* of the Panthéon in the fifth arrondissement just up from the Jardin du Luxembourg, which was changing fast; Marville's photograph defined it in terms of contrast and incommensurability.[21] The new street's very pavement appears unstable. The broad triangular swath of rutted fresh roadway has an oddly indistinct surface, to the degree that its materials are not easily defined: dirt? macadam? gravel? cobblestone? brick? The road seems to hover between horizontal spatial strata. The causes of the new pavement's vagueness include (but are not exhausted by) the ghostly *pentimenti* of pedestrians and carriages (in motion when the shutter was opened) concentrated at the rear center, not far from the façade of the somewhat blurred yet imposing Panthéon. The prevailing structures in this capacious view of a sector undergoing extensive change

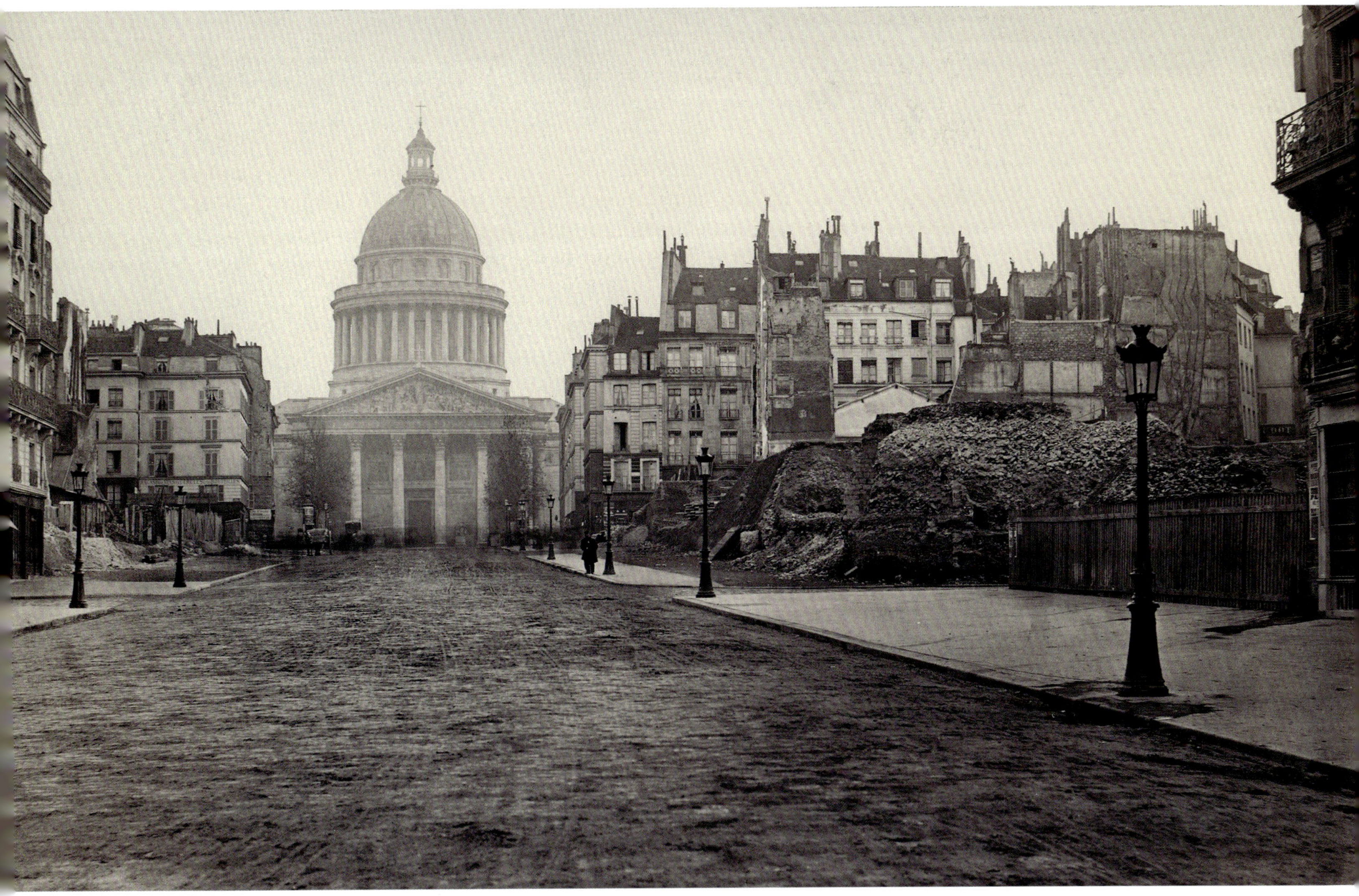

are paradoxically the most diminutive elements of the built environment: the street furniture composed of two rows of spiky, up-to-date gas streetlamps spaced along the rims of the two facing, perspectivally splayed orthogonal sidewalks flanking the new street. (One in-focus, dark-clad pedestrian, though tiny, shares visual standing with the third lamp at right.)

1.8 Charles Marville, *La rue Soufflot et le Panthéon*, 1876–77. Albumen print from collodion negative. Musée Carnavalet.

This is the only case of a Marville photo examined so far that contains more than one lamp: six identical posts are fully legible here (maybe there are two more), but the largest lamp, in the right foreground, holds sway. The lampposts constitute a team, as they should: arrayed and coordinated as identical sentinels, as soldierly, sharply focused examples of up-to-date construction; alert (on duty), ready for illumination, unmindful of (oblivious to) the human blur close by.

Cherchez la lampe

Why does this streetlamp flaunt its unpleasant perpendicular right in the middle of the picture?

PAUL SÉBILLOT, 1877[22]

The importance of the lamps in Marville's photograph of the rue Soufflot (fig. 1.8) echoes the role played by the prominent daytime gas lamp at the center of Gustave Caillebotte's 1877 *Temps de pluie* (fig. 1.9), a venerated demonstration (or performance) of social disaffection. Caillebotte's canvas reminds us that all the canonical vanguard images of the 1870s and earlier 1880s of the Paris boulevard or square (*place*) are daytime pictures. Rummage through your mental Impressionism archive—Monet, Renoir, Pissarro, Degas, Caillebotte—and this is quickly verified. No Impressionist Paris nocturnes were made in these years. Camille Pissarro's *The Boulevard Montmartre at Night* (1897; National Gallery, London) is the belated exception that proves the rule. That Claude Monet never painted an outdoor night picture in Paris, but did indeed paint a handful of electric nocturnes set in London

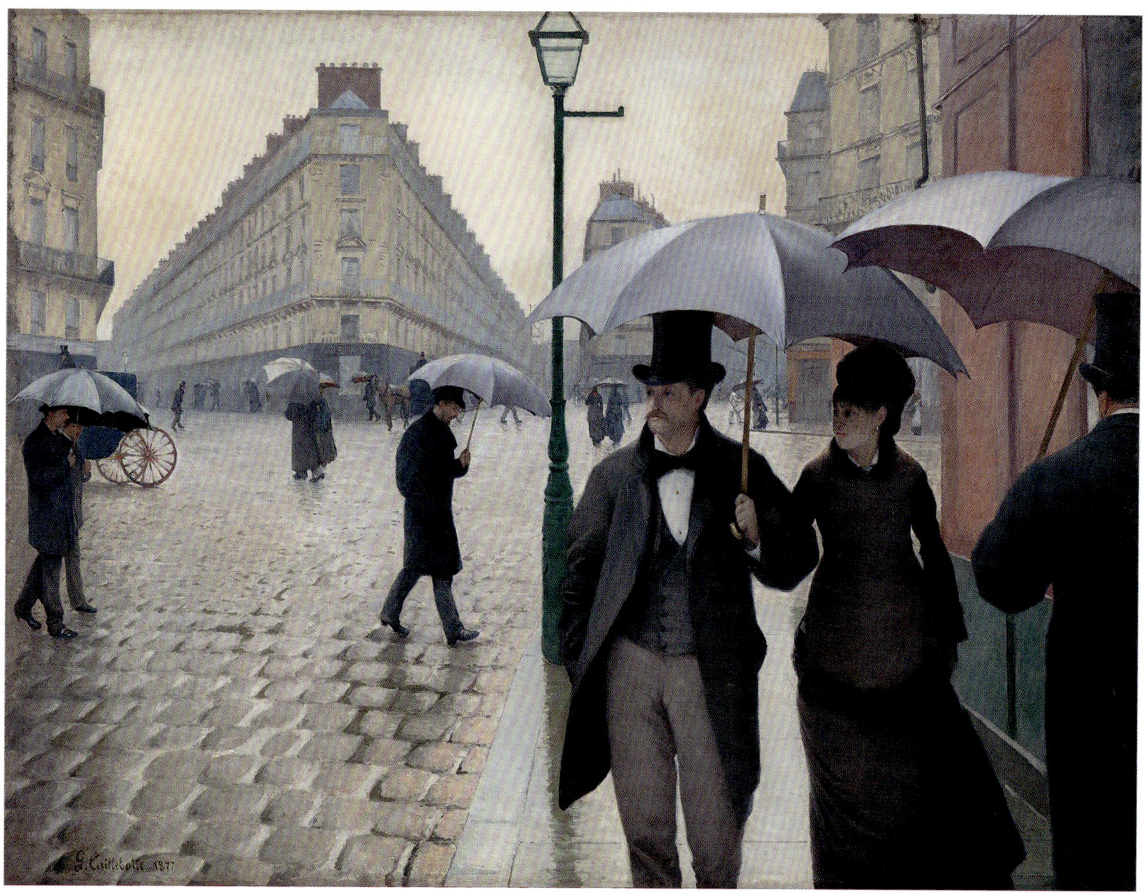

1.9 Gustave Caillebotte, *Paris Street: Rainy Day*, 1877. Oil on canvas. 212.2 × 276.2 cm (83½ × 108¾ in.). Charles H. and Mary F. S. Worcester Collection, Art Institute of Chicago. Photo: Art Institute of Chicago/Art Resource, New York.

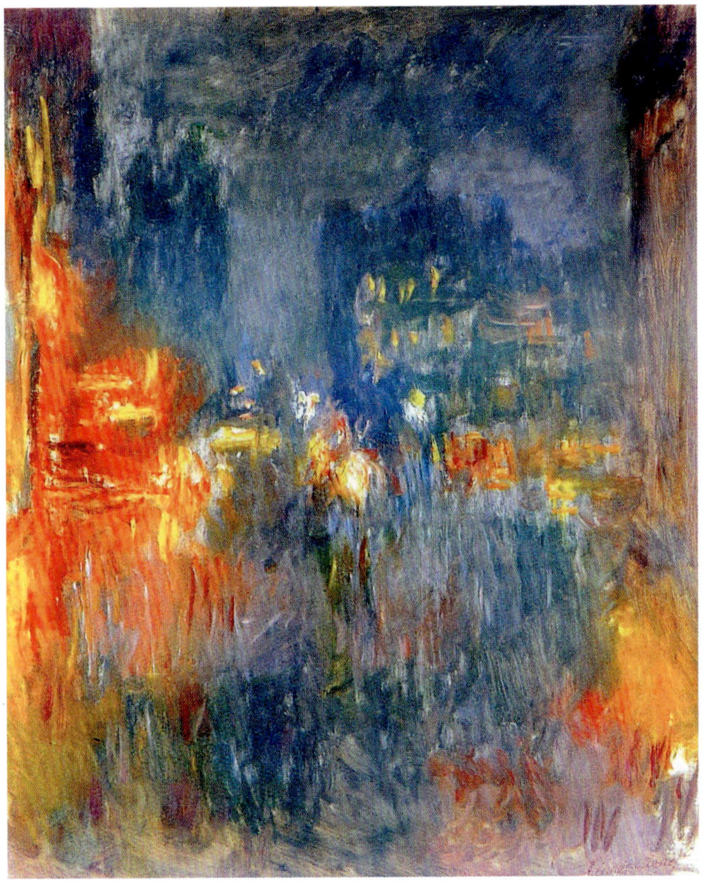

1.10 Claude Monet, *Leicester Square, La Nuit*, 1900–1901. Private collection. 80 × 64.2 cm (31½ × 25¼ in.)

(fig. 1.10) at the turn of the twentieth century should set us wondering about the very particular connotations of the nocturnal City of Light that prevented the Impressionist gaze from falling upon it.[23] The Impressionist painters did not choose the artificially illuminated street, boulevard, or square as a fitting setting for their records of the Parisian everyday. None of these nighttime outdoor locales was evidently reckoned appropriate for a modernist account of the fraught characteristics of the intersection of social and infrastructural modernity. A related point: while the Parisian modernist record of the city's *ur*-spaces of modernity has been well mapped by art history, the privileged time of day has not previously been much worried over.[24] For Parisian Impressionists, the telltale practices of urban modernity were lived and made out of doors by day, and indoors by night. That the case was decidedly otherwise for nonindigenous Painters of Modern Life, energetic creators of outsider Paris nocturnes, is the subject of chapter 5.

With this context in mind, we home in on the green lamp at the center of Caillebotte's much-studied *Paris Street: Rainy Day*, 1877 (fig. 1.9), a painting that is widely taken to be—thanks in large measure to interpretations made and influenced by the late Kirk Varnedoe—the visualization par excellence of urban *anomie* in the recently

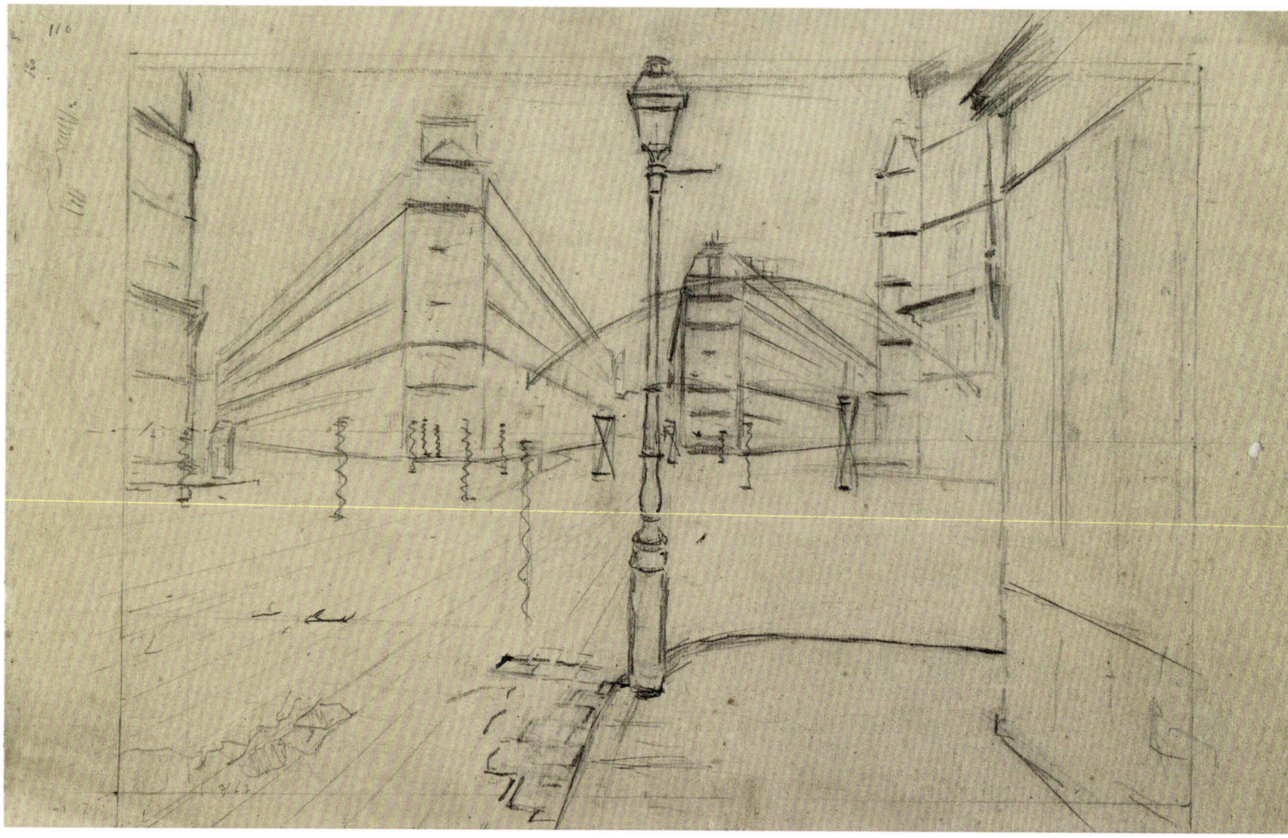

1.11 Gustave Caillebotte, *Study for Paris Street: Rainy Day*, 1877. Graphite, with touches of erasing and touches of charcoal, on tan, moderately thick, moderately textured, handmade laid paper. 30.2 × 46.5 cm (12 × 18⅜ in.) Restricted gift of the Jentes Family Foundation, 2011.420, Art Institute of Chicago. Photo: Art Institute of Chicago/Art Resource, New York.

modernized and patched-up capital city.[25] The structure of the picture's illusion of the built environment, and the placement, choreography, and identity of its major and minor figures have been studied over and over again, but little (*pace* Paul Sébillot) has been said about the green, unlit gas streetlight dead center in the picture. Its force and centrality in the preliminary drawing (fig. 1.11) leave no doubt that it was central to Caillebotte's conception of the space and its eventual inhabitants. The rugged handling and thickness of the white paint in this central upper portion of the canvas (fig. 1.12) render the lamp's glass surfaces emphatically blank and opaque. Despite its complete inutility during the day, the prominence of this paradigmatically networked device makes it nonetheless a key "figure" in the ambitious and large tableau. But while this weighty streetlight is a kind of actor in the painting's narrative, its daytime uselessness and blindness make it a mere bystander to the picture's stagey image of the bourgeois social everyday in modernity. What remains of interest—albeit a question that cannot easily be answered—is why Caillebotte bothered to endow a useless, and indeed old-fashioned if urbanistically integral, piece of street furniture with such consequentiality?[26] The questions do not stop there. Why paint the lamp and its painstakingly limned reflection in green? Period streetlamps were not green. Why did so many French Impressionists and their successors cast daytime streetlamps as actors in the theater of modern life?[27] Our burgeoning acquaintance with the intensity of the period's fascination with city lighting will help to address these questions.

In Caillebotte's painting, the conspicuous and somewhat mysterious glaucous eye of the solitary, unlit streetlamp inflects the picture's otherwise largely consistent depictive and connotative messages, despite the complexity of the painting's spaces. The situation is somewhat similar in Marville's *rue Soufflot* (fig. 1.8). Thanks to the photograph's perspectival structure and the camera's normative shutter speeds that disallowed the freezing of moving subjects, the elegant cast iron lamps carry a lot of weight in the picture—visually if not syntactically. They perform an important portion of the photograph's signifying work by unambiguously indexing new and completed construction: the installation of the lamps is complete, and their style, unlike that of the prominent green *réverbère* in Caillebotte's painting, is the very latest.[28] They are visually arresting as well, as saturated black forms in sharp focus (along with the solitary man on the sidewalk). The lamps are unique among the elements of the built environment shown by the photograph in that they are both brand new and completely stable. Elsewhere, as our inventory stressed, some of the structures undergoing transformation are blurred, and thus the process appears inchoate. Witness,

1.12 Gustave Caillebotte, *Paris Street: Rainy Day*, 1877, detail. Oil on canvas, 212.2 × 276.2 cm (83½ × 108¾ in.). Charles H. and Mary F. S. Worcester Collection, Art Institute of Chicago. Photo: Art Institute of Chicago/ Art Resource, New York.

too, the partially demolished buildings adjoining the Panthéon at right, directly behind a huge pile of debris. The similarity between Marville's rue Soufflot lamps and Caillebotte's *réverbère* is therefore only partial. The painted lamp is old, "a relic of time past" in Michael Marrinan's words, a remnant of the adjacent unimproved district including the rue de Turin; it is an outsider to the trendy *quartier* of the painting's setting, beyond and behind the lamp-structured near foreground.[29] The lamps photographed by Marville are, by contrast, up to date; the unmistakable insignia of the newness of a first wave of the neighborhood's ongoing transformation.

The layout of Marville's *Rue d'Hautpoul et entrée des carrières de la rue Compans* (*Rue d'Hautpoul and Entrance to the Quarries from the rue Compans*; nineteenth arrondissement), a work of around 1877 (fig. 1.13), echoes the lamp-centricity of Caillebotte's painting as well as the *mise-en-scène* of a later Paris canvas by Vincent van Gogh, *The Outskirts of Paris* (1886; fig. 1.14). T. J. Clark has argued that Van Gogh's painting is distinctive by dint of its "very emptiness, and the literalness with which

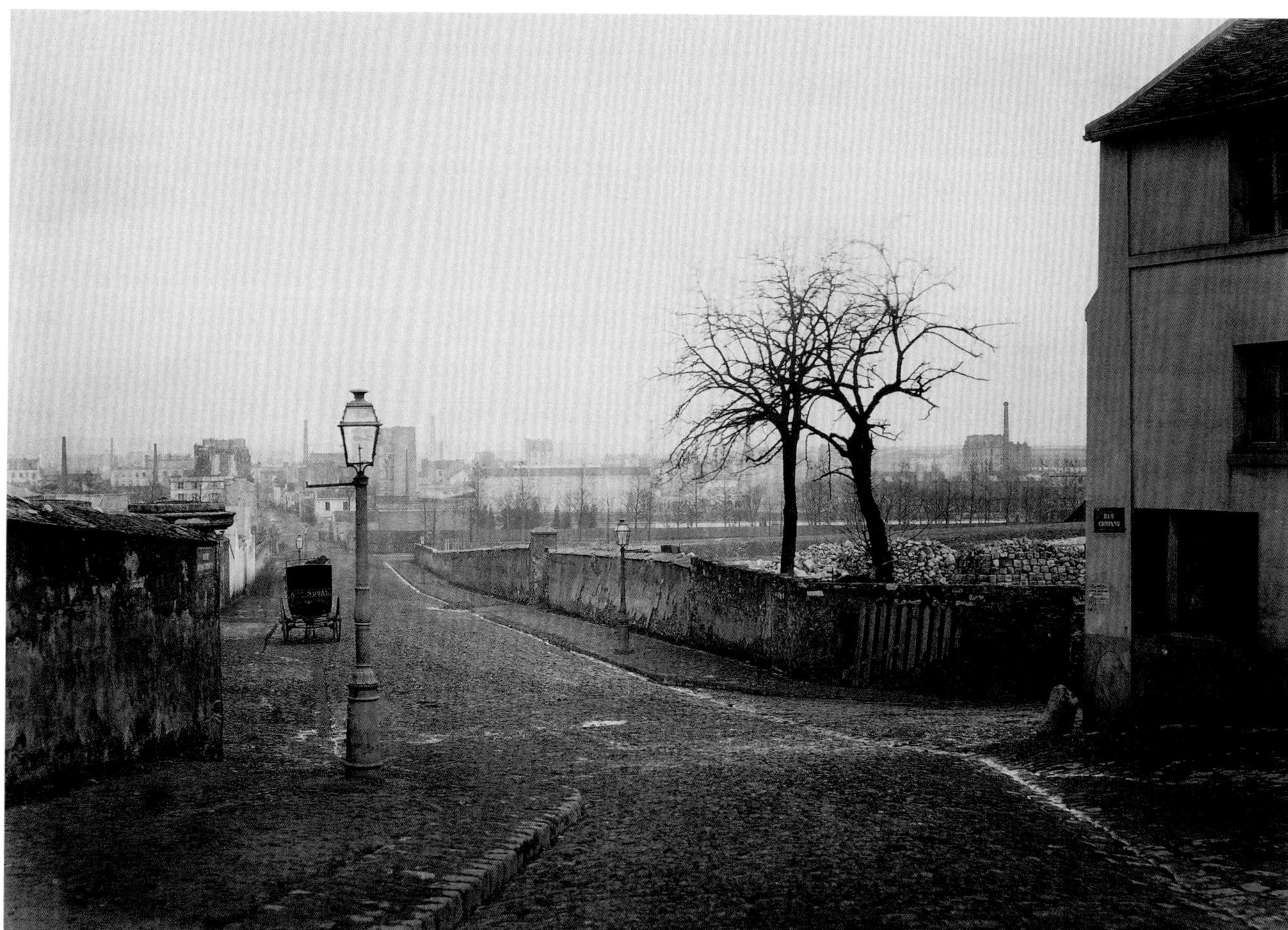

1.13 Charles Marville, *Rue d'Hautpoul et entrée des carrières de la rue Compans (Rue d'Hautpoul and Entrance to the Quarries from the rue Compans;* nineteenth arrondissement), c. 1877. Albumen print from collodion negative. Image © Charles Marville/Musée Carnavalet/Roger-Viollet/The Image Works.

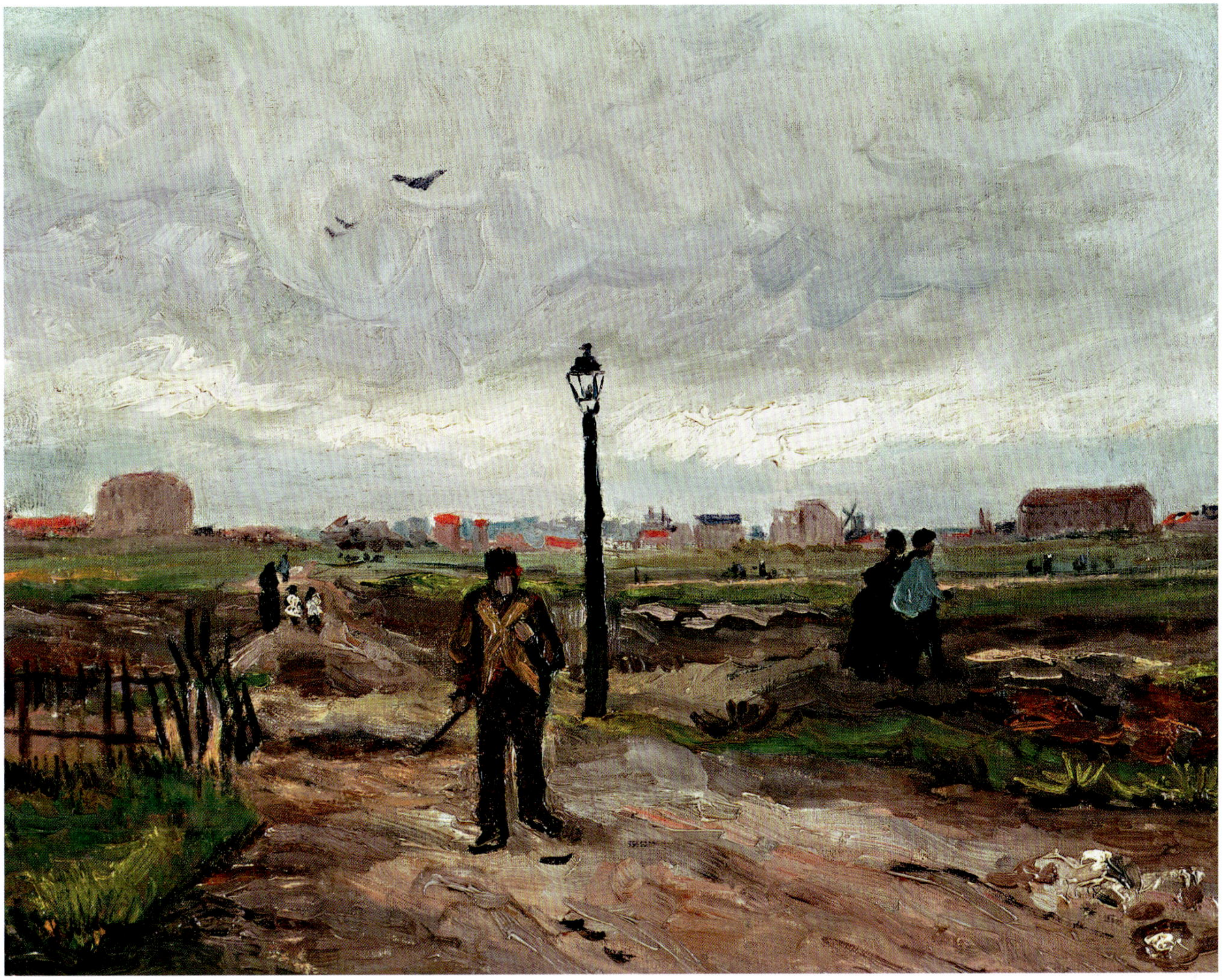

1.14 Vincent van Gogh, *The Outskirts of Paris*, 1886. Private collection. Photo © Christie's Images/Bridgeman Images.

signs of change are spelt out in it unobtrusively. . . . He draws up a kind of inventory of the edge of Paris."[30] The artist's stock-taking includes a prominent gaslight that Clark calls "the premonitory gas standard."[31] It figures prominently among the things Haussmann's critics blamed him for: "the factories, the mess of fields and paths and stranded gaslights."[32] Marville's photograph of a gloomy and desolate spot in the nineteenth arrondissement, a northern city district packed with quarries, is in many ways of a piece with the setting of Van Gogh's painting, except for the contrast between the prominent pieces of lighting hardware in both. Marville used a solitary lamp, as did the Dutch painter, to stress the homeliness and the neglect of the pictured environment, but the lamps differ in date, style, and impact. These are critical dissimilarities between otherwise analogous pictures.

Van Gogh's gaslight-punctuated *banlieue* marks the painting's terrain as the outer limit of the still-developing burgeoning city close by. It is a zone at the northern periphery, almost beyond the reach of the French capital's technologized urban modernity. The lamppost thus qualifies as Clark's cautionary and baleful object, and is in many ways the mirror image of the 1830s-era Oudry lamp dramatically bisecting Caillebotte's 1877 canvas. It served as a marker of an earlier era of the city's material fabric, standing on the very edge of, and indeed abutting, a recently updated *quartier* of Right Bank Paris. Caillebotte's intrusive green lamp functions as the return of the repressed in the register of street furniture. Van Gogh's lamp is the harbinger of a future, while Caillebotte's *réverbère* is a lone survivor of a past quickly becoming a ruin overtaken by thrusting new development.

The plot thickens. The domineering lamppost of Caillebotte's 1877 painting is *identical* to the gas lamps in Marville's 1877 photograph (fig. 1.13), containing one prominent post plus three amply spaced and blurry *confrères*. Marville's dominant old-fashioned light telegraphically signals the grimness of this section of northern working-class Paris, whose bleakness exceeds that of the marginal district summoned up by Van Gogh's painting (fig. 1.14). The chimney- (at least ten) and quarry-packed locale of Marville's photo is the site of labor and poverty; it is not new, let alone prosperous or chic. One sign of its identity is that the apparatuses of technologized modernization are absent, have stayed far away.[33] In Van Gogh's scenario, the up-to-dateness of the lamp strikes a different note; it introduces the temporality of unfolding change and heralds a dubious urban future. In Caillebotte's and Marville's 1877 representations, the prominent 1830s-style lamps are unambiguous markers of the past, of the old: what survives on the edge of a prosperous and otherwise modernized district of Right Bank Paris, on the one hand, and the sure identifier of a poor industrialized section of far northern Paris toward the end of the 1870s, on the other. *Cherchez la lampe.*

· 2 ·

Losing the Moon

JOHN SINGER SARGENT IN THE JARDIN DU LUXEMBOURG

Dusk begins to fall. The shadow flows from trees that rise above one another in dark masses.... Shadow expands still more, drowning the whole terrace, erasing the white outlines of the marbles.

J.-K. HUYSMANS, 1881[1]

John Singer Sargent's 1879 nocturnes (figs. 2.1 and 2.2) are subtle picturings of dusk in the Luxembourg Gardens; renderings of *l'heure bleue*, the redolent French term for the time and color of twilight.[2] Both paintings, whose order of creation is not known, call attention to a bright summer full moon rising above the silhouette of the chestnut trees densely clustered on the east edge of the gardens.[3] Discrepancies between Sargent's gardens and J.-K. Huysmans's written description in the epigraph point to the singularities of the paintings. The young painter's avoidance of pools of shadow sets his subtle tonal system apart from Huysmans's chiaroscuro, and Sargent included people: an isolated pair of well-lit modern wanderers dominates the left foreground, while numerous tiny, black-pigmented figures in silhouette busy themselves at mid-picture.

Two of Sargent's early commentators perceptively accentuated the paintings' delicate mood and human occupants.

> *In the Luxembourg Gardens* [fig. 2.2] is a splendid impression, broad and summary, showing the Luxembourg Gardens in a fog, with glinting broken reflections in the basin of the fountain, and forms wandering through the bosquets surrounded with aureoles of mist and exhalation. *Edward Stranahan, January 1880*[4]

> In this canvas [fig. 2.1], Sargent has shown the hushed pensiveness of mood, the perfection of sensitively modulated color of which his many-sided genius is capable, but which it has, in fact, rarely attained to such a degree as here. . . . Various groups of figures lend the scene a measured animation.
>
> *Handbook of the Minneapolis Institute of Art, 1917*[5]

But neither Stranahan nor the 1917 handbook author noted another feature of consequence: the multiplicity of lights—as many as five different kinds—that glow in the pictures. My interest in the two park paintings derives from those lights. My reading strives to pinpoint the role played by the diverse illuminations and the quirky narrative ingredients of Sargent's strangely empty and disconsolate summer Luxembourg, painted at the crepuscular moment of moonrise, haunted by the specter of recent modernizations nearby, including glaring gas and electric lights outside the eastern gates.

The two paintings are very close in size: 73.6 × 92.7 cm (29 × 36½ in.; fig. 2.1) and 65.7 × 92.4 cm (25⅞ × 36⅜ in.; fig. 2.2). Compositionally they are also virtually

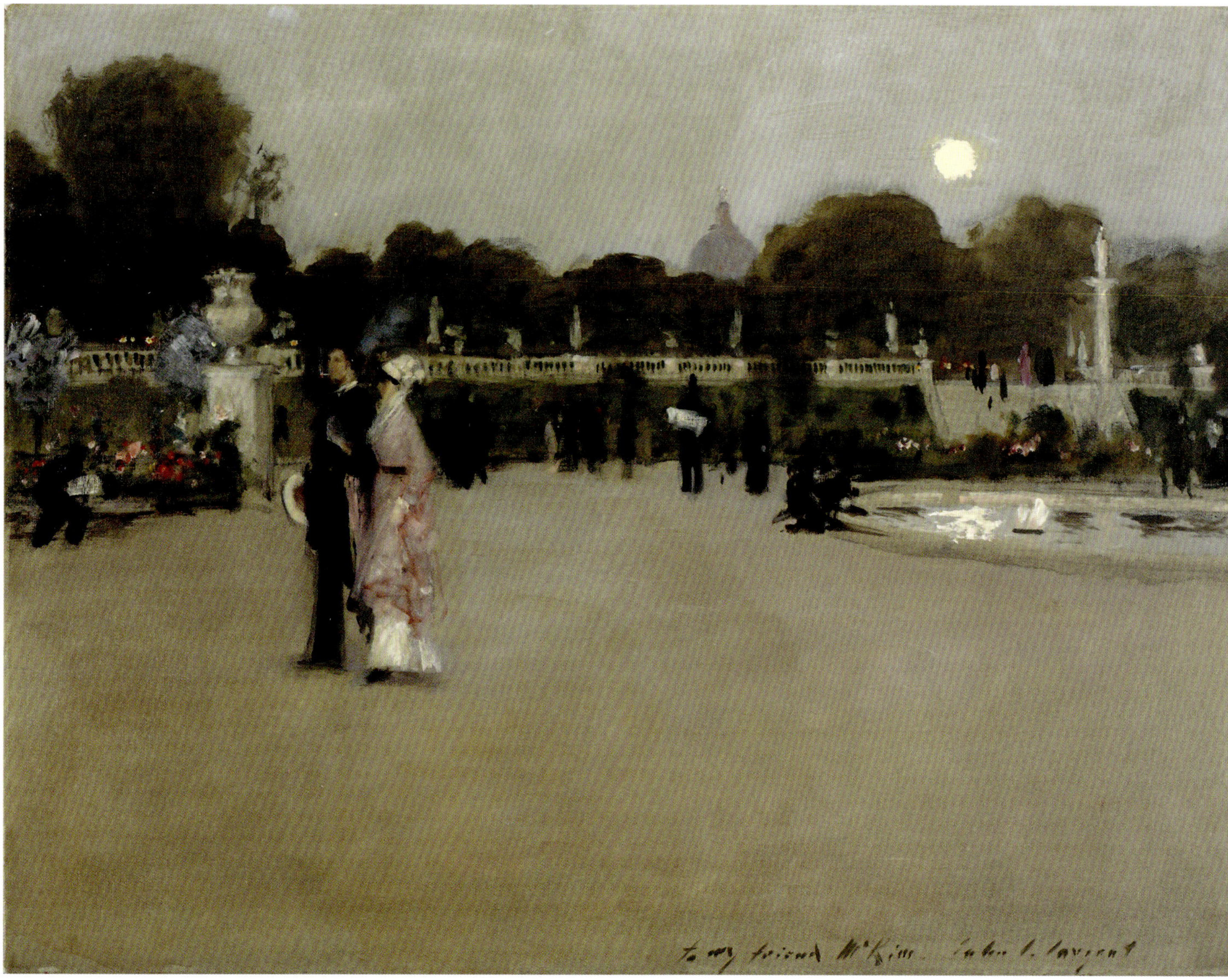

2.1 John Singer Sargent, *Luxembourg Gardens at Twilight*, 1879. Oil on canvas. 73.7 × 92.7 cm (29 × 36½ in.). Minneapolis Institute of Art, Gift of Mrs. C. C. Bovey and Mrs. C. D. Velie 16.20. Photo: Minneapolis Institute of Art.

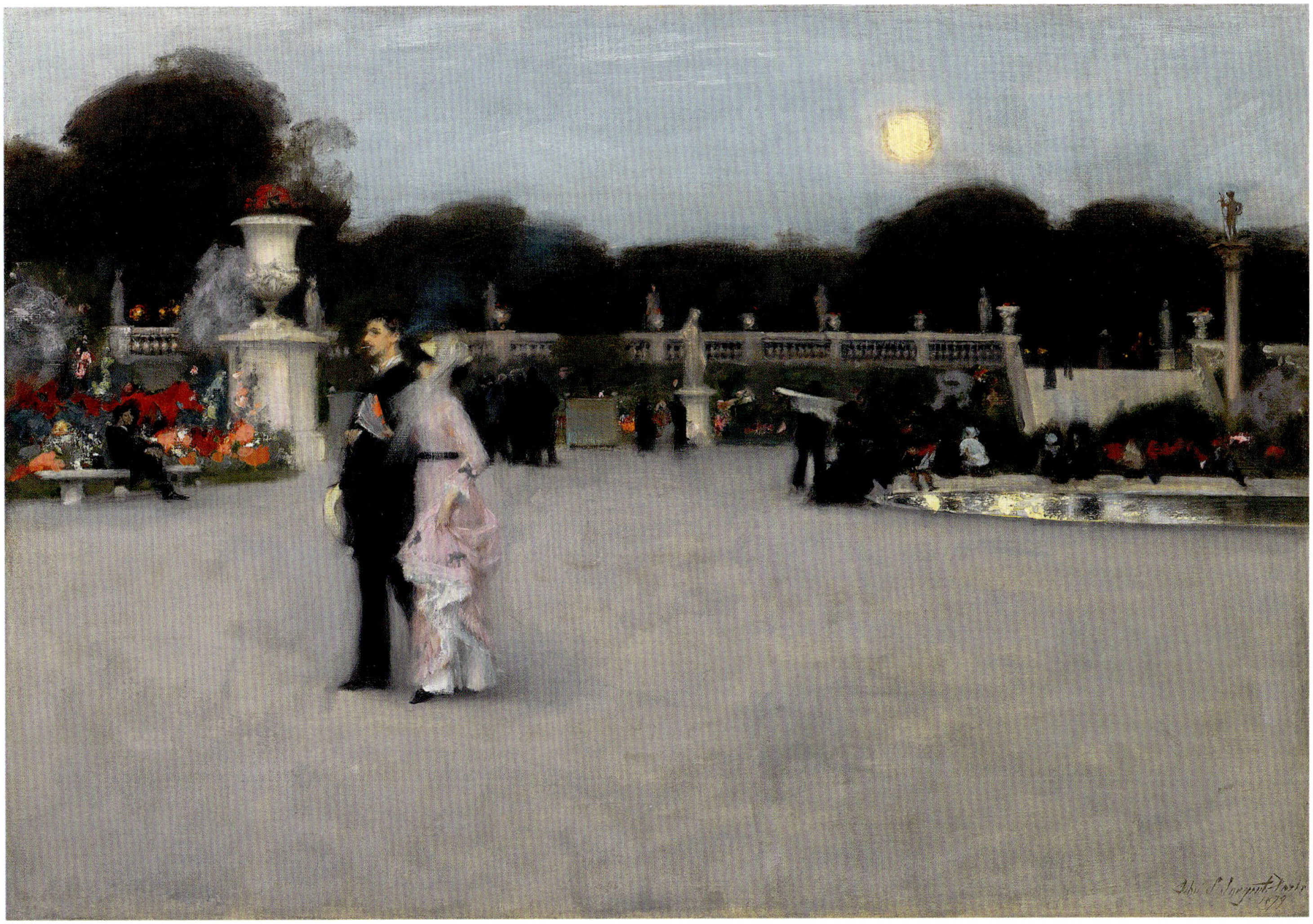

identical, with some shifting of the small-scale silhouetted figures (the newspaper-reading man near the boat basin and the fellow on the bench at far left, for example), and some adjustment of the natural lighting conditions (the Philadelphia picture [fig. 2.2] is comparatively foggy). Both canvases are set facing due east toward the rising full moon (the directional lodestone, the dome of the Panthéon, is visible in the Minneapolis version [fig. 2.1]) and thus more or less on axis with rue Soufflot, seen to advantage in Marville's 1876–77 photograph (fig. 1.8). In the Minneapolis picture the moon is toward the southeast, while in the Philadelphia picture it is more fully east-southeast. (Perhaps the moon and fuzzy sky "block" the Panthéon in the Philadelphia version.) Sargent's acknowledgment of the moon's transit across the sky suggests that the making of the pictures was to a degree rooted in observation—a surviving drawing done *sur place* seems to support that (fig. 2.3)—and shows some alertness to celestial mechanics.[6] (See *quartier* map [fig. 2.4].)

2.2 John Singer Sargent, *In the Luxembourg Gardens*, 1879. Oil on canvas. Image: 65.7 × 92.4 cm (25⅞ × 36⅜ in.); framed: 96.5 × 123.8 cm (38 × 48¾ in.). Philadelphia Museum of Art, John G. Johnson Collection, Cat. 1080.

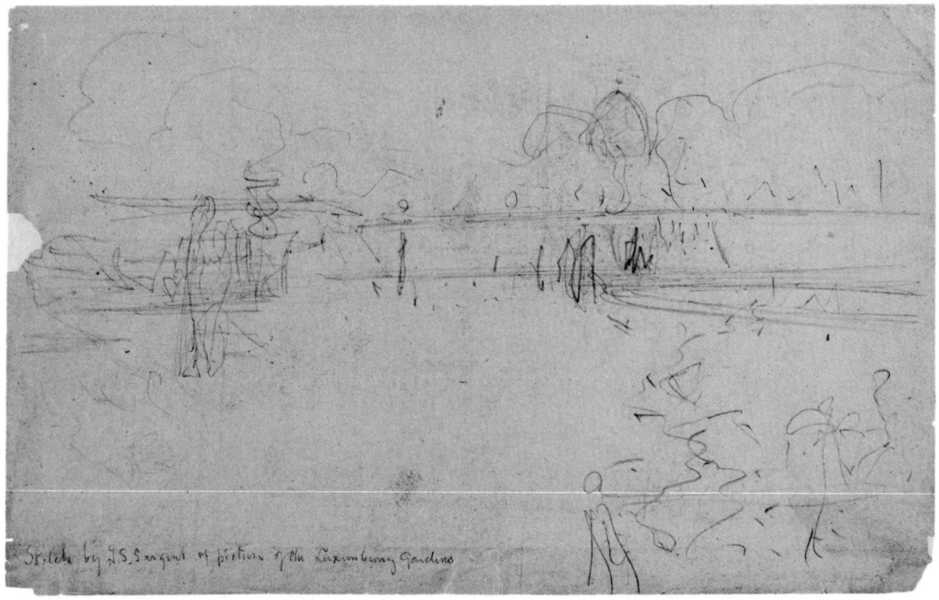

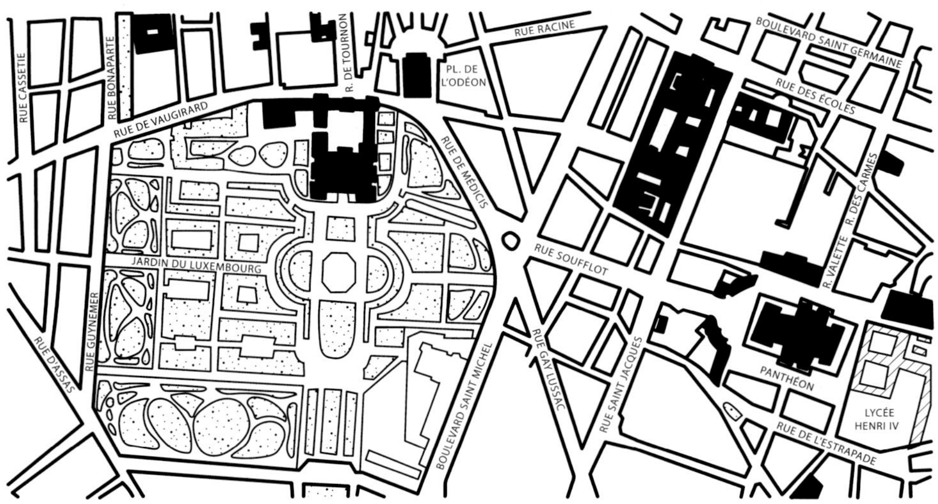

2.3 John Singer Sargent, *Study for "The Luxembourg Gardens at Twilight,"* c. 1876–78. Graphite on ivory paper, formerly mounted on a page of James Carroll Beckwith. Album 3. Sheet: 19.8 × 31 cm (7 13/16 × 12 3/16 in.). Gift of the National Academy of Design, 1935.85.3.153, New-York Historical Society. Photo © New-York Historical Society.

2.4 Map of the 6th arrondissement, Paris, adjoining the Luxembourg Gardens. Drawn by Mel Keiser.

If we wanted them to mark a sequence—despite Sargent's forgetfulness about the order in which he painted them—the differential moon placement makes the Philadelphia sky "earlier" than the Panthéon-containing Minneapolis sky. I state this for the astronomical record despite two things: the small moon orb's position was likely determined primarily by compositional rather than scientific deliberations, and an opposite sequence is strongly implied by the contrasting surfaces of the paintings. The demonstrably looser facture of the Minneapolis canvas has led to informed speculation that it was the first picture, and may even have been done, in part, *en plein air,* inasmuch as Sargent himself used the words "preliminary sketch" and "replica" in his own epistolary memory of them.[7]

To complete the cosmological record, though I have no intention of tying the works to a specific day, during the summer months of 1879 that Sargent was in town, there were two dates of coinciding sunset and rising full moon: Thursday, July 3 (at 8:56 p.m.), and Saturday, August 2 (at 8:27 and 8:28).[8] These particulars are adduced to demonstrate that a coincident twilight and rising full moon in the summertime is factual. Then as now the park's closing time was pegged to sunset: each fortnight the time changed so that it fell between thirty minutes and one hour following sundown. So the "moment" referenced by Sargent's pictures is a quite particular one, aligned with the concurrence of a rising full moon in the evanescent twilight conditions of a simultaneous sunset, and with the time when the streetlights would have been in full blaze. Having the closing time pegged to sunset in the era of artificial light could be seen as an anachronism, despite its survival to the present day. Sargent's choice of "moment" could be understood to draw attention to that anachronism, and to the threshold nature of the time of day and season he selected: day at the brink of night in the gardens, not long at all before closing time in summer, about 9:00 p.m. or so. How his selection of that moment tallies with the narratives he crafted in the paintings is taken up in what follows. But let me insist again that it does not make sense to tie the evanescent imaged light conditions of the paintings literally to a "real" time or "actual" light conditions. Charles Baudelaire's 1859 observation about the emptiness of depictive truth is a tonic prompt. "These things [diorama and theater] because they are false, are infinitely closer to the truth; whereas the majority of our landscape painters are liars, precisely because they have neglected to lie."[9]

Sargent's two garden pictures are eccentric tableaux in several respects, not least of which is their unusualness with regard to his practice. The mere fact of their production belied the artist's habit of only making "subject pictures" while away on trips.[10] In the summer of 1879, however, Sargent was mostly in Paris, where he had been living since 1874 and would remain until 1884.[11] Paris motifs were also rather rare for him, and the two Luxembourg canvases stand out in sharp relief against constellations of related pictures of the same park, in regard to both rhetoric and iconography.[12] They are distinct, for example, from the many child-centric depictions of Luxembourg sociability by other artists.[13] Furthermore, the contrast between the scale, handling, and lighting of the prominent promenading couple and those of the middle-ground staffage (the tiny black-pigmented figures) in Sargent's pictures is arresting and discordant, even when measured against the disposition of figures in avant-garde genre paintings.

There were two further exceptions to the young artist's modus operandi in 1879. His oblique and tonally jarring interiors of the *Rehearsal of the Pasdeloup Orchestra at the Cirque d'Hiver* (figs. 2.5 and 2.6) are also emphatically city paintings.[14] Sargent was otherwise painting portraits and regionalist genre pictures (set in Spain and Italy in the summers) during his studies with Carolus-Duran, as the portraitist and art teacher Charles Auguste Émile Durand was called. Sargent's pictures of social life in a city park at twilight, combined with the daring orchestra rehearsal paintings, add

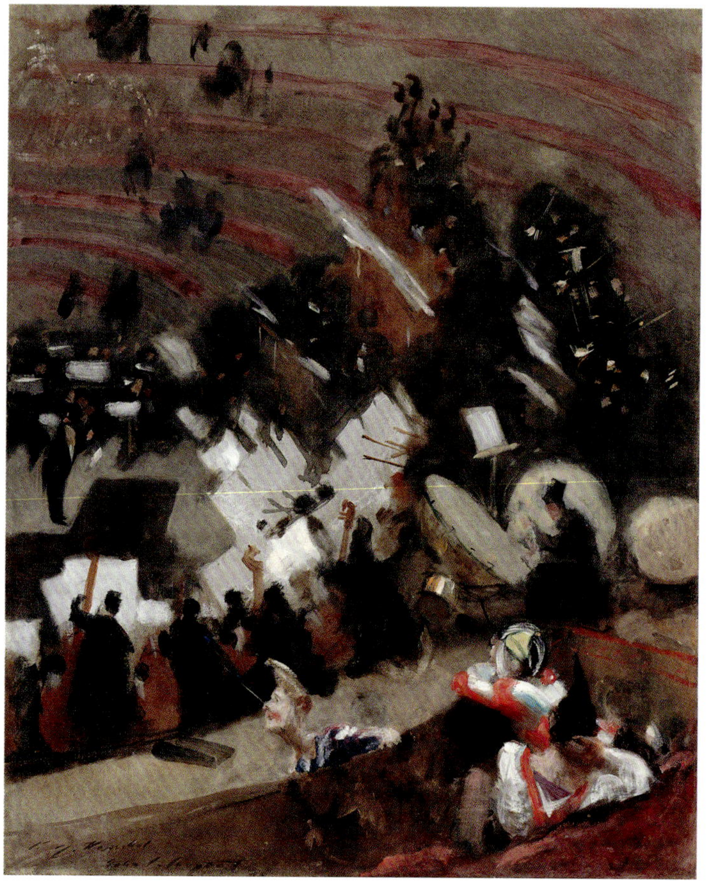

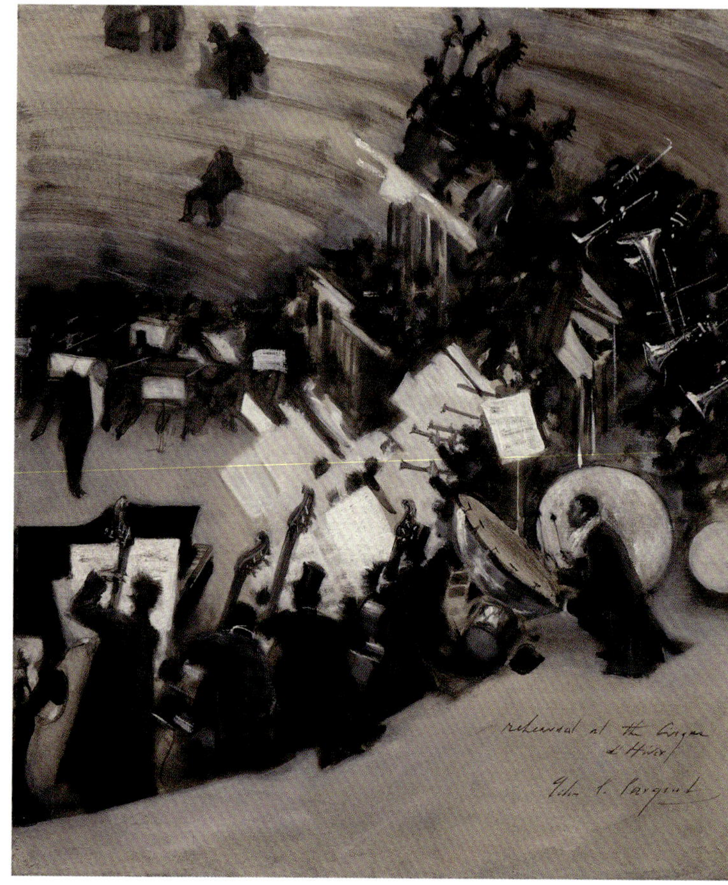

2.5 John Singer Sargent, *Rehearsal of the Pasdeloup Orchestra at the Cirque d'Hiver*, c. 1879. Oil on canvas, 93 × 73 cm (36⅝ × 28¾ in.). Signed, lower left: "To G Henschel/John S. Sargent." Anonymous loan, 81.1972, Art Institute of Chicago. Photo: Art Institute of Chicago/Art Resource, New York.

2.6 John Singer Sargent, *Rehearsal of the Pasdeloup Orchestra at the Cirque d'Hiver*, c. 1879–80. Oil on canvas. Signed lower right "rehearsal at the Cirque d'Hiver John S. Sargent." No. 22.598, Museum of Fine Arts, Boston.

up to a distinctive small corpus insofar as each of the summer pictures instanced thematic originality and striking illumination. First, a word on the defining attributes of the Jardin du Luxembourg.

Previous accounts of the two garden paintings have understood Sargent's park site to have been a "shining emblem of modernization" that epitomized "the French capital's new energy," but I will argue that it is the opposite of an upbeat achievement and laudable identity that informed Sargent's 1879 garden paintings.[15] Rather than a uniformly cheery emblem of the best of public park facilities and happy social life in the new Paris, the Luxembourg was a fraught place. The garden was a recently dismembered, verdant public property whose beloved tree nurseries had been cut down. It was a park on whose parterres Communards had only recently spilled their blood. It was still, at 60 acres, the largest green zone on the Left Bank, but its edges had recently been pruned and newly bordered on the east side by brightly gaslit new streets, the rue de Médicis and the even more imposing boulevard Saint-Michel. Another new construction lay close at hand: the prolongation of the rue Soufflot to the boulevard Saint-Michel, the street just east of the garden running up to the Panthéon, had been completed only in 1876. When photographed twice by Charles

Marville in 1877 (see, e.g., fig. 1.8), it was clearly newly paved, thronging with life (indexed by multiple ghost images) and lined with gas streetlamps likely mirroring those on the adjacent new streets it now met, namely, the rue de Médicis and the boulevard Montmartre.[16] The paintings thus look in the direction not only of the Panthéon and the rising moon, but toward a rare concentration of artificial light on three adjoining streets. (See fig. 2.4.)

The Luxembourg quickly comes into focus as an exemplary later nineteenth-century big city public space: a product of contentious urbanism and state-of-the-art public works interventions, on one hand, and a *lieu de mémoire* of the most violent and divisive political clash in recent memory, on the other. Added to the events that had already fashioned a dramatic identity for the garden was the emplacement in March 1879 of Jablochkoff candles along the newly extended rue Soufflot. The new arc lamps, located near the east entrance to the garden, were vivid examples of the topical and controversial coexistence of brighter light and eye-piercing glare—convenience plus discomfort—brought into being by the city government's irregular flirtation with technologized enhancements of the city's illumination. Only weeks before Sargent's painting campaign, the park was newly subject of an evening to an unprecedented kind of glare through the trees on its eastern perimeter cast by the newly implanted experimental arc lights (Jablochkoff candles) installed sometime during the spring of 1879.[17]

As suggested in the introduction, they were the shock lights of their day. A Chicago newspaperman, Charles Barnard, weighed in on the effect of the lamps on the avenue de l'Opéra in 1878: "The dusky street flashes into sudden glare, white, and intense. . . . The electric light is as cruel as the sun."[18] Starting in August 1875, Sargent shared a rented studio within walking distance of the garden.[19] He would have been keenly aware of the change in the visual atmosphere of a park he traversed frequently. His familiarity with the garden as a preserve of natural light and greenery may have enhanced his sensitivity to its evening penetration by yellow-orange gaslight through the trees and eventual cold white electric light from a close-by street. He would have been accustomed, in other words, to the experience of a green space that shared its crepuscular starry and moonlit evenings with diverse mechanical luminosities. As the ensuing analysis proposes, the blaze of the lamps and lights shining upon the Luxembourg left its traces in Sargent's pictures. Small wonder he devised idiosyncratic, even contradictory ways to represent such a garden, which he endowed with a gingerly compromised guise. Sargent's pictures appropriately regard the garden as a locus of subtly disenchanted twilit repose.

An explicit link between the two pairs of Paris paintings is that their settings, the Cirque d'Hiver and the Jardin du Luxembourg, were both experiencing the glare of the city's new electric arc lamps just at the time that Sargent elected to paint them. The newness of those lamps and their startling visual effects, I argue, helped draw Sargent's attention to the two sites. The Cirque d'Hiver (Winter Circus) is a round polygonal building designed by Jacques-Ignace Hittorff, the German-born French architect, and built in 1852 as the winter home of a circus company; it also hosted orchestra rehearsals, among other events.[20] Jules Étienne Pasdeloup was determined to bring contemporary music, including that of Richard Wagner and Gabriel Fauré, to a broad swath of the Paris public.[21] His modestly priced (some sources call them free) "Concerts Populaires" were held on Sunday afternoons from 1861 to 1887. Sargent was a passionate music fan, who can easily be imagined—indeed, he was—in the audience of Pasdeloup's performances.[22] Like the Hippodrome—a large setting for horse racing in Paris, refurbished in 1879 and lit with Jablochkoff candles in 1880, and familiar to art historians through the work of James Tissot— the Cirque d'Hiver was undoubtedly lit electrically.[23] The piercing lights must have inspired the strident vigor of the blacks and whites he used in the two orchestra rehearsal canvases, especially in the Boston version. As with the Luxembourg boat basins, discussed below, I argue that in the Winter Circus interiors Sargent was fascinated by the contrasts and the brutally customized chiaroscuro of Paris under new lights. But while he orchestrated two complex, almost checkerboard patterns of blacks and whites in the Winter Circus paintings, industrialized lights as such are mere pinpricks in the gray-lavender atmosphere of the garden pictures.

In the Chicago version (fig. 2.5), the sheets of music are conspicuously white and the musicians' generalized treatment in exclusively black echoes the approach to figure painting that Sargent would put to use in the garden paintings.[24] Two other details are notable: the exceptional brightness of the whites used for the three impish clown figures in the near foreground (up in the balcony) and the transparent ghost of a circular, unlit light fixture hanging from the ceiling at the far left. Regarding the former, the emphasis upon the circus clowns at the orchestra rehearsal may reference Pasdeloup's democratic and inclusive audience ambitions as well as the dominant identity of the building, and the ghostly light fixture, hitherto unremarked, may have been intended to say farewell to earlier gas illumination in the Cirque d'Hiver.

The slightly smaller but more finished version in Boston (fig. 2.6) possesses even brighter whites in the zones of sheet music and drum surfaces. Here Sargent has added certain formal male dress details, as well as the outlines of spiky trombones and certain other instrument highlights (such as the keys on the bass viols) to the repertoire of bleached-out white shapes. As in the case of the Chicago picture, the plunging perspective engineered for the striking circus interior scene, viewed from the position of an unseen onlooker, pushes the picture in the direction of schematization, even abstraction—visual effects often summoned up by those whose eyes were dazzled by bright lights. Also notable is that the Boston picture echoes another effect of bright illumination in being an almost complete monochrome. Only curving strokes of brown modulate the picture's otherwise exclusively black-and-white palette.

Sargent also revealed his interest in painting night lights in a Spanish setting in the out-of-doors at about the same date in *The Spanish Dance* (fig. 2.7), one of a

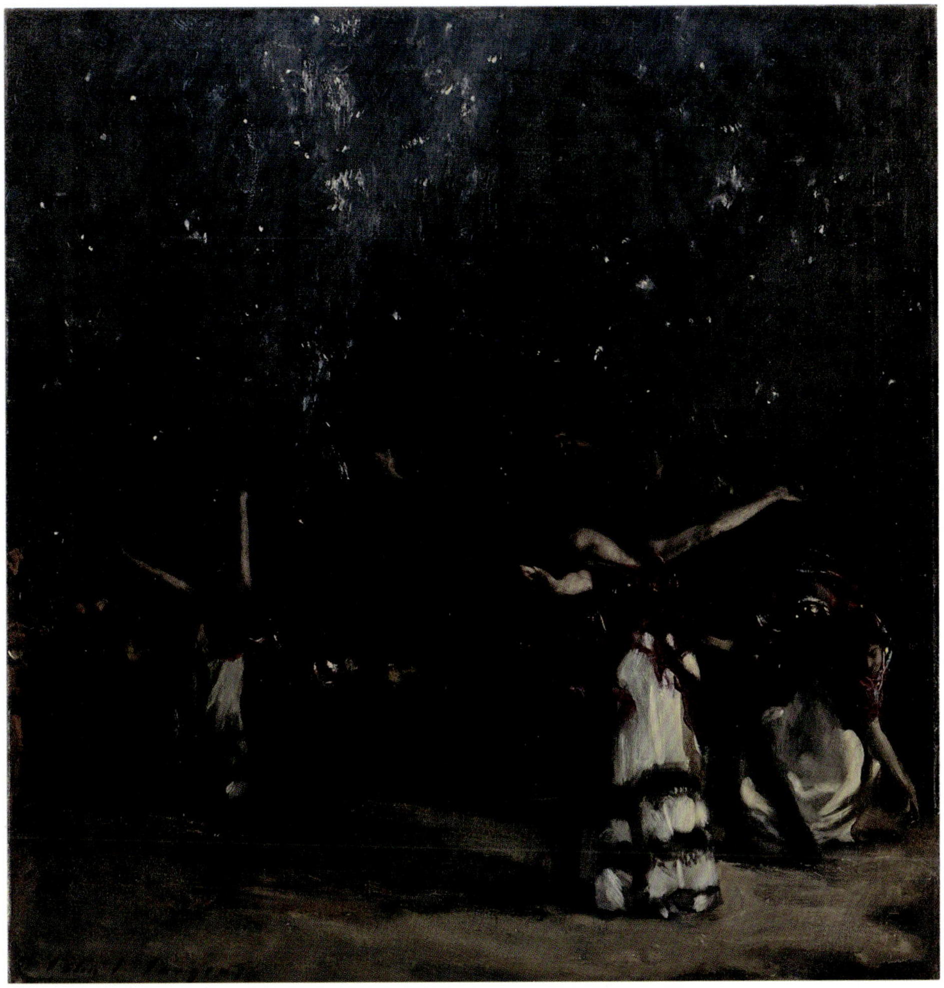

2.7 John Singer Sargent, *The Spanish Dance*, c. 1879–82. Oil on canvas, 88.2 × 83.7 cm (34¾ × 33 in.) Courtesy of Hispanic Society of America, New York.

sequence of related canvases, dated to around 1879–82.[25] They are unmistakably Whistlerian. *The Spanish Dance* has been specifically and closely connected to Whistler's famous *Nocturne in Black and Gold: The Falling Rocket* (1875), object of Ruskin's critique in the notorious 1878 libel suit. Though Sargent would not have seen Whistler's canvas, he would have known about it.[26] Certain of the Spanish picture's characteristics, a very rare nighttime image for Sargent, point in that direction. Its very dark blacks and a suggestion of fireworks commingled with stars in the inky night sky suggest the impress of a Whistler kind of nocturne. In any case, the interest in multiple kinds of light against a nighttime sky is of a piece, if in a different register of illumination, with the exploration of a multiplicity of nights that characterizes the roughly contemporaneous Luxembourg Garden pictures. Multiple lights—even when set against a Spanish sky—further instance the heyday of illumination discourse in Paris.

We return now to the Luxembourg Garden canvases (figs. 2.1 and 2.2) to examine their intricate tissues of illumination. A graveled expanse of path surrounding the Luxembourg's signature octagonal boat basin dominates the depicted garden spaces. Reflections in the basin's water do a lot of work in both canvases, so it is not surprising that the artist plotted its size and placement with care. The painted basin is a mere oval slice of shimmering water at the right middle ground, while in the actual garden the basin is a comparatively colossal feature and strong focal point.[27] Sargent conspicuously rounded off and diminished the thick stone lip of the faceted basin in both paintings. His handling and placement of the pools reduced their size as well, and pulled them down to ground level, to the degree that the pools appear almost flush with the ground in both versions. In the Minneapolis painting, this effect is pronounced: strokes of opaque pearl-gray cover the basin's forward edge, resting on top of the dazzle marks and making a small rectangle. It both flattens the pool and enhances the luminosity of the reflected light on the water.

Though they are only about one-third of the vertical dimension of each canvas, a solitary young couple (about 8 inches tall) walking across the gravel of the sunken garden surrounding the great basin is the central narrative event in each picture.[28] The young people disclose little. They do not interact with one another, though they are very close together, nor do they connect to any onlookers inside or outside the pictures. The profile view of a person is always the least expressive, and the members of this promenading couple do not rupture their dual quasi-Egyptian poses. To the degree that the couple's presence and implied movement instance sociability at all, it is of a muted, even abstract kind. Because they are prominent, we take their identity and social status seriously. It turns out that their clothing is not fashionable; they are more bohemian than bourgeois.[29] But what matters most is their separation from everyone else. They are shown heading, without explanation and in lockstep, due north in the direction of the palace, the long way to an exit. The couple is at the same time taking its distance from the bustle of activity still unfolding around the north end of the boat basin. What interests me more is what else they are shown walking away from: the lights, all of them. Their backs are turned away from the moon and its reflection in the basin, and away from the gas and arc lights beyond the trees to the east. Their trajectory plots an axis parallel to but ignorant of the north-south axis of light that Sargent engineers, a line that more or less follows the strong, straight horizontal line of the balustrade at about the horizontal midline. They embody a lack of interest in, almost to the brink of disdain for, the contest unfolding behind them in the garden between celestial and artificial light.

The idiosyncrasies of the silhouetted figures in the murk in both works have gone largely unnoticed. Despite their animation, they are ignored by or at least totally unconnected to the "world" of the ambling duo moving left. It is certain that the shadowy black figures strewn anonymously across the middle ground are not molded by the ostentatious requirements of social ritual, save perhaps the practices of toy boating and newspaper reading. Their primary *raison d'être* seems to be to denote shadowiness as a tonal condition, as the paradoxical consequence of both being in the dark and being struck by bright light. As Nancy Forgione observes in her impressive discussion of the rise of the silhouette in late nineteenth-century Paris: "Although the artistic use of the silhouette did not depend on actual lighting conditions, the availability of those conditions could stimulate the experience of seeing in terms of silhouette and high contrast rather than in terms of chiaroscuro."[30] Indeed, in this connection, she remarks the silhouetting effect of the electric arc lights, installed on a trial basis in Paris starting in 1878. She defines the silhouette as "a type of shadow, an occluding of light."[31] Her comments help me to connect the shadowy black figures to the visual effects of the bellwether rue Soufflot lights. I shall argue that the inclusion of these figures helped Sargent, on the one hand, to establish in a shorthand way the budding darkness of a crepuscular moment (one star shines in the sky and an impasto spot dazzles at the top of the Panthéon in figure 2.1) and, on the other hand, to register the starkly silhouetting effects of the Jablochkoff candles near the garden, even though the painted figures are blurry and painterly rather than sharply edged—the paradoxical effect of being unable to see detail in the gloaming.

Other qualities of the nocturne genre are relevant to Sargent's canvases, and return us to the men struggling to read their newspapers in the crepuscular light. Literary scholar Susan Stewart notes, "The history of the nocturne itself has a lyric, and not merely a narrative, dimension."[32] She points to the genre's engagement with the expansion of sense impressions at night. In the dark, new modes of moving, touching, smelling, and seeing are sparked and enlarged. The struggle and mood of Sargent's murky human figures speak to the altered sense environment of night. It is harder for them to see; it is more difficult for others to see them.

"Il s'argent"

The silvery reflection of the moon in the pool of water gives sparkle to the grayed lavender palette.

PATRICIA HILLS, 1986[33]

Modernity, contrary to its popular connotations, is not the world in a sweepingly transformed state. . . . It is the hybrid and dissonant experience of living intermittently within modernized spaces and speeds, and yet simultaneously inhabiting the remnants of pre-capitalist life-worlds, whether social or natural.

JONATHAN CRARY, 2013[34]

Sargent painted five different lights in each picture, some more easily perceived than others. The more noticeable are the natural effects: (1) The dusky gray-blue sky is crepuscular (one star twinkles in the MIA picture); and (2) the orb of the silver-yellow, rising full moon glows with reflected sunlight. Look carefully among the

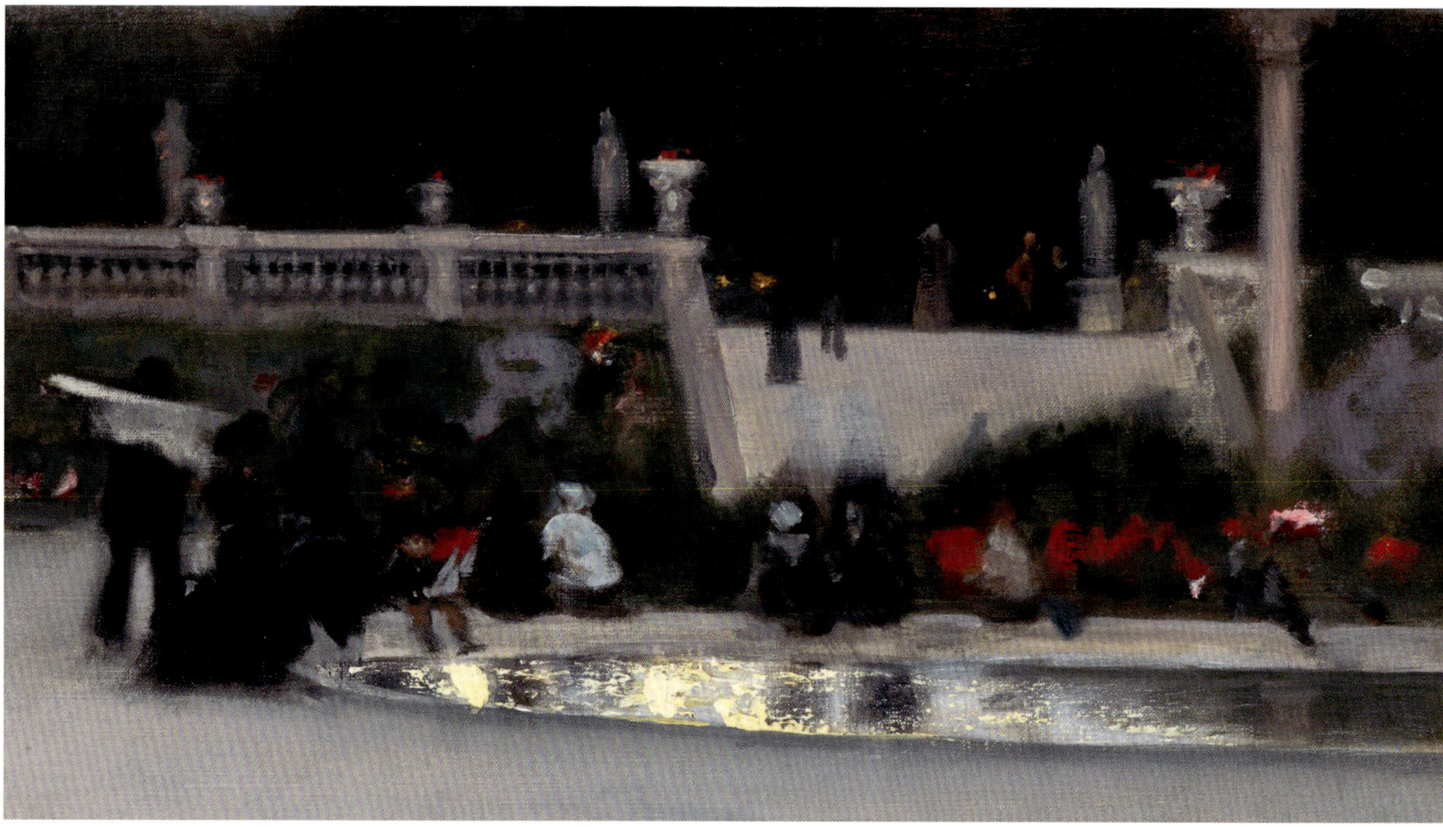

2.8 John Singer Sargent, *In the Luxembourg Gardens*, 1879, detail. Oil on canvas. Image: 65.7 × 92.4 cm (25⅞ × 36⅜ in.); framed: 96.5 × 123.8 cm (38 × 48¾ in.). Philadelphia Museum of Art, John G. Johnson Collection, cat. 1080.

shrubbery, scanning patiently just above the balustrade at left over to the top of the stairs at right, to find another (figs. 2.8 and 2.9): (3) Balls of orange, referencing the gas streetlights twinkling along the boulevard St. Michel and the rue de Médicis, penetrate the dark greenery on the park's eastern fringe.[35] A quotation from Patricia Hills heralds the most elusive illumination: (4) "The detail of the glowing cigarette interrupts the scene with the discreteness of a firefly on a July night."[36] See the glowing ash at the end of the strolling man's smoke. The last, though easily spotted, is the most complex: (5) Whitish yellow light dazzles via encrusted bright paint on the surface of the boat basins. I argue that these thick patches do double duty as patches of reflected moonlight and as the displaced glare of the nearby Jablochkoff lights. The two garden nocturnes thus disclose Sargent's interest in artificial lights concatenating in the fragile atmosphere of simultaneous sunset and moonrise, a scenario that illustrates the hybrid and dissonant experience of modernity as defined by Jonathan Crary.

Sargent, an emergent Whistlerian, was attracted to painting scenes set in conditions of gathering darkness, and undeniably attuned to the profusion of lights punctuating this particular garden landscape at twilight. The painter's likely familiarity with Whistler's nocturnes and his reputation also bears upon Sargent's attraction to a telltale confluence of visual effects in the Luxembourg Garden district. Whistler's

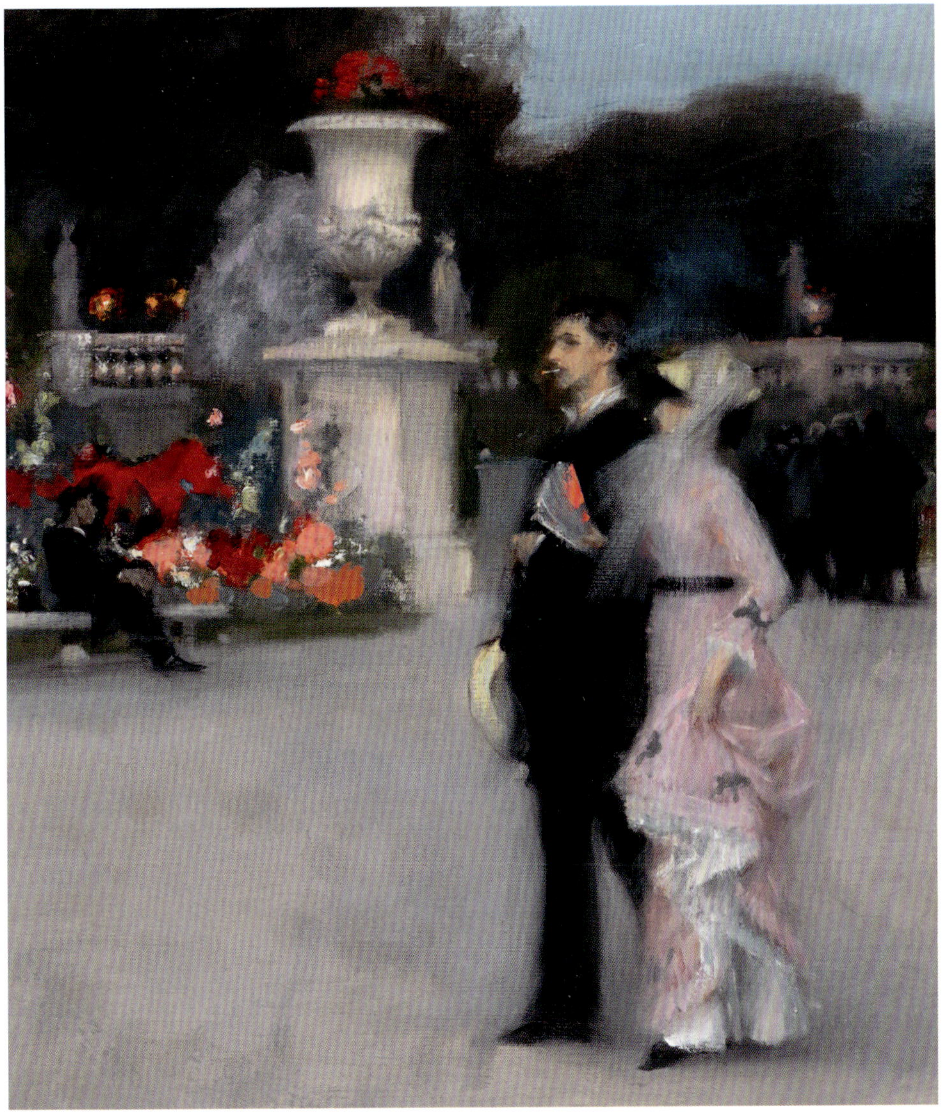

2.9 John Singer Sargent, *In the Luxembourg Gardens*, 1879, detail. Oil on canvas. Image: 65.7 × 92.4 cm (25⅞ × 36⅜ in.); framed: 96.5 × 123.8 cm (38 × 48¾ in.). Philadelphia Museum of Art, John G. Johnson Collection, cat. 1080.

nocturnes were urban; thirty-two of his night canvases were set in London.[37] In the eyes of some viewers, according to John Siewert, "the public notoriety of the Nocturnes in the 1870s and beyond generated a close association between the artist and the enigmatic, perhaps unsavory nature of night."[38] This reading was possible to project onto Whistler's paintings and the artist himself because the darkness and vagueness of his artworks courted illegibility.

That an urban nocturne fashioned in the 1870s might more or less automatically be understood to bear connotations of untoward or disagreeable properties of night is worth considering in relation to Sargent's city twilight campaign, but it does not hold as a stable characterization of his two comparatively pale and generally benign canvases. The silhouetted figures at middle ground do court unreadability, however, and their indistinctness tentatively hints that furtive or unseemly people are present in the Luxembourg near closing time. Assessing their presence from another angle, is their contrast to the strolling duo a metaphor for class difference in modernized Paris? Do the shadowy people speak to anxieties about the invasion of respectable territory by the other, by the disreputable? Does the latent social difference between the two cohorts of garden users mirror or even reinforce the dissonance between the two orders of light in the park, which Sargent negotiated so subtly in both pictures? And vice versa?

The deliberately painted glimpses of gaslights on the adjoining streets helped to define clashing orders of illumination, inasmuch as their effects differ from and go beyond thc delicate illuminations supplied exclusively by celestial bodies. The outrageous arc lights on the rue Soufflot gave rise to even more dissonant visual conditions. While it would not have been possible in subtle genre paintings like these to paint a patch of brilliant glare in the sky at the center of the painting in order to register the ocular ignominy of Mr. Jablochkoff's lamps, the hyper-glare on the basin's water, an unlikely effect of a moon not exactly overhead, helped to create the impression of an extra-bright, one might say deviant, spot of light in the park in the summer of 1879.[39]

Sargent chose the primordially poetic moment of twilight complete with full moon, a staple of Romanticism, exemplified by the great early nineteenth-century moonlights of Caspar David Friedrich. But Sargent moved to disenchant the moment of full moonrise at twilight by his attention to social and environmental modernities, including elements of the built environment framing the park's prominent well-lit strollers and furtive middle-ground denizens.[40] It is instructive to acknowledge analogous combinations of nature and culture in the city of Paris investigated by Sargent's distinguished predecessor, Johan Barthold Jongkind (1819–91).[41] Like Jongkind's preferred territory, Sargent's modern terrain, the Luxembourg, was characterized by the diversity and simultaneity of technologized and natural lighting systems.

In Jongkind's 1854 nocturne, *Notre Dame de Paris au clair de lune* (fig. 2.10), a luminous white full moon shines at the center of a dramatic and irregularly clouded sky: the blues are cut through by diagonal swaths of scudding gray and white clouds. The clouds at the right and left margins of the sky are deep black, aligning with the industrial valences of the smoke-belching boat afloat on the Seine at the lower center. The light of the bright moon falls from behind and well above the central mass of buildings to silhouette the distinctive façade of the dark cathedral and hulking hospital viewed from the southwest across the river, and to produce a shadow cast by the bridge. As in Sargent's canvases, reflected light shines brightly upon the surface of water. Also as in Sargent's paintings, the dark scene is punctuated by small patches of orange flame burning in the glass lanterns of a sequence of paired streetlamps on the bridge over the Seine (two pairs) and along the street (four more) that continues north across the island. There are also oil lamps burning orange in some of the windows of the massive, old, and decrepit Hôtel Dieu (pulled down in 1878), adding another ten luminous small rectangles of light to the delicate but irregular checker-

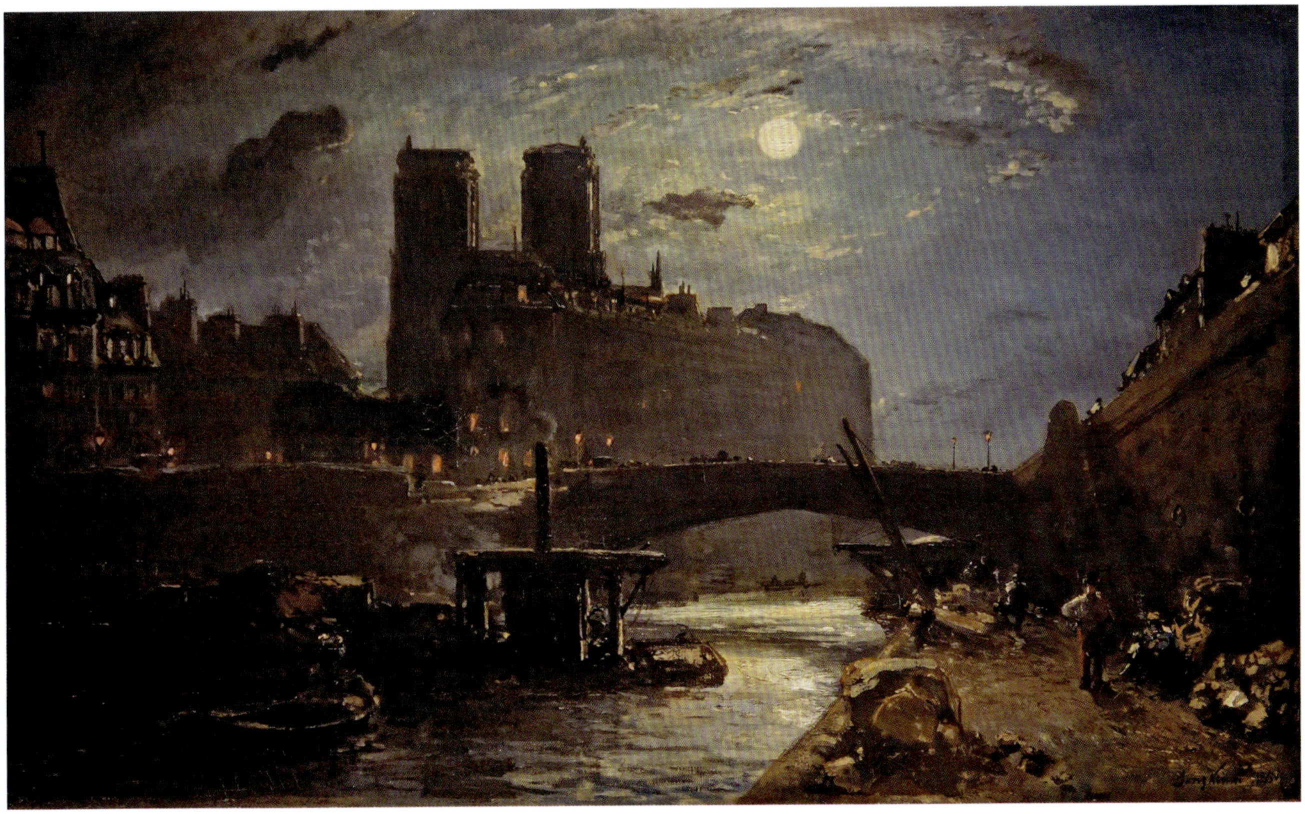

2.10 Johan Barthold Jongkind, *Notre Dame de Paris*, 1854. Reims, Musée des Beaux-Arts.

board of lights at mid-picture, punctuating and enlivening the otherwise gloomy, even sinister "natural" urban nocturne. The canvas is an ambitious orchestration of the coexistence of human and celestial light in moonlit Paris, complete with barely legible manual laborers on the *quai*, a forbidding setting.

In Sargent's nocturnes, the "battle" between *lumière* and *éclairage* is less pitched than the contest staged within Jongkind's canvas. Despite their commonalities, the evanescent tones of Sargent's garden pictures are worlds away from the undeniably grim chiaroscuro of Jongkind's nocturnal and largely premodern central Paris.[42] There is, nonetheless and unmistakably, a match, but the competing parties are delineated by means other than the strictly descriptive. Sargent's pictures provide the fundamental ingredients of a romantic twilight, but delicately disallow and disenchant the romance via both subtle and blatant means. Most explicit is the artist's attention to gravel and streetlights, but his handling of reflected moonbeams works against romance in a different way. I argue that it is the cleverly displaced glare of the nearby Jablochkoff candles that masquerades as excess moon glow on the boat basin.[43] It is a displacement; a metaphorical maneuver that muffles and pastoralizes a locus under assault by blazing experimental light machines during the summer of 1879. The argument is that the multiple instances in the park of hyper- and vulgar modernity seem to have intrigued him, and provoked him to devise a highly sophisticated metaphorics of reference. Or rather that it was the odd and composite modernity of this particular park that attracted Sargent and pushed him to devise a highly sophisticated metaphorics of reference.

Sargent's response to the multiply illumined twilight in the park also puts his paintings in line with jokes that were appearing in caricature during the era of maximum light volatility, gags that bear witness to the topicality of the contest between moonlight and industrialized light. In a caricature by Cham (Amédée Charles Henri de Noé, 1818–79) (fig. 2.11), for example, the competition between the full moon and Jablochkoff's light in particular is the crux. The caption reads: "La lune se réargentant elle-même pour paraître plus brillante que la lumière de M. Jablochkoff" ("The moon resilvering herself to appear more brilliant than Mr. Jablochkoff's light").[44] The moon is female, and she paints her round and smiling face from a pail of "solution d'argent" ("silver solution"). She employs a brush, evoking both the work of an artist and the womanly labor of "making up" a face.[45] Sargent's paintings could have been installed flanking Cham's caricature in a triptych entitled *The Moon Attempts to Outshine the Lights of Modernity*. There could not have been a more modern subtext for an urban nocturne in 1879 than the full moon rising at sunset on a perfect summer night attempting to outshine brash new lights on the street.

The competition between moonlight, always poetic, soft, and sad, and streetlight, invariably brash and upstart, was indeed a chestnut of journalistic humor. *Les Becs de gaz* by G. Lorin (fig. 2.12), from an 1883 issue of *La Vie moderne,* is a fantasy that shows a sorrowful moon kidnapped, and unhappily relegated to the glass lantern of a colossal riverside streetlamp to provide the light. The multiple vignettes of *Paris la nuit* by Loys (fig. 2.13), a centerfold from *La Caricature* (1881), also depend upon the difference between the sweetness of a benign moon (smiling down at upper right) and a glaring *réverbère* (dead center) that affords the conditions in which Parisians have the enhanced visibility in which to behave badly or to be vulnerable

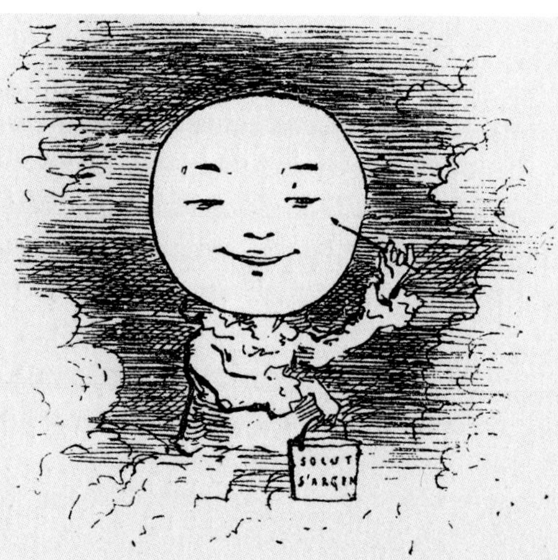

Fig. 33. — La lune se réargentant elle-même pour paraître plus brillante que la lumière de M. Jablockhoff.
(Dessin de Cham, 1880; collection de l'auteur.)

2.11 Cham, "La lune se réargentant elle-même . . ." (1880). Figure 33 from M. Eugène Defrance, *Histoire de l'éclairage des rues de Paris* (1904). Courtesy of University of Wisconsin Library.

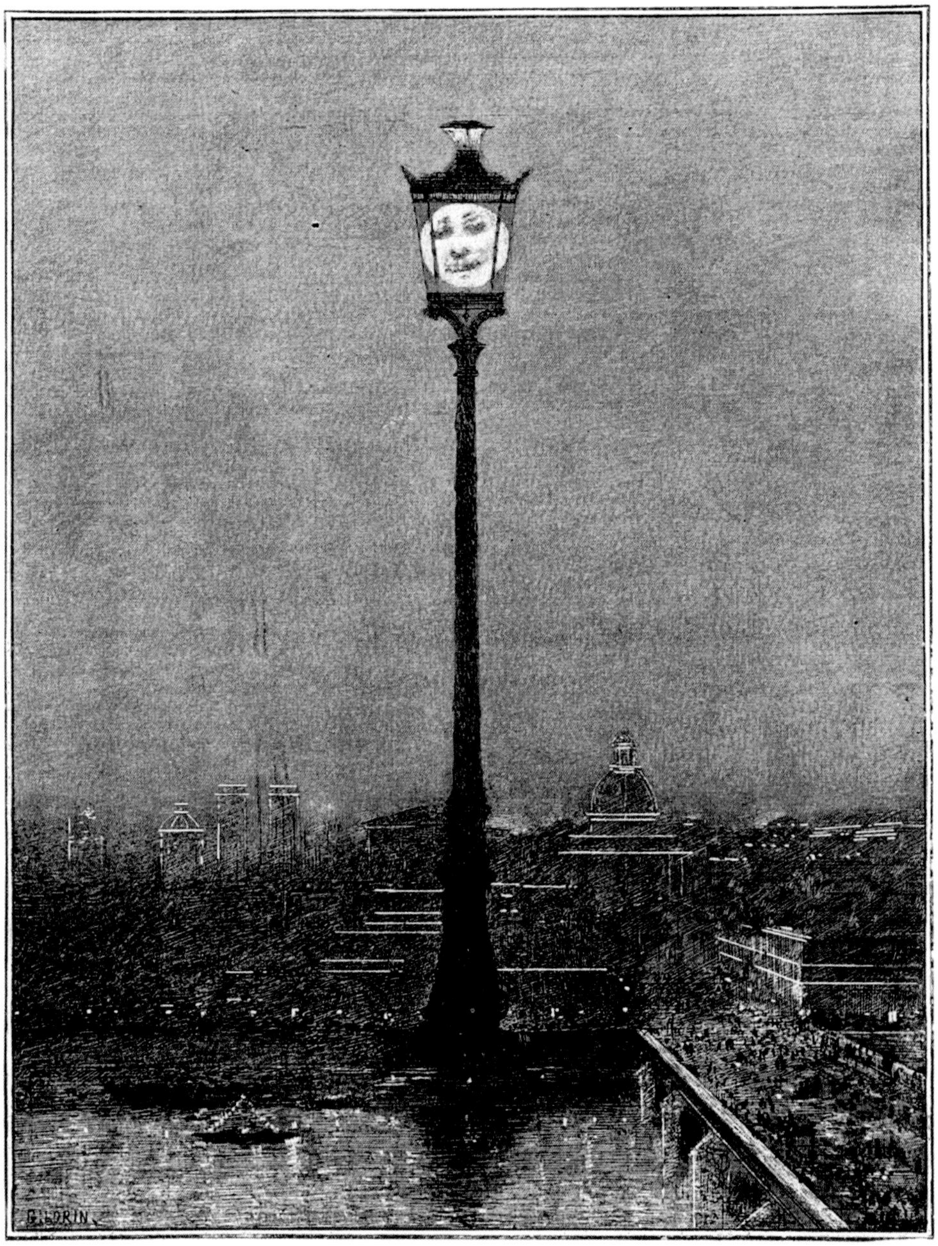

2.12 Drawing by G. Lorin, engraving by Michelet, *Les Becs de gaz*, *La Vie moderne* 16 (April 21, 1883): 261. Photo © Musée Carnavalet/Roger-Viollet/The Image Works.

to risks. As the print puts it with regard to the moon, "Chaste Phoebe," at upper right: "Why chaste? More numerous than the stars in the sky are the criminals to whom she has lent her obliging light since the world has been the world and Phoebe has illuminated it."[46]

It was a kind of competition bruited in somewhat different terms when the first arc lights shone in Paris during the July Monarchy. In response to the evidently more benign electric shafts of light of 1840, an unnamed commentator for *L'Illustration* wrote in terms that continued to resonate, and to inform the practices of both Sargent and Cham thirty-nine years later: "Electric light, which has a pallid color,

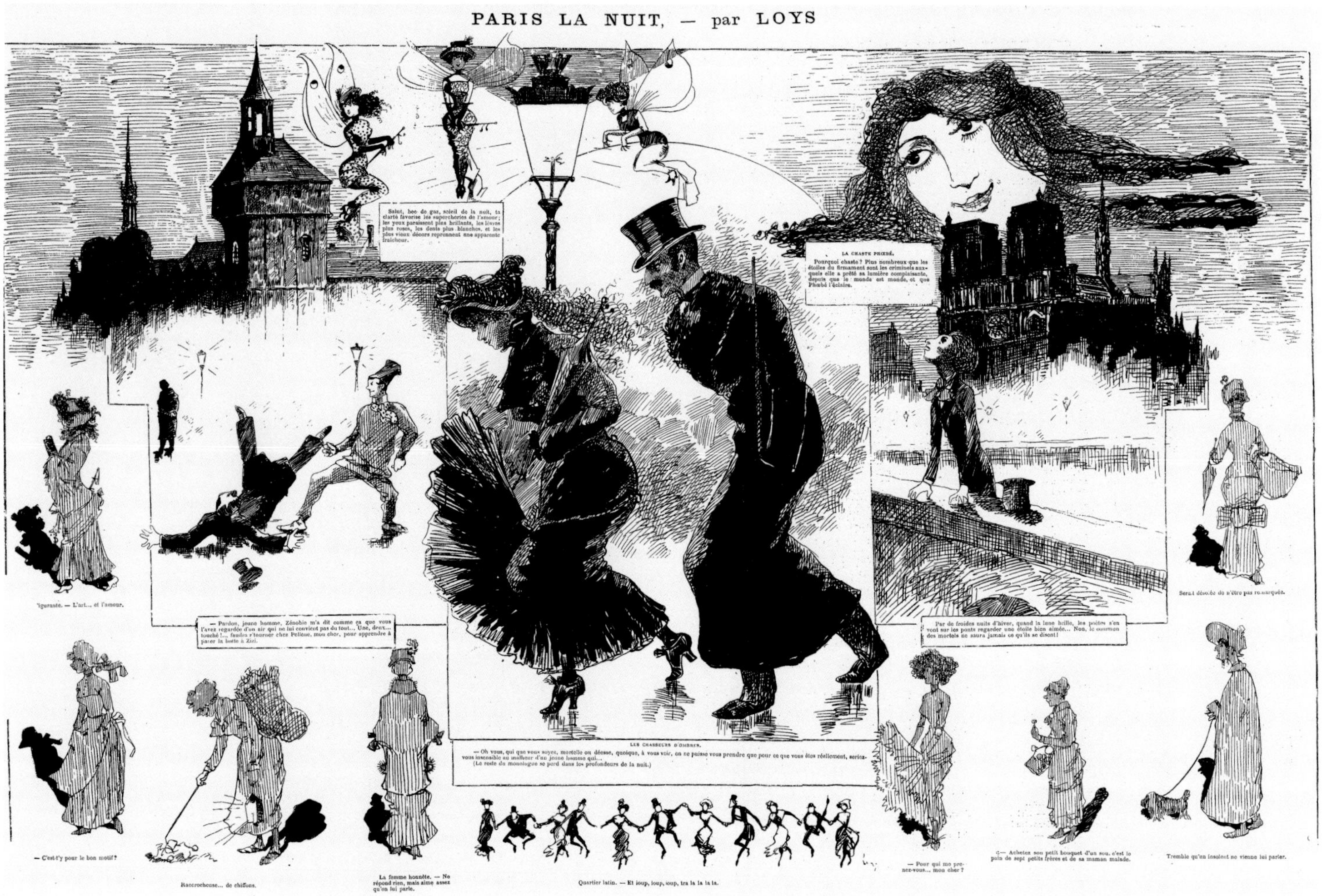

2.13 Loys, *Paris la nuit, La Caricature*, no. 63 (March 12, 1881): 84–85.

like that of the moon, projected its beams that by no means tired the eyesight. . . . You could read very easily thanks to this luminous beam, at a distance of 150 meters."[47]

Even if my readers accept the portion of my argument that links Sargent's canvases to the thematics of Cham's caricature, they may nevertheless resist a reading of the painted dazzle on the water of the two boat basins as displaced Jablochkoff brilliance. Indeed, the topographically perspicuous might balk in view of the distance between the sidewalks of the rue Soufflot and the garden's boat basin. (Consult the *quartier* map, fig. 2.4.) Another way to understand Sargent's delicate network of lights, both transmitted and reflected, requires putting even more pressure on Cham's witty contest. His example has emboldened me to imagine a cheeky pronouncement by the twenty-three-year-old painter.

> You want an up-to-date Paris nightscape? One that does not shy away from recording natural light and artificial lights? Twilight *and* gaslight? Moonlight *and* arc light? Mr. Jablochkoff's blaze almost jeopardized my campaign in the garden, but I

fought back by heightening the intensity of the moon's reflection in the water. Now there's your modern nocturne.

Sargent's imaginary speech opens another path to the glimmering boat basins: one that places less emphasis upon the putative operations of displacement, and more upon the brightness of the clotted, white-yellow patches of dazzle as a blow struck on behalf of the moon and her light. This line of argumentation would mesh well with a certain reading of the paintings' middle-ground, newspaper-reading men. Reading a newspaper after the sun has set by struggling to catch rays reflected by the moon is both somewhat pitiable and comical. There is thus something of the caricatural in these broadly brushed, generalized, black-clad characters endeavoring to pore over their papers in the gloaming.[48] One way to understand the stressed perusers is to read them as blows struck on behalf of the viability of the illuminating force of moonlight. Why exit the park to read in the abundant illumination of streetlight close by, when moonlight, the old light, the time-honored source, still functions well enough? This synchs up with an approach to the boat basin reflections that would hold that paint was laid on thickly to shore up the effectiveness of the moon's struggle to excel even in the discouraging conditions of industrial disenchantment, to outshine the technologized *éclairage* of modernity—whether gas or electric. Even Cham's moon, after all, had to resort to a technologized chemical enhancement (brushed-on silver) to glow with sufficient brilliance in the modern world. The cartoon moon's silver solution resolves the contest between natural and artificial light by bringing them into convergence; by collapsing them into one another.

In the Luxembourg canvases, Sargent grapples with yet another critical component of urban modernity: after about 1850 in Paris, up-to-date leisure primarily happened at night because of its infrastructural requirements.[49] In the two garden scenes, Sargent succeeds, against the odds, in figuring metropolitan leisure in an entirely innovative way. It is set neither in daylight nor in the expected spaces of nocturnal leisure—neither the commercial interior nor the street. We have already observed the contrast between the cramped, middle-ground zones of substantial darkness and the brighter, clearer, and sprawling foreground. The dissimilarity between the inhabited spatial strata of the paintings defines multiple orders of time and light in the two works. He thus made astute choices in the ways he fashioned both the place and the time of day: a brutally modernized park filled with incongruous Parisians depicted on the cusp of night is the setting for a rather melancholic evocation of a world being lost as a darkening but moonlit park is invaded by shadowy inhabitants and threatened by new lights.

"Il s'argent" (he silvers himself) indeed.[50] Cham's "solution d'argent" and the moon "qui se réargent" (who resilvers herself) inspired the overdetermined and irresistible pun on the artist's surname. The pun functions effectively to acknowledge the painter's francophone cosmopolitanism, but, more pointedly, the linguistic jest goes straight to the heart of the paintings' evanescent and multiple luminosities, diagnostic features of modernity that informed Sargent's two Paris moonlights.

·3·

Bright Lights, Brilliant Wit

ELECTRIC LIGHT CARICATURED

Caricature is a double thing;
it is both drawing and idea—
the drawing violent, the idea
caustic and veiled.

CHARLES BAUDELAIRE, 1855[1]

New Luminosities and Comic Art

The new and varied *éclairage* that altered the city's visual environment in the final decades of the nineteenth century gave birth to lively social scenarios. This state of affairs was a godsend to graphic artists, especially those working in a comic vein. Inspiration was ubiquitous; there seemed to be new lights blazing every day. Umbrella-like electric lamps sprang up in the place du Carrousel of the Louvre (fig. 3.1). Four enormous electric light fixtures were strung on wires over the boulevard des Italiens (fig. 3.2).[2] Jablochkoff globes burned in the interior of the department store Les Grands Magasins du Louvre on the rue de Rivoli, whose lamps cast beams onto the adjoining sidewalk through the generous windows of the Porte Marengo (fig. 3.3). Who could keep up? What graphic artist could resist the visual dazzle and human bewilderment?

Such newly illumined commercial interiors and streets, and recently illuminated exhibition spaces were significant sources of humorous imagery. The spectacular one-off exhibition showcasing electric lights held in 1881, and a range of artificially illuminated art displays in museums and galleries, had a galvanic effect on caricature. The coincidence of novel illuminations and the polyvalent celebrity of electric light's metonym, Thomas A. Edison, was also providential: the confusing effects of the new lighting and the risibility of the legends attaching to Edison sparked new lines of wit. Humor provoked by dazzling environments, on one hand, and the Genius of Menlo Park, on the other, revivified two of the mainstays of earlier Parisian caricature: opportunities for sexual mischief in Paris and the cluelessness of Americans.

The concurrence of the era of nonstop innovation in electric light and the 1880 revival of *La Caricature* by Albert Robida (1848–1926) was a boon for Parisian visual culture.[3] Robida's journal is our point of entry into the thematics of electricity and electric light in the comic visual arts. The publication's razor-sharp visuals shine a piercing light on some of the signature beliefs and preoccupations of the Parisian sociocultural scene around 1880, at least in the eyes of the subscribers to Robida's eight-page weekly.

Hilarious and ludicrous situations fostered by new lights as well as other electric contraptions may have motivated Robida's publishing venture. The journal's specialization under his direction was *la caricature des mœurs* (the caricature of morals and manners), an intentionally less political program than that followed by the 1830s publication of the same name edited by Charles Philipon (1800–1861).[4] The front page of the June 19, 1880, issue, an amalgamation of picture and text by Robida himself, showcased the journal's prowess in the realm of social caricature. "Nouvelle et Merveilleuse Invention d'Edison" ("New and Marvelous Invention by Edison";

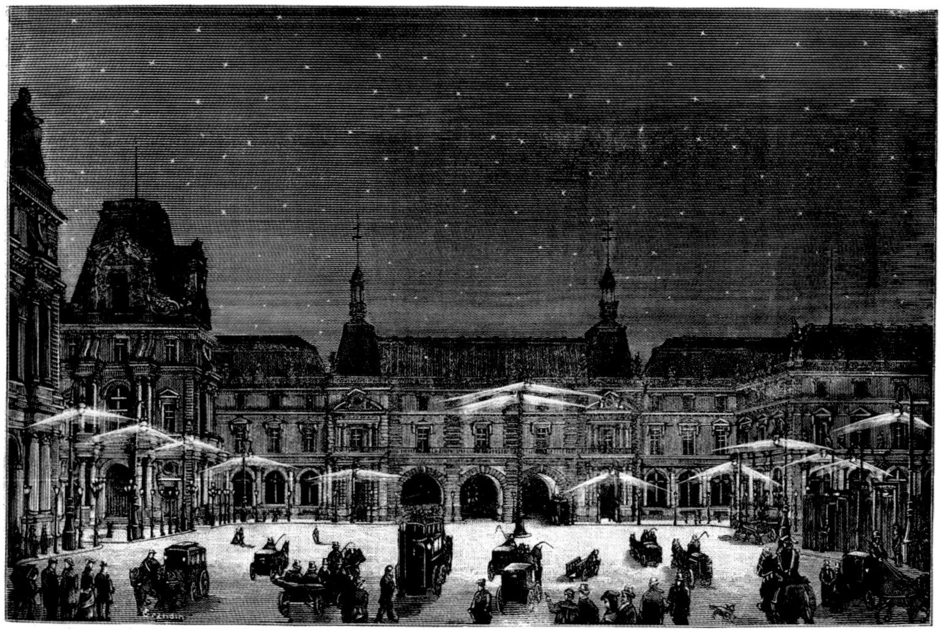

3.1 "Éclairage électrique de la place du Carrousel." *La Lumière Électrique: Journal universel d'électricité* 66 (November 16, 1881): 239. University of Chicago, John Crerar Library.

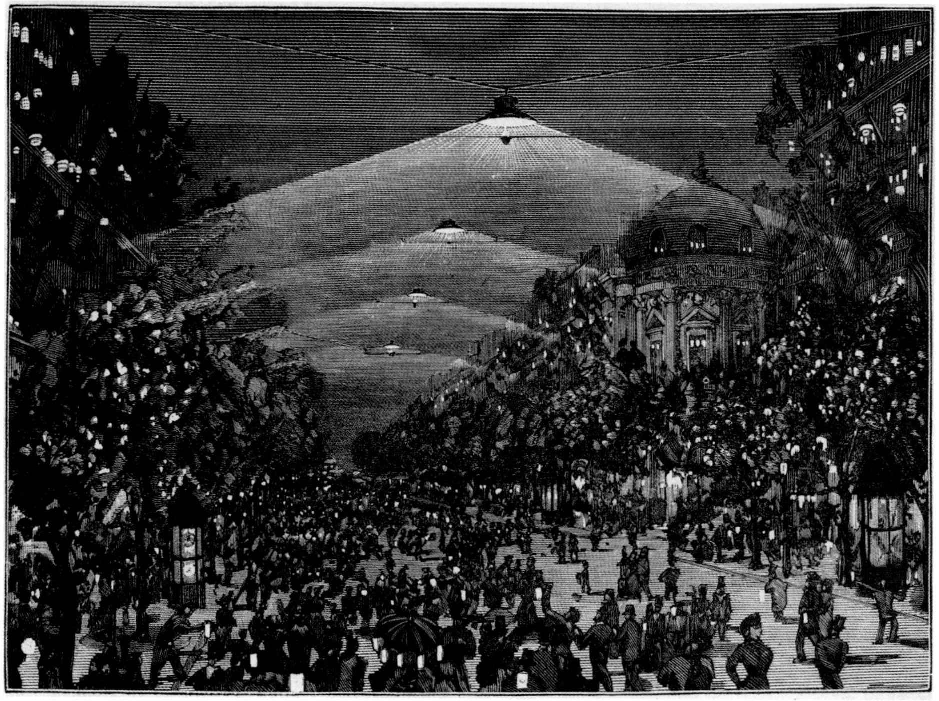

3.2 "Éclairage électrique du boulevard des Italiens par les Lampes Million." *La Lumière électrique: Journal universel* 36 (August 3, 1881): 153. University of Chicago, John Crerar Library.

fig. 3.4), a tour de force of the humorous imagination rooted in actuality, starred the wildly charged-up Thomas Edison in his laboratory. The date of June 1880 was well into the flowering of the American inventor's transatlantic reputation as the Wizard of Menlo Park, a term used famously by Villiers de l'Isle Adam in his novel *L'Ève future*, begun in 1878. The successful test of the incandescent Edison bulb in New

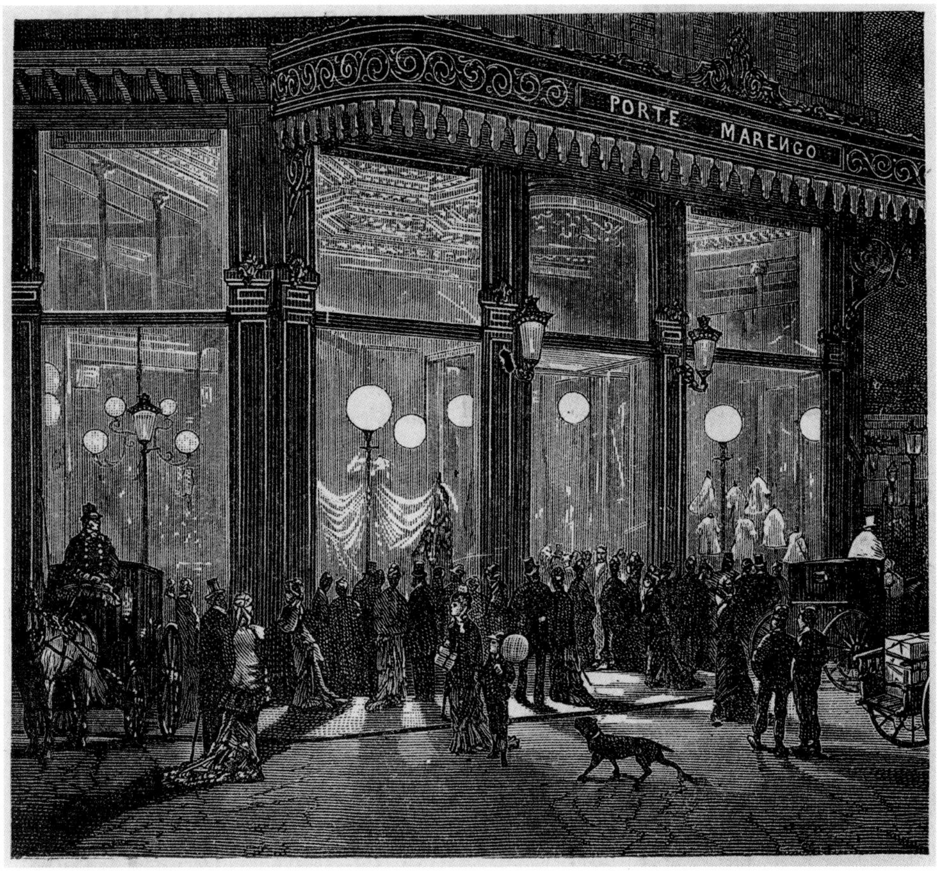

3.3 "Les bougies Jablochkoff au (Magasins du) Louvre." Figure 118 from Henri de Parville, *L'Électricité et ses applications. Exposition de Paris* (1882). University of Chicago, John Crerar Library.

Jersey in late 1879 secured Edison's reputation as electric light's flashiest prodigy, and dispelled most of the doubts that had governed the thinking about him in the French *électricien* crowd.[5] His standing in the French capital rose even higher when his contributions to the 1881 International Electricity Exhibition grabbed the headlines, even though Edison himself did not travel to Paris until 1889.

In Robida's front-page scenario (fig. 3.4), a fiercely determined Edison has invented a clever contraption that harnesses the omnipotent force par excellence, electricity, to satisfy the husbandly need to track and control the romantic meanderings of unsupervised wives. The device, brilliantly named, was the Fidèlimètre designed "to indicate the degree of faithfulness of women" ("à indiquer le degré de fidélité des dames") by tracing the ups and downs of perpetually erratic young women, the temperamental twins, albeit married, of Verdi's "La donna è mobile" (*Rigoletto*).[6] A woman would sport the gadget as if it were a flattened pocket watch, pinned like a brooch or dangling from a fob. The dial registers her actions and intentions between two extremes, from PARFAITE (perfect) to DÉSASTRE (disaster).

The women tricksters modeling the device are Americans, exemplary feral women. A joke defining Edison as the *ne plus ultra* of technological wizardry unsurprisingly derides American mores at the same time. Admiration for Edison's gadgets

Numéro 25. PRIX : 30 CENTIMES 19 Juin 1880.

A. ROBIDA
RÉDACTEUR EN CHEF

La Caricature

PUBLICATION
DE LA
LIBRAIRIE ILLUSTRÉE

Abonnements d'un an, Paris : 16 francs. — Départements : 18 francs. — Union postale : 20 francs. — Bureaux, 7, rue du Croissant.

NOUVELLE ET MERVEILLEUSE INVENTION D'ÉDISON
LE FIDÉLIMÈTRE

L'homme extraordinaire, l'immense savant auquel nous devons déjà nombre d'étonnantes inventions, M. Edison, vient encore de se signaler par un coup d'éclat foudroyant. Après le téléphone et le phonographe, il s'agit aujourd'hui d'une découverte destinée à révolutionner le monde tant par son éclatante nouveauté que par ses incroyables et incommensurables conséquences *scientifiques*, SOCIALES et surtout MORALES.

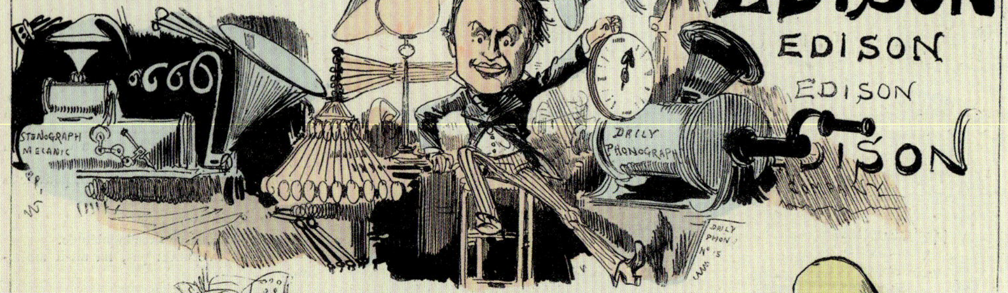

Cette nouvelle invention c'est le FIDÉLIMÈTRE, servant à indiquer le degré de fidélité des dames. On sait que jusqu'à présent ceci avait échappé aux investigations de la science. Edison est venu ; brisant les entraves de l'antique routine, il a cherché et il a trouvé !

Le *Fidélimètre* est un cylindre plat, ressemblant beaucoup à une montre en or, argent ou nickel. Sa face supérieure est garnie d'un verre de montre, sous lequel oscille une aiguille d'une extrême sensibilité.

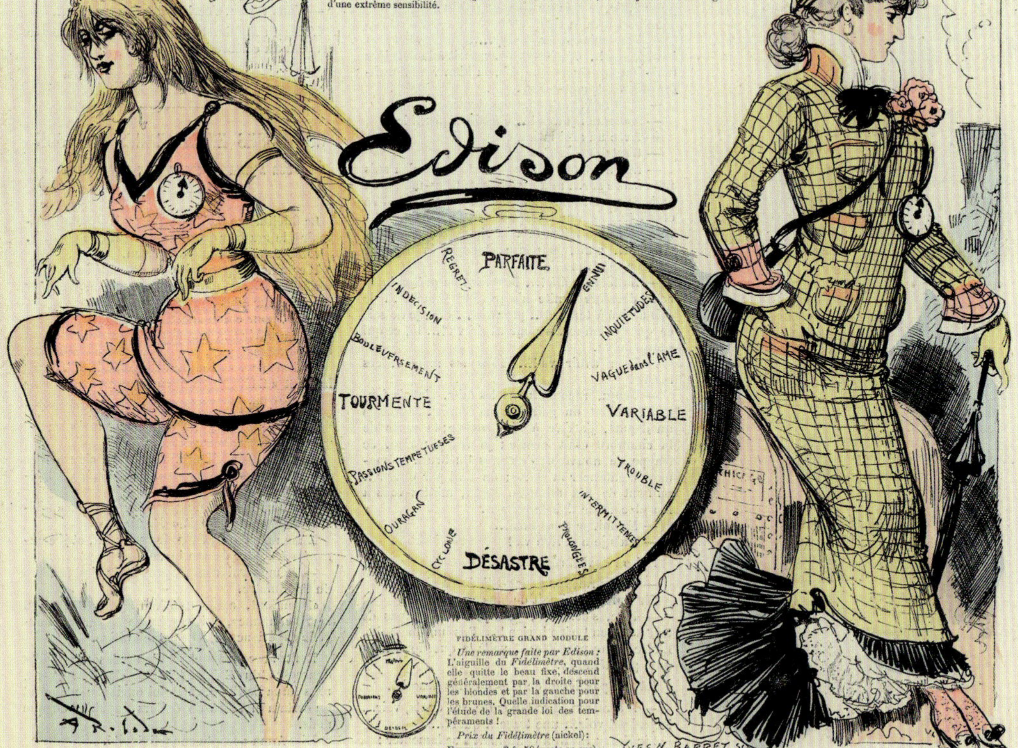

FIDÉLIMÈTRE GRAND MODULE

Une remarque faite par Edison : L'aiguille du *Fidélimètre*, quand elle quitte le beau fixe, descend généralement par la droite pour les blondes et par la gauche pour les brunes. Quelle indication pour l'étude de la grande loi des tempéraments !

Prix du Fidélimètre (nickel) :

France. . . 2 fr. 50 (poste en sus).
Amérique. 12 dollars.

UTILITÉ. — BAINS DE MER

L'oncle Sam est un malin, chacun sait ça. Dès la première apparition du *Fidélimètre*, toutes les épouses de l'Union en ont été pourvues. Quel objet d'ailleurs est d'une utilité plus incontestable ?

A l'ouverture de la dangereuse saison des bains de mer, qui font courir tant de risques aux maris retenus à la ville par les affaires, le *Fidélimètre* a été adopté avec enthousiasme par toutes les épouses de la fashion américaine.

L'utilité incontestable du *Fidélimètre* n'échappera, nous l'espérons, à personne. Nous sommes heureux d'être les premiers à le faire connaître à Paris. Une Compagnie financière se forme pour l'exploitation, en France, du brevet de l'invention Edison ; les actions, non encore émises, font déjà 128 fr. 50 de prime.

VOYAGES

La jeune américaine voyage beaucoup ; pendant que son mari siège à son comptoir ou dans les assemblées politiques, elle court le monde en toute liberté.

Mais le *Fidélimètre* est un témoin et une garantie. Toutes les semaines, une photographie du *Fidélimètre* est envoyée au mari, et, à la moindre déviation, celui-ci envoie, par télégraphe ou téléphone, l'ordre de rentrer *at home* !

had, moreover, to be undercut by the frequently fierce transatlantic economic and cultural struggle over the ownership of electricity. The gag is leavened by another trope of Parisian boulevard humor: American women are cheeky and insolent compared to *les françaises*.

The corpus of caricatural responses to the new lighting scenarios in *La Caricature* and elsewhere tests some of the leading theories of the distinguishing achievements of the modern idiom of caricature. On one end of the theory axis, caricature is defined as an inherently democratic and potentially subversive genre, and thus a potent tool of counter-discourse and ridicule.[7] At the opposite end, caricature is construed as an irretrievably conservative mode whose purpose is classification and the recycling of types.[8] Attentive to Baudelaire's caveat that caricature is characterized by doublings (*dédoublements*), both in its comic form and as a spur to laughter, this interpretation navigates between the poles of caricature theory, but often gravitates toward the latter, defining caricature as a mode that reinforces traits of preestablished types.

Returning to Robida's busy front page: at right is the urbanite with a Chicago sticker on her steamer trunk; a train belching steam is close by. Under the heading VOYAGES, she is dressed sharply to roam freely.

3.4 Albert Robida, "Nouvelle et merveilleuse invention d'Edison." *La Caricature* 25 (June 19, 1880). Courtesy of Northwestern University Libraries.

> The young American travels a lot; while her husband sits at his counter or in political assemblies, she runs around the world in complete freedom. But the Fidèlimètre is a witness and a guarantee. Every week, a photograph of the Fidèlimètre is sent to the husband, and, if there is the least deviation, he sends, by telegraph or telephone, the order to return home.[9]

The other young American, bounding into the surf in bathing costume with long hair loosely streaming, demonstrates the benefits of the device at the seashore under the heading, UTILITÉ—BAINS DE MER (Usefulness—Seaside Resorts).

> Uncle Sam is cunning, everyone knows that. As soon as the Fidèlimètre was available, all the spouses in the States were provided with them. Besides what object has a more incontestable usefulness? At the opening of the dangerous season at the seashore, which runs so many risks for the husbands stuck in the city by their business affairs, the Fidèlimètre was adopted with enthusiasm by all the spouses in the American fashion.[10]

The social and technological modernity of the contraption was thus all-inclusive. Nothing less could be expected from the capacious mind of Robida, an artist whose ability to imagine the future was unparalleled.[11] The Fidelity Meter was a model specimen of the American inventor's outsized talent seen through the lens of Robida's comic dexterity, inasmuch as this mechanical brainchild commingled the resources of an electric sensor, photography, telegraphy, and the telephone. Robida nonetheless trivializes Edison's futuristic intermedial contraption. Indeed, a clear sign of Robida's comic dexterity is his ability to lionize and belittle Edison's resourcefulness simultaneously, all the while keeping the laughs flowing. All that know-how, the jest starring two American women suggests, is marshaled merely to perform a housekeeping task: keeping irksome *Américaines* in line. Its hilarity and originality drink deeply from the wellspring of Parisian stereotypes—about American men (credulous cuckolds) and women (morally lax), and the upstart inventor himself (a wily genius solving a trivial problem). Taxonomization is in operation.

Female viewers who may admire the agency of the Americans who are on the verge of outfoxing authoritarian spouses are not allowed the pleasure of positive identification and recognition because the mobile women are imagined to be under the control of an electric behavior tracker. Looked at by women similarly constrained, the scenario may foster the humor of painful recognition, which is an inherently conservative social form, according to Laura Kipnis. If cathartic laughter follows, "the laughter depends on our recognizing the world as it is, and leaving it the way we found it. Like all cathartic laughter, it questions nothing."[12] The jokes are funny nevertheless, and Edison's demonic look may foretell a malfunctioning Fidelity Meter.

Laughing at Lights in the Street

The piercing glare of the electric street lights that shone in prominent locations between 1877 and 1882 was a rich source of inspiration for both Draner (Jules Jean Georges Renard, 1833–1926) and the indefatigable Cham (Amédée Charles Henri de Noé, 1818–79).[13] In an apparently unpublished drawing (c. 1879), Draner explored a complaint about brightly glowing clusters of ball-globed arc lights on the traffic islands of a large boulevard (fig. 3.5). Two men dressed as *voyous* (thugs) on a third traffic island discuss how annoying "they" are with "their" electric light. (Pointedly, the lights are not "ours.") The fellow at left likely speaks the second line. "[O]n ne sait plus où aller travailler!" ("Due to the glare, now I can't even see where I'm supposed to work.") The caption aligns electrical brilliance in the newly lamped City of Light with diminished rather than improved visibility. But it is an exchange between *voyous* (hooligans; petty criminals), not workers.[14] They cannot see where to pull off the next job thanks to "their" lights. The laments of two urban outlaws convert the usual sober discussion of enhanced productivity under bright light into a cheeky insider joke.

Cham exploited the comic potential of blinding glare in numerous lithographs, most of them set in spaces adjoining Charles Garnier's new Opéra, opened in 1875. The avenue de l'Opéra was inaugurated on September 19, 1877, and the very same evening, the façade of Garnier's Opéra was illuminated by Jablochkoff carbon arc lights.[15] As we have already learned, the avenue itself was lined with Jablochkoff candles in 1878.[16] Garnier was a passionate advocate of electric light, determined to light the interior of his lavish building accordingly.[17] Indeed, on October 18, 1881 (during the run of the International Electricity Exhibition across town), according to

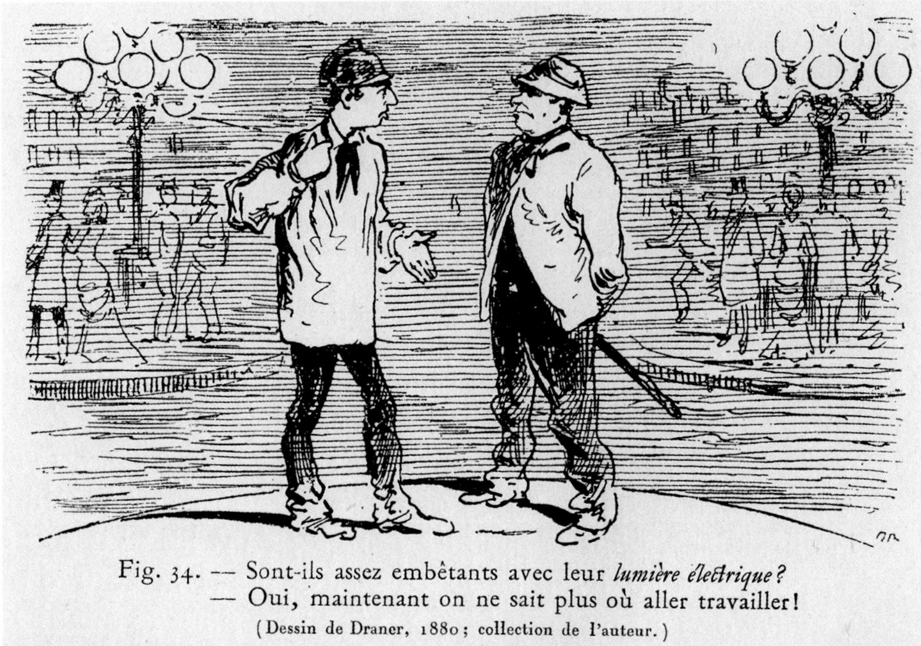

3.5 Draner, "Sont-ils assez embêtants . . ." (1880). Figure 34 from M. Eugène Defrance, *Histoire de l'éclairage des rues de Paris* (1904). Courtesy of University of Wisconsin Library.

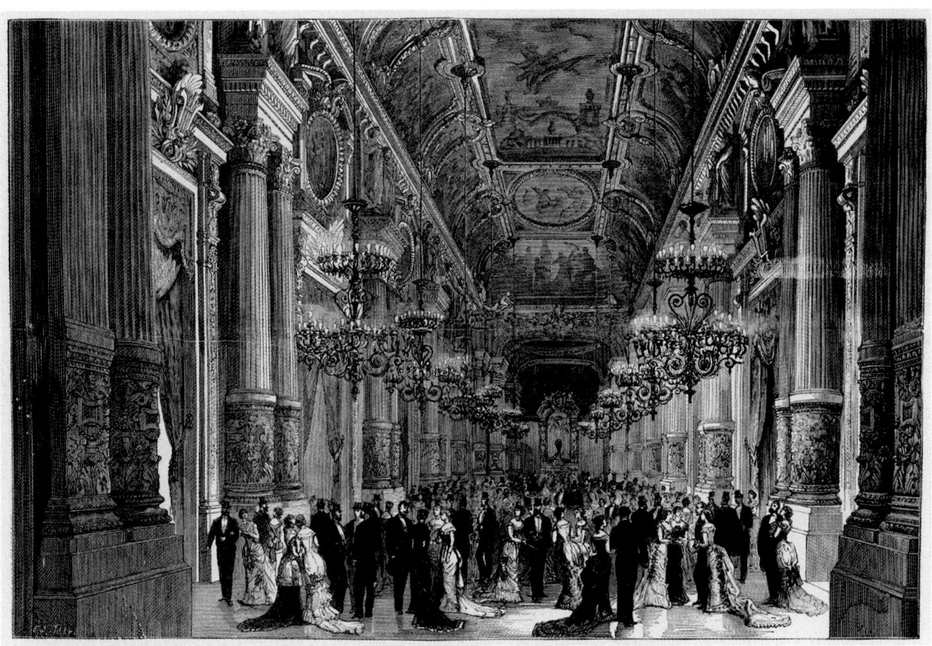

3.6 "Le Foyer de l'Opéra éclairé à la lumière électrique (Système Edison)." *L'Illustration* 2021 (November 1881): 341. Courtesy of Northwestern University Libraries.

architectural historian Christopher Mead, "the entire public half of the Opéra, exterior as well as interior, was illuminated by systems invented by Jablochkoff, Edison, Swan, and others" (fig. 3.6).[18]

Cham's attraction to the Opéra building and its surroundings had two bases: first, the location was a high-profile site of electric arc lighting, and second, the comte de Noé was a close personal friend of Charles Garnier.[19] In Cham's c. 1879 drawing (fig. 3.7), a multiglobe electric arc streetlamp has taken a heavy toll on passersby. The nighttime glare in the *place* has blinded all three men in top hats walking dogs. But the animals are not pets. The leashed dogs gloss the predicament of the impaired strollers. The increasing numbers of blind Parisians unleashed a surge in the numbers of seeing-eye dogs. Cham cast service dogs in another comic scenario (fig. 3.8), set directly in front of the Opéra (labeled Académie de Musique). Indeed, a full-fledged "station de caniches" ("poodle station") adjoins the *candélabre* in the left middle ground. The joke turns on a nocturnal stroller's preemptive need for a specialized dog as he was—or inevitably would be—blinded by the plaza's lamps.

"Une ombrelle le soir pour traverser la place de l'Opéra" ("An umbrella to cross the place de l'Opéra in the evening"; fig. 3.9) revolves around the quintessential deracinated urban eye, the umbrella deployed at night when the weather is dry and the sun is down. In the trio of struggling walkers, the torsion in one man's stance and the severe tilt of the woman's body signal the physical exertion required to endure the night's penetrating rays. The postures also amusingly emulate bodies beset by heavy rain or strong winds. "Aveuglement de la place de l'Opéra" ("Blindness in the place de l'Opéra"; fig. 3.10) references eye specialists often imagined to benefit directly from the wounding intensity of the new lights. The witty caption identifies the damaging light as a scheme by unemployed oculists: "Éclairage électrique fondé par une société d'oculistes sans ouvrage" ("Electric light established by a society of unemployed oculists").

Cham's drawings pose incisive questions about audience, not defined sociologically, but in Baudelaire's sense of concern for the locus of laughter. For Baudelaire, the wellspring of humor is laughing at another's misfortune.[20] In Cham's scenarios of light-caused inconvenience, who is the laugher and who is laughed at? And to what effect? Are the bourgeois pedestrians battling the environment the laughers or the targets of the joke? Is electric light under assault in these drawings, or are its opponents made to appear ridiculous? The answer is that the bourgeois lampooned in Cham's work is not literally caricatured; that is, he is not given exaggerated features. In the context of Cham's gentle humor, the bourgeois *dramatis personae* are encouraged and expected to laugh at themselves.[21]

Doctors of the era understood the correlation between bright light and eye damage or compromised vision—a standby of both comic art and popular journalism—to be less certain. Dr. N. Théodore Klein, Parisian opthamologist, for example, posed the key question in 1873: "Must we admit, with all those who have written on this subject, that visual acuity diminishes when the light is too strong?" His initial answer seems to assert that exposure to intense light produced ocular damage. "Very intense lighting, like electric light, produces a painful excitation that few people can even bear." But it quickly becomes clear that in Klein's judgment the case against electric light was not the open-and-shut one delivered by the caricature and anti-electricity crowd. Klein, in fact, perceived a reversal: "But we have always observed, following the first moment of dazzle, an augmentation of sharpness."[22]

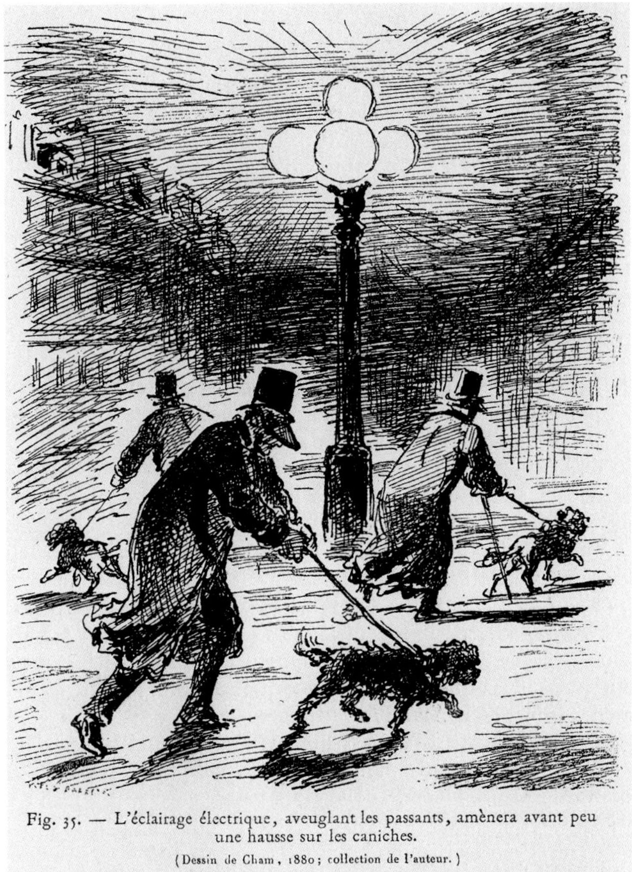

3.7

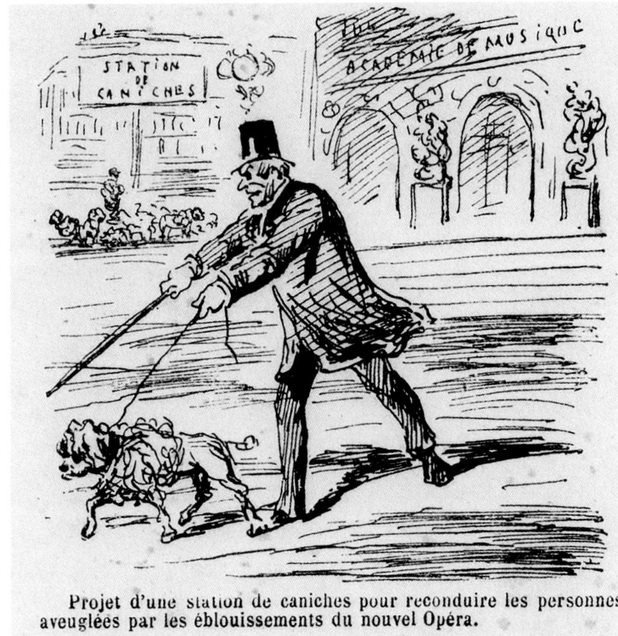

3.8

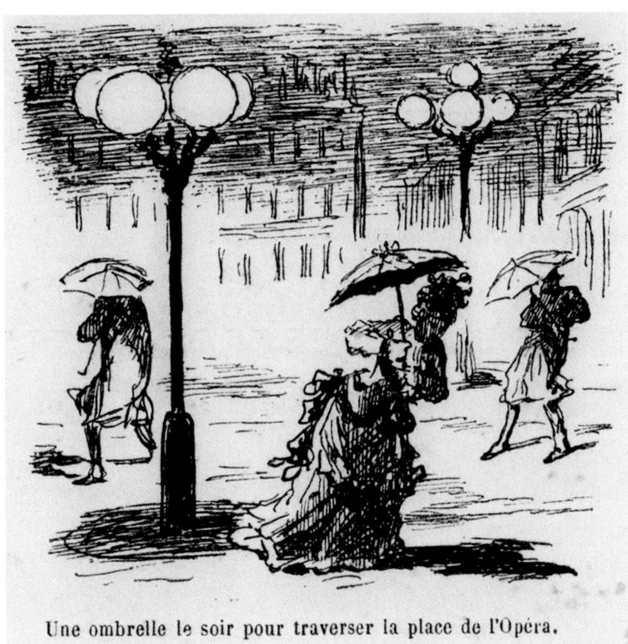

3.9

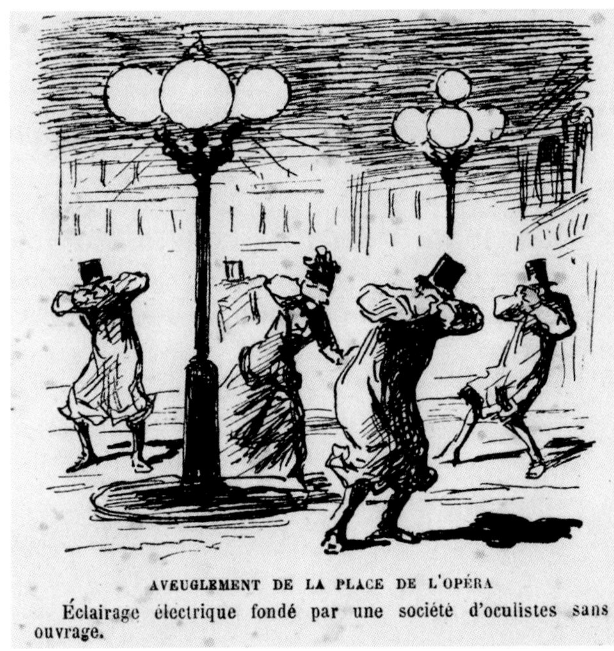

3.10

3.7 Cham, "L'Éclairage électrique . . ." (c. 1879). Figure 35 from M. Eugène Defrance, *Histoire de l'éclairage des rues de Paris* (1904). Courtesy of University of Wisconsin Library.

3.8–3.11 Cham, illustrations from "De tout et partout." In *Les Folies parisiennes par Cham: Quinze années comiques 1864–1879* (1883).

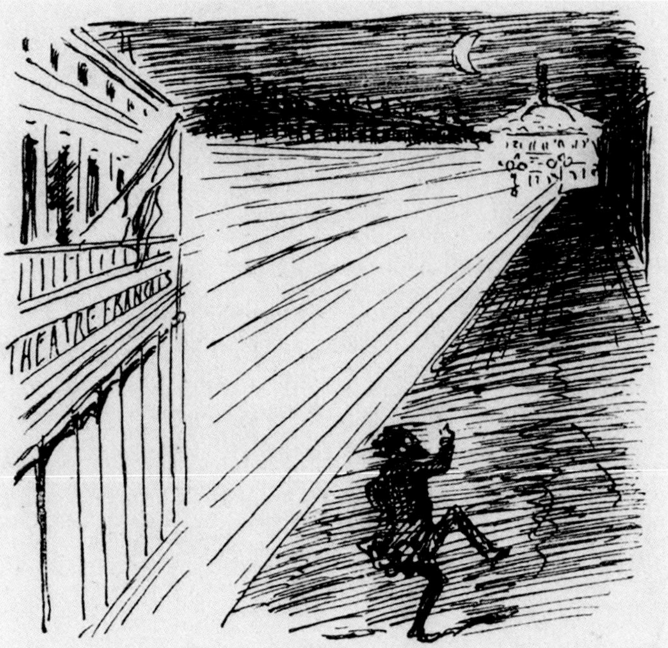

Joie de M. Perrin, le directeur des Français, en voyant son théâtre éclairé aux frais de l'Opéra, en face.

3.11

"Joie de M. Perrin" (fig. 3.11), addresses the theme of glare once more, but in a different key. Rather than joking about the somatic price paid by Parisians down on the ground in the immediate vicinity of prestigious musical culture, this print envisions the Opéra district seen from above occupied by a singular Parisian, the director of the Théâtre Français. Cham dissolves the avenue into a blazing river of electric light, a luminous ribbon that shoots down the street's full length. Like a ramrod-straight tail of a comet, it cuts a broad swath across the second and first arrondissements to connect two celebrated cultural institutions. The shimmering electric candelabras in the *place* before the façade of Garnier's building appear to have generated a luminous stream that obliterates the avenue, erasing all its fixtures and features. What's more, as a result, the Théâtre Français is ablaze for free: "éclairé aux frais de l'Opéra" ("lighting courtesy of the Opera"). The only competing light is the crescent moon shining in the dark sky—artificial light's perpetual antagonist and the image's sole marker of natural night. M. Perrin, the director of the Théâtre Français, dances for joy just outside the stream of light, exuberantly alone alongside the radiant avenue. His energetic dance combines delight, surprise, and chagrin. Cham's clever concoction enables M. Perrin to appear both discomfited by the miracle of light and intoxicated by it simultaneously. The observer would smile, I argue, but not laugh. By fancifully dissolving an avenue in light, Cham pictured the inglorious optical traits of a blaze of Jablochkoff streetlights in the realm of aesthetic form in exceptionally inventive terms.

Éclairage in the Spaces of Art

Although one may be tempted to mock the experiments of the nineteenth century, this would be more than a little unfair: even today, complaints about the poor lighting in museums can be heard all over the world.

ANDREAS BLÜHM AND LOUISE LIPPINCOTT, 2000[23]

"Le Salon Nocturne," drawn by Draner, was the eye-catching front page of the May 29, 1880, issue of *La Caricature* (fig. 3.12). It addressed the effects of one of the newest public uses of electric light, the evening illumination by arc lights of the Paris Salon, the enormous annual state art exhibition held in the Palais de l'Industrie (the 1855 structure also known as the Palais des Champs-Elysées). The 1880 Salon was only the second illuminated by electricity—the first was the previous year—but it clearly remained newsworthy enough to deserve comic treatment in 1880, the inaugural year of Robida's paper.[24] Tracing the history of artificially lit art spaces, including the electric Salons, is necessary in order to do justice to Draner's complicated nocturnal Salon.

To begin at the beginning: what is meant by artificial light? The definition of "artificial" in the *Nouveau Larousse illustré* demonstrates that in the later nineteenth century the link between the artificial and art, on one hand, was closely connected to the bond between the artificial and light, on the other. To wit, the first entry under "artificiel, elle" (artificial) is "fait avec art" (made with art). The second is "lumière artificielle" (artificial light), or rather light mediated or enabled by a contraption or transformative process.[25] Both of the primary definitions of "the artificial" stress the transformation of nature. Keep this in mind as we encounter the widespread opinion that electric light was inimical to art.

It is certain that there were art exhibitions in Paris illuminated for evening viewing before the electric Salons, and gas very probably fueled their lights. The first Impressionist exhibition, for example, held in Nadar's former studios at 35, boulevard des Capucines in 1874, was lit artificially to prolong opening hours beyond the day into the evening (from 8:00 until 10:00 p.m.).[26] Accounts of the exhibition do not specify as much, but the light must surely have been gas or even oil, because while those flame-based illuminations were relatively common indoors in 1874, electric interior lighting would not make inroads until 1878–79 at the very earliest, and then only in commercial spaces. It remains difficult, nonetheless, to believe that art history does not know this for certain. Following directly upon a discussion of electric arc light, Pierre Pinchon writes suggestively: "Thus, the first Impressionist event benefited from artificial light in 1874, which enabled the viewing of artworks after sundown."[27] Without further evidence, the significance that the reader should assign to his opening "ainsi" (thus) remains in doubt.

Numéro 22. PRIX : 30 CENTIMES 29 Mai 1880.

A. ROBIDA
RÉDACTEUR EN CHEF

La Caricature

PUBLICATION
DE LA
LIBRAIRIE ILLUSTRÉE

Abonnements d'un an, Paris : 16 francs. — Départements : 18 francs. — Union postale : 20 francs. — Bureaux, 7, rue du Croissant

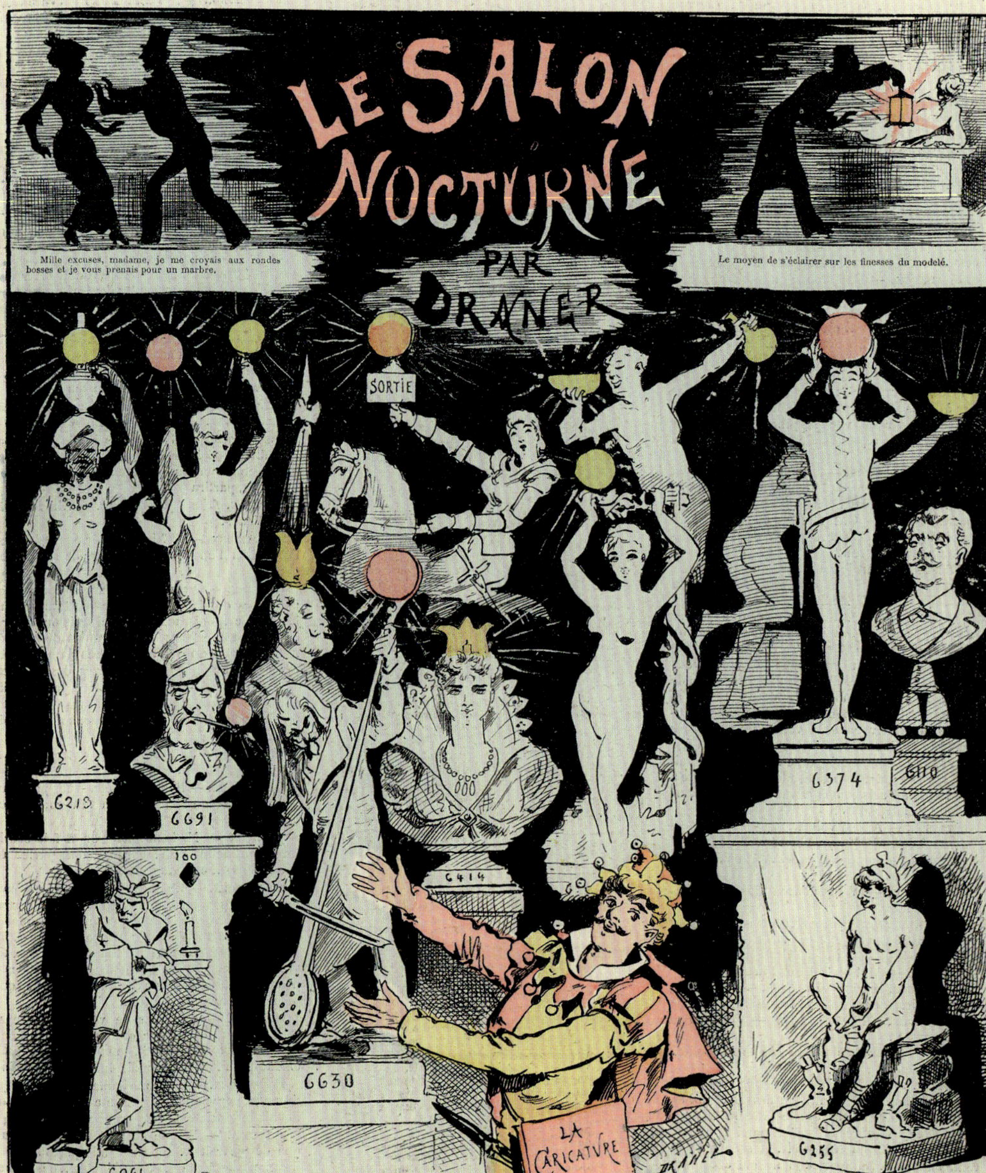

Ce pauvre Dante, il ne s'est pas méfié des effets du cidre.

Il fallait y aller carrément; comme cela, tout le monde eut été ébloui et l'on n'aurait pas pu reprocher à l'administration son absence de lumières.

Onze heures sonnent! Enfin je vais donc pouvoir aller me coucher.

Though, to repeat, the exact lighting technologies used in the eight Impressionist exhibitions have yet to be mapped in the otherwise abundant scholarship that treats the criticism garnered by the shows, Edgar Degas left a tantalizing demonstration of his acquaintance with current illumination technologies. A postscript to an early 1879 letter to his friend and colleague, Félix Bracquemond, about plans for the upcoming Impressionist show (the fourth, held in 1879) displays the artist's familiarity with the electric lighting innovations in his environs: "The Company Jablochkoff proposes to do the lighting with electric light. Etc. Etc."[28] Art historian Ronald Pickvance surmises, based only on Degas' remark, that "the Compagnie Jablochkoff may have installed electric lighting—though no evening openings were advertised."[29] I must conclude the opposite: that the absence of evening hours suggests instead that artificial lighting was not installed in the very grand premises secured for the 1879 display at 28, avenue de l'Opéra.

Scant mention is made of the lighting conditions at the venue for the 1880 exhibition. A centrally located space in a fashionable address on the Right Bank, 10, rue des Pyramides (à l'angle de la rue Saint-Honoré) was secured, but it was a building under construction. Period critic Gustave Goetschy found it noisy and badly lit. Interior decor expert and critic Henry Havard agreed: "In other words, they have gained in neither lighting nor comfort."[30] No reference is made either to artificial light or to evening hours.

3.12 Draner, "Le Salon Nocturne." *La Caricature* 22 (May 29, 1880). Courtesy of Northwestern University Libraries.

We can, however, write a history of earlier nineteenth-century gaslit art spaces. Gaslight was apparently first used to brighten indoor displays, especially in museums, in Britain in the early decades of the nineteenth century.[31] Paris dealer galleries used indoor gas lighting in the 1870s and 1880s. The exhibitions of the Cercle de l'Union Artistique (known as the Mirlitons) took place in a building on the place Vendôme in the 1870s. Their exhibitions ("petits salons") stayed open as late as 11:00 p.m., made possible by the room's ample gaslight chandeliers that were, according to art historian Martha Ward, "carefully distanced from the walls and ceilings to avoid shadowing and equipped with cup-shaped reflectors."[32] Georges Petit was also widely praised for elegant and sensitive lighting when he opened his gallery in 1882.[33] Ward observes that in addition to the other striking features of Petit's premises, most notably the space's grandiosity, blue-chip location (the rue de Sèze), and chestnut-red walls: "For evening viewing, the regiment of twenty or so regularly spaced chandeliers, equipped with gas bulbs [*sic*] and copper reflectors, added still more warmth to the *salle*, effects that reviewers found marvelous."[34] Pierre Pinchon adds that the evening lighting chez Georges Petit "made his luxurious and worldly gallery situated near the Madeleine an obligatory place for Parisian night life."[35]

Was the Paris Salon alone among larger exhibitions to experiment with an electric arc–lit art interior? Not entirely, and small wonder, since, *pace* Georges Petit, there was abundant discontent with the quality and effects of gas lighting. Overlapping with the first electric Salon, an exhibition of Old Master drawings at the École des Beaux-Arts, opened on May 1, 1879, was lit by electric lights installed by the

Compagnie Jablochkoff. Evening hours (8:00 to 11:00 p.m.) were able to be held as a consequence.[36]

The rise of incandescent bulb technology ushered in yet another era of electrically lit art interiors.[37] As with gas lighting, British institutions were ahead, but a bulb-based electric lighting system for illuminating artwork was introduced in Paris in 1881; a high-profile feature of the International Electrical Exhibition, discussed below, was an exhibit of paintings lit by *la lampe Soleil*.[38] The first commercial gallery lit by electric light was across the Channel: the Grosvenor Gallery in 1882, opened by Sir Coutts Lindsay in 1877. There is evidence to suggest that a lordly friend recommended it to him after visiting the Paris International Electricity Exhibition in 1881.

In 1881, the specialized Paris press illustrated examples of art-clad interiors illuminated by incandescent bulb lighting, the newest thing, and not exclusively in Paris. The October 1, 1881, number of *La Lumière Éélectrique*, an opulent issue of the pacesetting journal, illustrated a "Galerie de tableaux à New York" (fig. 3.13) grandly illuminated by Edison bulbs, and observed in the process the advanced state of electric interior lighting across the Atlantic.[39] The September 8, 1883, issue of *The Electrical World*, an American paper, showed off "An Art Room" (fig. 3.14) lit by an incandescent bulb chandelier. Because neither of these illustrational pictures made for the specialized electricity press was conceived in a caricatural mode, there is no critique of art environments lit by bulbs. Indeed, this sort of picture works against the messages sent by caricatures that signaled the perils of electric light. With regard to

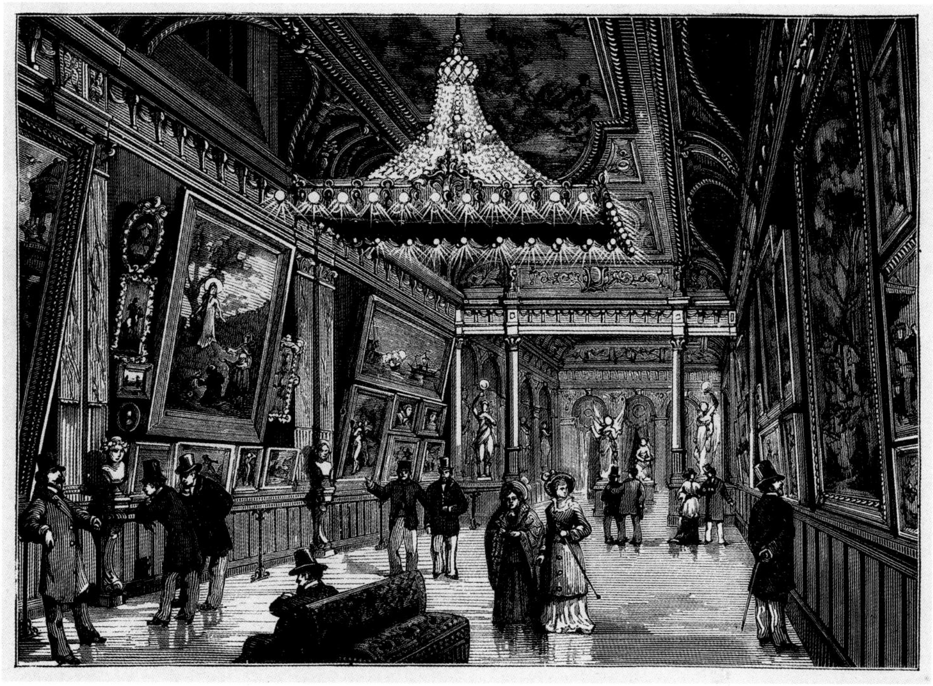

3.13 "Galerie de tableaux à New York éclairée par des lampes d'incandescence du système Edison." Figure 150 from Ém. Alglave and J. Boulard, *La Lumière électrique: Son histoire, sa production et son emploi* (1882). University of Chicago, John Crerar Library.

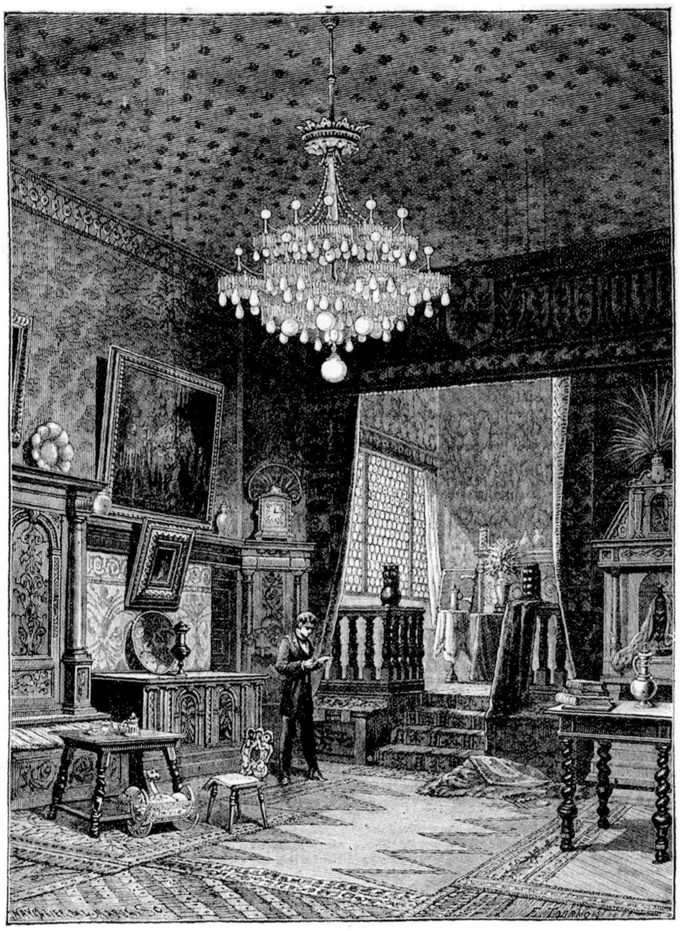

3.14 "An Art Room Lighted by a Chandelier of Incandescent Lights." From *Electrical World* 2, no. 2 (September 8, 1883): 17.

both human comfort and technological effectiveness, such depictions declare that interiors thus illuminated by and for the benefit of the elite were not just up to date, but elegant and easy on the eyes. It is a pity for historians and art historians alike that verbal estimations of the effects of electric light on art in galleries and in the home are rare.[40] Discussions of the impact of electric light on work shown in the Paris Salon do exist, however.

The Electric Salon

Frank Géraldy, "le secrétaire de rédaction" (copy editor or editorial assistant) of *La Lumière électrique*, who published as "F. G.," expressed his warm appreciation—and presumably that of the entire editorial staff—of the quality of the light in the first electric Salon. His account appeared in the journal's third issue, dated 15 June 1879.[41] I quote it in full since it is a savvy tribute to the positive properties of arc light, and is a benchmark text against which to assess the opinions of the light's detractors.

> The lighting of the Palais de l'Industrie is a success, the public is already showing up in great numbers and everything indicates that the crowd will do nothing but increase. This eagerness strikes us as entirely justified. The appearance of the palace thus illuminated is agreeable.
>
> The general sensation is that of an unobtrusive and slightly subdued daylight. In the paintings galleries the effect is completely satisfying; the globes are placed high enough, their light is concentrated by reflectors on the walls; one gladly forgets the time and one gives oneself over to the examination of the paintings with the same quietude as during the day. In the garden, where the sculptures are, the light is not quite abundant enough; it is true that the problem was more complicated; an immense nave, no ceilings or reflective surfaces, light to distribute everywhere into every nook and cranny; the result is not perfect, but it is certainly remarkable.
>
> Several specific remarks may be made about this large-scale experiment.
>
> It seems to us to demonstrate that the light does not have the dazzling and exaggerated flash for which it is sometimes criticized. The globes in the garden are of medium height and are not bothersome. We particularly point out the lamps [*foyers*] on the entrance landing on the first floor, they are almost the height of a man; one looks them in the face very close up and no one appeared bothered by them. As compensation, the scintillations and changes in the color of the light were very apparent. In the painting galleries, this effect is little felt, either because the colors of the paintings themselves prevent it from being perceived, or because the globes that are closest blend their lights in a more equal fashion; but in the garden, the implacable whiteness of the statues does not let even the smallest disparities go unnoticed, and we must admit there are plenty of them.
>
> Despite this small reservation, we hasten to recognize that this illumination, the largest yet attempted in an enclosed space, does the greatest honor to those who carried it out; it shows yet again what we can expect from electric light, and it cannot fail to accelerate its progress.

Géraldy's meticulous assessment was no doubt an influential one in the French electricity community. The subtle distinctions he draws between the effectiveness and *douceur* (softness) of the lighting in the painting galleries versus the cacophonous coloristic effects of arc light shining on white stones are both discerning and instructive. Only two weeks later, a brief unsigned piece in *Le Monde illustré* drew nearly identical distinctions between the pleasant quality of the light falling on the paintings up in the higher galleries and the problematic impact of the arc light on the sculptures down in the nave. "In the case of sculpture, the lighting produces pearly shadows, sometimes a bit harsh." Overall conclusion: "In sum, this innovation has obtained a great success this year."[42] The two accounts help us to grasp the optical essentials of arc light. Written descriptions are especially valuable in this era when unadulterated photography was not up to recording the appearance of an illuminated exhibition interior at night, especially not when the space had a glass ceiling.[43]

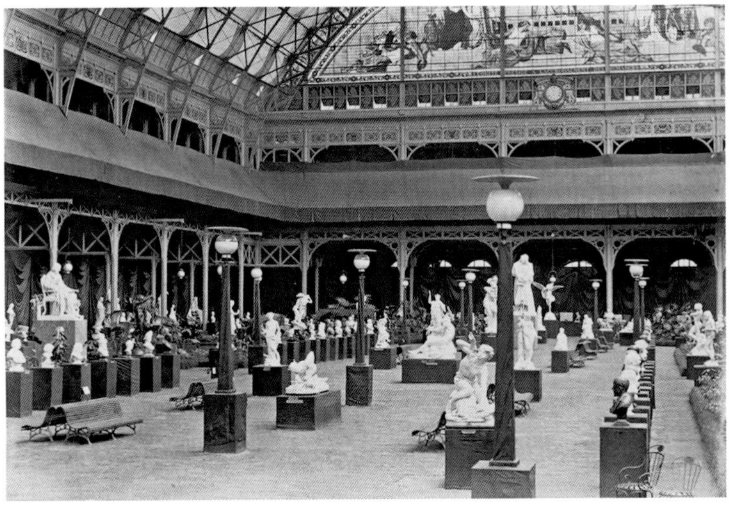

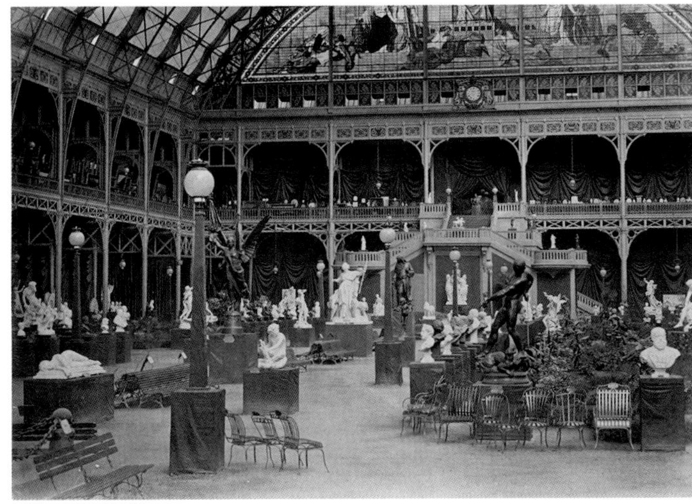

3.15 Salon of 1879. View of the sculpture garden. Photo: G. Michelez. Archives Nationales, France.

3.16 Salon of 1879. View of the sculpture garden. Photo: G. Michelez. Archives Nationales, France.

There are, however, photographic records of the Salon sculpture nave in the daytime dotted by its unlit light fixtures (figs. 3.15 and 3.16). The lamps, up on their blocky plinths, are much taller than the "medium height" or "the height of a man" described by Géraldy, and it is fascinating to discover that the lamp hardware, clearly taken seriously, was changed from one year to the next. Circular reflector discs surmounted the 1879 lights. They were gone in 1880. The 1879 lamps also appear more densely packed than those used in 1880. Léon Du Paty's line drawing or etching (fig. 3.17) of the illuminated sculpture court in 1879 confirms the considerable height and concentration of the lamps. Their rays shone well over the heads of the exhibition visitors. At the same time, his unprepossessing representation is at pains to invoke a glare-free and refined art-viewing environment. Highlighting three poised female visitors enhanced the impression of well-being and comfort experienced by visitors to the nave. Indeed, the glowing lamps that punctuate the ground floor of the lofty palace resemble streetlamps out of doors, though the ease of the unchaperoned women delineates a benign and protected environment. There is no indictment of glare here, nor any traces—how could there be?—of Géraldy's troubling "scintillations and changes in the color of the light."

M. Vierge's wood engraving (fig. 3.18), from a summer 1879 issue of *Le Monde illustré*, builds upon analogous standards of social etiquette, but registers the piercing luminosity of the Jablochkoff globes with greater success. That the two murky halftone figural sculptures in the main nave could easily be mistaken for smock-wearing attendants replacing carbon rods in the lamps cleverly showcases, I suggest, the artist's familiarity with the signature traits of this challenging lighting equipment. The irony of the depiction, of course, is that while the evening Salon was illuminated to allow workers to attend the exhibition, all of the visitors in Vierge's picture are bourgeois. In the near foreground, shoring up the respectability of the place, is a double example of night reading in public under the new lights: a woman has the *livret*

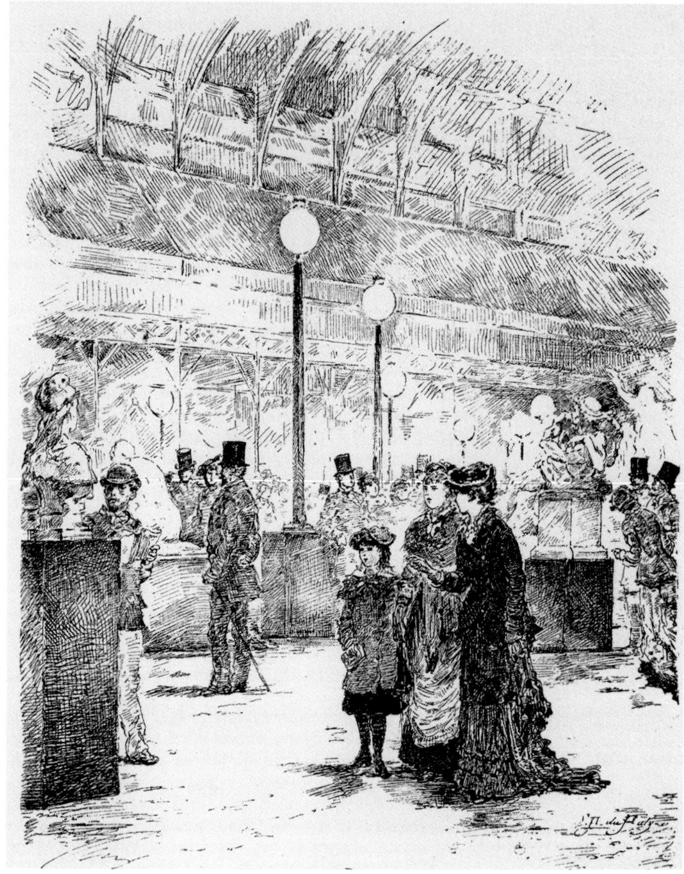

3.17 Léon Du Paty, *At the Salon, Evening*, 1879.

open (shared with her daughter), while a man peruses *Le Moniteur Universel*. Only one person, the man standing at left, looks out at "us," shading his eyes with his hand, a small but unmistakable suggestion of the uncomfortable glare in the nave of the palace. An amusingly gendered activity unfolds at right. A circle of four young women (accompanied by an older male chaperone figure) gathers around to ogle the attractive naked stone body of the sculpted *Génie gardant le secret de la tombe* (*Genius Guarding the Secret of the Tomb*). They make smart use of the opportunity afforded by enhanced luminosity. In sum, Vierge's handsome print completely sidesteps the population of art seekers for whose benefit the night illuminations were primarily intended.

3.18 M. Vierge, "Paris—Le Salon éclairé par la lumière électrique.—Les visiteurs devant la statue de M. de Saint-Marceaux, *Génie gardant le secret de la tombe*." *Le Monde illustré,* July 5, 1879, 12. Bibliothèque Nationale de France.

Searching for comments on the lighting in the electric Salon, a remark by J.-K. Huysmans about "éclairage" with respect to a picture in the 1879 Salon caught my attention. His appraisal of Léon Bonnat's *Portrait de Victor Hugo* includes this observation: "The illumination is, as usual, crazy [*dément*]. It is neither daylight, nor twilight, nor Jablochkoff light."[44] It is puzzling until we realize that he is not discussing the actual lighting in the gallery, but is instead complaining about the unsatisfactory illusion of light devised by the painter within the artwork. The critic's familiarity

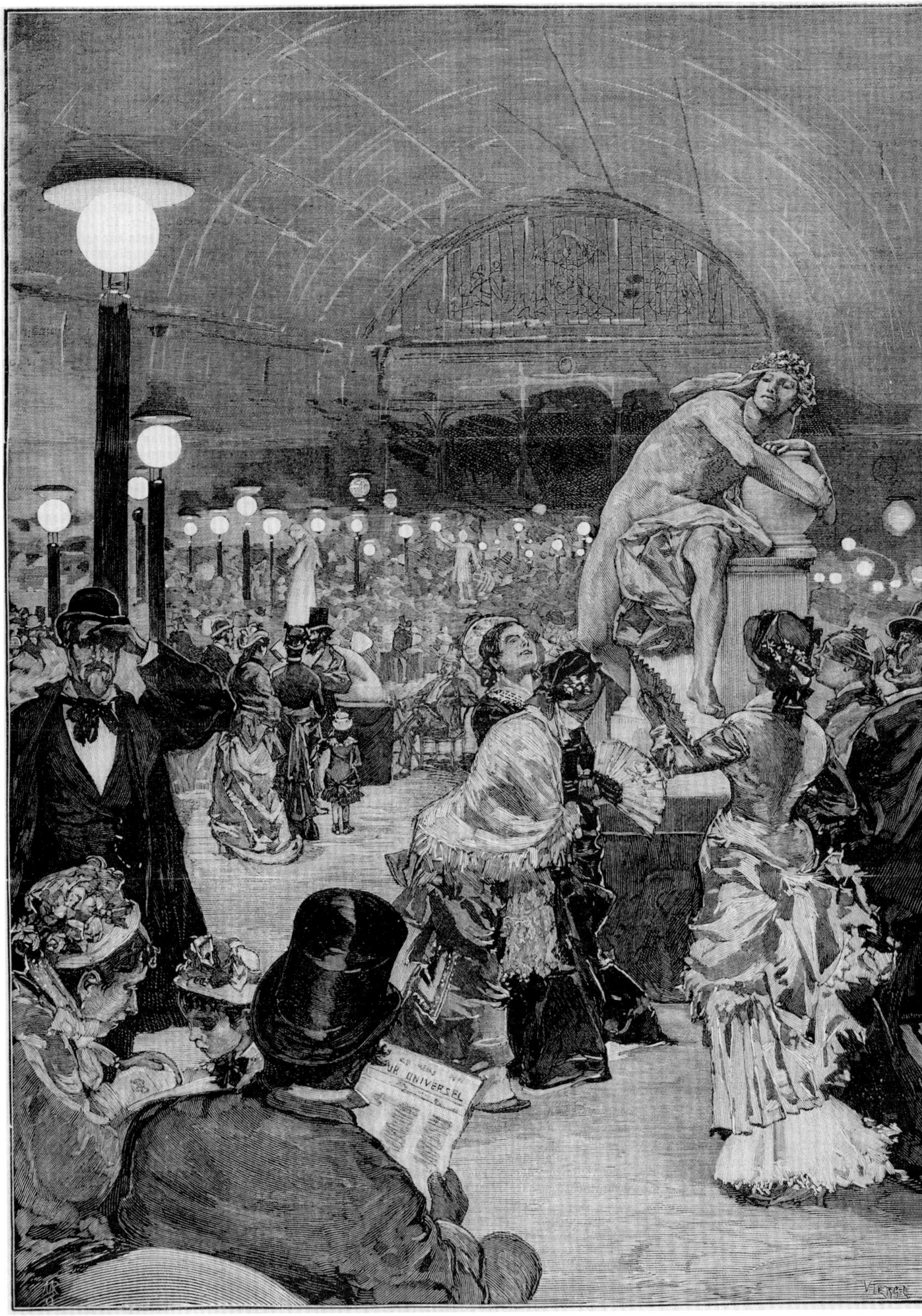
UR UNIVERSEL

with the effects of Jablochkoff light served his denunciatory purpose well. He was able to mobilize an impeccably up-to-date metaphor.

Jules Claretie (1840–1913), theater critic and acerbic social observer, was, like Huysmans, *au courant* concerning modes of lighting, but while Huysmans was silent on the subject, Claretie expressed his dim view of the electric Salon in writing. His indictment of the shoddy connotations of electric *éclairage* appeared in a discussion of the second electric Salon (1880), included in the first installment of his eventual seventeen-volume *La Vie à Paris* (1880–1910). (His chatty weekly columns were collected in book form at the end of the year.) He took the opportunity to critique the exhibition more broadly by observing sarcastically that visiting a Salon was nothing but a pretext for a promenade, a chance to shake as many hands as possible and catch up with the maximum number of people in the shortest time. "The tableaux hanging on the wall are completely unimportant. They are merely backdrop [*tapisserie*]. You'll look at them later. The important thing is the opportunity to show yourself off."[45] Cynicism colored all of his comments on the exhibition. For example, he declares that soon, with more and more artists ("un peu sauvages") disdaining public exhibitions in order to close themselves up alone in their studios with their dreams, "these exhibitions, in a few years, won't have any connection with art at all."[46]

He then plunges his critical dagger: what could be less about art or rather more anti-art than electric light?

> They have already announced, in huge red letters, on the official posters, that the Salon will be, every evening, illuminated by electric light: it is the first step toward an inevitable transformation. The evening when, to enhance the leisurely atmosphere for the visitors, they will add the seduction of music to the charms of painting—of painting passed through the gas of Jablochkoff—the exhibition of the work of living artists will agreeably replace the Folies-Bergère. . . . This would be, for the painters of modernity, as one says, of a charming originality.[47]

His disdain for the beams of arc light was all-inclusive, but his knowledge of lighting technology was not. He refers to the "gas of Jablochkoff," a lighting system that involved no gas whatsoever. Claretie was no *électricien*.

Claretie's visceral response of disdain and disgust regarding the large red letters announcing electric light is the mirror image of another mention of the new entanglement of letters and electricity rooted in the same moment, but that issues from an entirely different sphere. Taking their cue from Marcel Proust and glossing the dominant attitudes of the *fin de siècle*, historians Alain Beltran and Patrice A. Carré proclaim: "Désormais le Progrès s'écrira en lettres électriques!" (Henceforth Progress will be written in electric letters!).[48]

Claretie again brought up the problematic valences of the light—its commercialism and vulgarity—when he described the exodus of a crowd of fashionable attendees, among whom one visitor exclaims: "It's adorable! With that trail of light [*trainée de lumière*] falling on the paths and the chestnut trees, you would think that you were exiting les Italiens [trendy theater on the *grands boulevards*] in springtime!"[49] Claretie's conjoining of the vulgar and the commercial contrasts sharply with the effect of the soothing images of the 1879 Salon under the lights by Léon Du Paty and M. Vierge, in both of which an exclusively bourgeois crowd visits the glowing ground-floor displays (figs. 3.17 and 3.18). They contrived an atmosphere of social tranquility as well as homogeneity during the pro tem conversion of an indoor space into a street in the sculpture garden of the Palais de l'Industrie. Just as the Salon was frequently likened to a vast *magasin*, the lamp-filled nave of the Palace of Industry resembles a lamp shop in the two graphic works. Though the two images do not imply the hectic and grimacing crudity of the scene denounced by Claretie, the commercial aura of the display of art objects under the lights as they picture it is striking, much like the interior of the Grands Magasins du Louvre, an early department store adopter of electric arc light for its sales floors, mentioned at the beginning of this chapter (fig. 3.3).[50]

Frank Géraldy's review of the 1880 Salon, written from the perspective of a savvy *électricien*, contrasts with Claretie's assessment in every way. Géraldy gave a level-headed evaluation of the problems with the Jablochkoff system lights, and provided the specialized readership of *La Lumière Électrique* with a glimpse of tendentious cultural politics behind the scenes. After straightforwardly stating that the "salon de peinture" again opens its doors in the evening, illuminated by electric light, he noted: "It was not without difficulty. The jury opposed it, in the interest of art; it believed that the canvases were poorly seen and faultily appreciated in that light." The practical Géraldy made fast work of the jury's obstructionist aesthetic fastidiousness:[51]

> Those scruples are justified, it seems to me, if it is a question of only seeing the Salon in electric light and suppressing the light of the sun; but it is a matter of furnishing a supplement: the painters lose nothing and gain several hours. In any case, the most important judge is the public; people came in large numbers last year, they will come again this year; the public thus finds visiting the Salon in this fashion to its taste and convenience; are the painters interested in depriving them of it? It is clear that the answer is no.

His evaluation of the lights was no-nonsense and aware of the possibility that class prejudice lay behind the critique of electric arc light. He readily admitted that the technology was still experimental ("encore dans sa période de travail") and that the problems were simply too minor to cause excitement, let alone an uproar. "I know the defects that are attributed to electric lighting such as it is now: scintillations, color changes, reflections that are difficult to avoid, peculiar beams of light that modify coloring. I am the first to agree that there is some truth there, but of middling size."[52] He did aver, as he had with respect to the first electric Salon, that the only real problem was the impact of the lights on white marble or plaster sculpture.[53]

Which may explain why the most ambitious caricature of the electric Salon, that by Draner (fig. 3.12), focused exclusively upon a gallery of white sculpted figures, suggesting that the colors cast by arc lights on pale stone objects were *the* headline news in the domain of the event's telltale opticality.

Caricature in the Spotlight at the Electric Salon

Caricaturists, experts in ridicule, humor, exaggeration, and the novel reinscription of the already known, were ingenious interpreters of the multivalent electric lamp: nothing but a gimmick that dazzled as well as a cunning contraption that ushered socially and morally fraught scenarios into being that illuminated social and personal "truths" normally concealed by darkness. Robida, the editor of *La Caricature*, assigned his seasoned colleague, Draner, to the 1880 electric Salon. His full-page multicolor sculpture gallery (fig. 3.12) is an ingeniously funny exegesis of the event. Surely the artists of *La Caricature* could not have guessed that it would be both the paper's first foray into an electric Salon (the May 29 issue was only the journal's twenty-second) and its last. The lighting experiment lasted only two years: it was terminated before the 1881 Salon.

Draner's hilarious nighttime visit is rooted more in facts than fiction. All but one of the numbered statues that crowd the page are identifiable as pieces on display. That said, the haphazardness and small size of the ensemble are absurd: Draner shows only fourteen of the seven hundred pieces of sculpture on display in the lamp-lit nave that year. The effort I make in what follows to match up Draner's white statues and their object numbers to those listed in the Salon *Livret* is painstaking forensic art history, but also a way of certifying the ludicrousness of Draner's grouping despite the overall exactness of his references.[54]

The balls of glowing, brightly colored, red and yellow light stand in for the piercing white globes of arc light. Their fanciful colors may function as exaggerated registrations of the characteristic "scintillations" that distorted color, described vividly by Géraldy. Like buoyant toys rather than glass globes cut loose from poles and batteries, they bounce and float to animate the crowded tiers of white sculpture in a festive way. The coronas of linear rays that emanate from the spheres designate the high wattage of their radiance. That the round globes also appear to operate within some of the sculptures' narratives or to provide novel attributes is part of Draner's spirited game, rooted in a playful attitude toward the untouchable, buzzing, and glaring globes of the real Jablochkoff candles.

Draner's definition of the light sources in play is intricate and detailed. There are only four old-style (preindustrial) lights in this nocturnal Salon, and only two of the sculpted bodies cast shadows—choices that enhance the delineation of light source–specific effects. No. 6219, at left, is a fanciful, dark-turbaned oriental man who does not correspond to the sculpture shown under that number. He is the only figure in Draner's Salon with an old-fashioned light on his head, an oil lamp insolently linking the Orient to old-fashioned lighting.[55] The oriental man's lamp glows bright yellow, and the imposing female bust (no. 6414) at center also gives off a yellow light; she has a candle flame on her head in the regal shape of a three-pronged crown.[56] The unnumbered bust of a smug bearded man (behind the wild no. 6630, "le démon de Rakoczy") also dons a flame on his pate, two-lobed in his case, and the male bust at the far right (no. 6110) is singular in radiating its own globe-free rays of light, presumably those of genius, suitable for a professor.[57] Joan of Arc, the eminent French heroine, is reduced to an exit sign. Indeed, this kitsch reduction of her stature—nothing now but an instrumental lamp—powerfully conveys the message that arc light devalues sculpture. Draner perhaps shared Claretie's viewpoint: nothing is more anti-art than electric light.

Another breed of figure occupies the bottom register. The spindly robed figure casting a dark shadow (no. 6061) at bottom left is Dante holding a candle. His caption classifies him as a tipsy nightwalker. "Poor Dante, he was not on his guard against the effects of cider."[58] At bottom right the helmeted and muscular, seated naked man (no. 6255) removing his boots and socks can be identified as Mercury. He sits atop a caption that explains his unshod feet: "11 o'clock sounds! Finally I'll be able to go to bed."[59] Mischievous captioning puts Dante and Mercury, the grandest personages on the entire page, along with Joan of Arc, in all-too-human conditions: intoxication and impatient drowsiness. The two men retain a modicum of dignity by being spared the possession of an electric light. All the other figures have become lamps—not merely illuminated, but converted into machinic commodities by Jablochkoff candles.

More complex and consequential is the caption below center, positioned as if spoken by the colorful yellow-and-red-clad jester, our guide, who carries a portfolio across his body labeled *La Caricature*, and gestures toward the illumined congregation of artworks.[60] "There should have been decisive action, that way everyone would have been dazzled (blinded) and nobody could have complained about the administration's lack of enlightenment."[61] This statement is a puzzle in many ways, but it unquestionably addresses the administration's decision to illuminate the Salon with electric arc light. It certainly plays on the familiar discursive rhyming and standoff between bedazzlement and enlightenment. Does it mean that the lights should have been left on all night? Does it reference discussions under way during the month of May about the efficacy and politics of lighting the Salon with electricity? The questions pile up and point to caricature's essential aggravating contingencies.

The electric Salon of 1880 was not only the largest in history (showing 7,289 works), but the last overseen by the Ministère de l'Instruction Publique et des Beaux-Arts. In effect, the state Salon was abandoned after 1880. As art historian Patricia Mainardi has shown: "The period from 1880 to 1885 saw the termination and collapse of the official French Salon system."[62] The quality of the arc lighting in two Salons notwithstanding, by 1881 the Paris Salon, as noted, had dropped the technology altogether, presumably for both political and practical reasons.[63] With respect to the latter, recall that this kind of light was demanding and complicated to maintain; its rods had to be continually replaced at short intervals.[64] Regarding politics, evening lighting arose as a Republican policy. Mainardi reports that it was only with the Republican accession to power in 1879 that Edmond Turquet decided "to have the Salon electrically lit so that workers could attend in the evening." And numbers rose: there were 155,000 more attendees than in the previous year.[65] Politics became explicit in the debate over the lighting conducted in the Chamber of Deputies. The Right was against the electric illumination and opposed, quoting Mainardi again, to "any other innovation that threatened the 'aristocratic' tone of the exhibition, and a deputy on the Left reminded them that 'everybody comes to the Salon at night!'"[66] The deputy no doubt had the working majority in mind, rather than the select gathering of gentry included by the nonhumorist limners of the electric Salon (figs. 3.17 and 3.18).

In his 1880 front-page Nocturnal Salon (fig. 3.12), Draner combined flat figures in silhouette with bleached pieces of volumetric sculpture. The two such vignettes, flanking the title, narrate the high jinks of top-hatted men, and illustrate an ascendant theme in caricature in the era of illumination discourse: bedazzlement by electric light causes and excuses male sexual mischief. The scenes demonstrate caricature's finely tuned ability to recycle social types, in this case the naughty men and women of Second Empire boulevard culture.

Silhouette as both index and residue of glaring illumination, first raised in our discussion of Sargent's nocturnes (in chapter 2), served Draner's purpose well. The picture at left stages a happenstance encounter between two stark and jaunty adults in flat monochrome profile. The man makes a mistake at the woman's expense. "A thousand excuses, Madame, I thought I was feeling round lumps and I took you for a marble."[67] The man's outstretched groping arms stop within inches of the recoiling woman's bosom. He claims not to be a sexual marauder. He was merely enjoying the haptic pleasures of stone sculpture as well as he could in obscure conditions. Draner repeated the sight gag in a two-page, ten-drawing miniseries in early 1881, "Effets de Brouillard" (Effects of Fog), in which one scene featured a man and woman facing one another in murky light. He feels her chest: "A rounded surface. . . . That must be the column I am looking for."[68] Situational male blindness amusingly vindicates sexual effrontery. In the 1880 front page (fig. 3.12), the upper right cartoon, "The way to enlighten yourself on the cunning of a model," is of a piece.[69] The availability of a curvaceous stone woman in the dark justifies an impertinently close inspection of her forms using a portable lantern light. Arc light as alibi comes to the rescue.

Draner filled two subsequent pages of the May 29, 1880, issue with twelve small-scale black and white electric Salon jokes (figs. 3.19 and 3.20) on sundry topics. In this cluster, Jablochkoff candles make people sick. In one (fig. 3.20, left column), for example, an exhibition guard is shocked to see a well-dressed woman putting dark glasses and a visor on a sculpted bust. Two blazing arc light balls on narrow stems are close at hand. The woman replies: "I tell you again that my husband has always

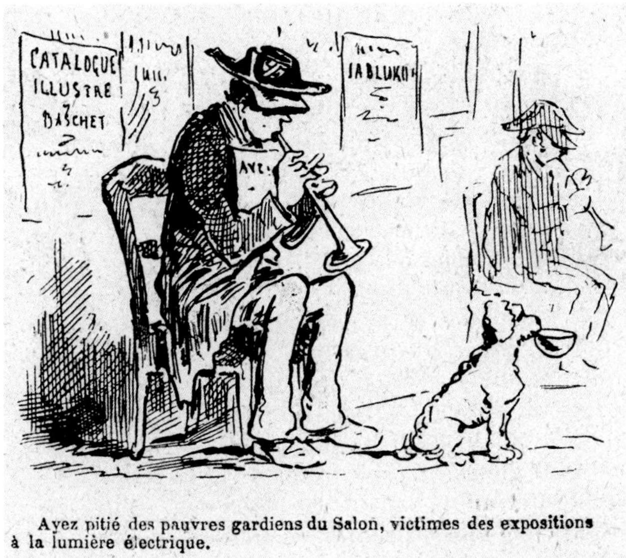

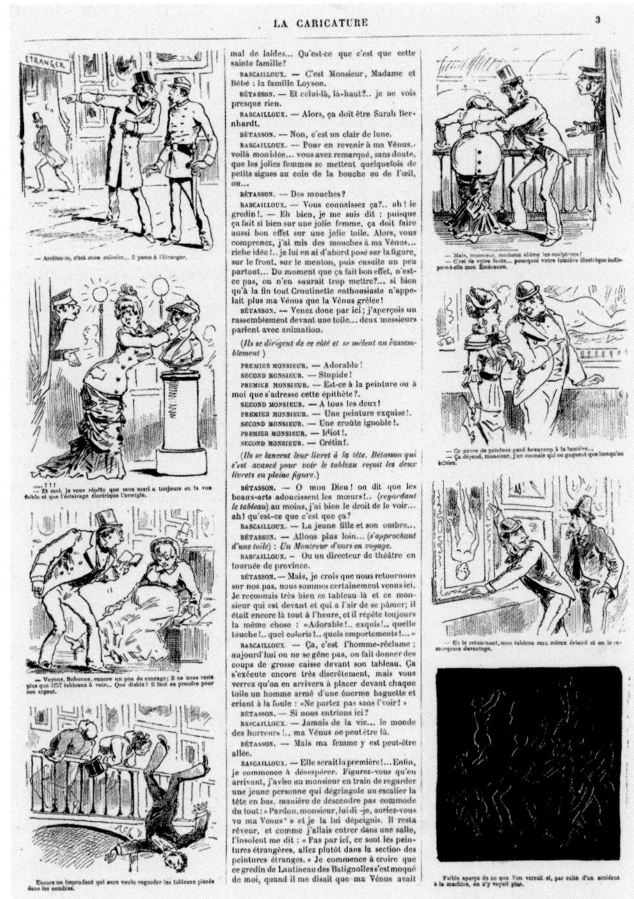

LA CARICATURE 3

mal de laides... Qu'est-ce que c'est que cette sainte famille?

BASCAILLOUX. — C'est Monsieur, Madame et Bébé : la famille Loyson.

BÉTASSON. — Et celui-là, là-haut?.. je ne vois presque rien.

BASCAILLOUX. — Alors, ça doit être Sarah Bernhardt.

BÉTASSON. — Non, c'est un clair de lune.

BASCAILLOUX. — Pour en revenir à ma Vénus. voilà mon idée... vous avez remarqué, sans doute, que les jolies femmes se mettent quelquefois de petits signes au coin de la bouche ou de l'œil, ou...

BÉTASSON. — Des mouches?

BASCAILLOUX. — Vous connaissez ça?.. ah! le gredin!. — Eh bien, je me suis dit : puisque ça fait si bien sur une jolie femme, ça doit faire aussi bon effet sur une jolie toile. Alors, vous comprenez, j'ai mis des mouches à ma Vénus... riche idée!.. je lui en ai d'abord posé sur la figure, sur le front, sur le menton, puis ensuite un peu partout... Du moment que ça fait bon effet, n'est-ce pas, on n'en saurait trop mettre?... si bien qu'à la fin tout Croutinette enthousiaste n'appelait plus ma Vénus que la Vénus grêlée!

BÉTASSON. — Venez donc par ici; j'aperçois un rassemblement devant une toile... deux messieurs parlent avec animation.

(*Ils se dirigent de ce côté et se mêlent au rassemblement*)

PREMIER MONSIEUR. — Adorable!

SECOND MONSIEUR. — Stupide!

PREMIER MONSIEUR. — Est-ce à la peinture ou à moi que s'adresse cette épithète?.

SECOND MONSIEUR. — A tous les deux!

PREMIER MONSIEUR. — Une peinture exquise!.

SECOND MONSIEUR. — Une croûte ignoble!.

PREMIER MONSIEUR. — Idiot!.

SECOND MONSIEUR. — Crétin!.

(*Ils se lancent leur livret à la tête. Bétasson qui s'est avancé pour voir le tableau reçoit les deux livrets en pleine figure.*)

BÉTASSON. — O mon Dieu! on dit que les beaux-arts adoucissent les mœurs!.. (*regardant le tableau*) au moins, j'ai bien le droit de le voir... ah! qu'est-ce que c'est que ça?

BASCAILLOUX. — La jeune fille et son ombre...

BÉTASSON. — Allons plus loin... (*s'approchant d'une toile*) : *Un Montreur d'ours en voyage.*

BASCAILLOUX. – Ou un directeur de théâtre en tournée de province.

BÉTASSON. — Mais, je crois que nous retournons sur nos pas, nous sommes certainement venus ici. Je reconnais très bien ce tableau-là et ce monsieur qui est devant et qui a l'air de se pâmer; il était encore là tout à l'heure, et il répète toujours la même chose : «Adorable!.. exquis!.. quelle touche!.. quel coloris!.. quels emportements!... »

BASCAILLOUX. — Ça, c'est l'homme-réclame; aujourd'hui on ne se gêne pas, on fait donner des coups de grosse caisse devant son tableau. Ça s'exécute encore très discrètement, mais vous verrez qu'on en arrivera à placer devant chaque toile un homme armé d'une énorme baguette et criant à la foule : «Ne partez pas sans l'voir! »

BÉTASSON. — Si nous entrions ici?

BASCAILLOUX. — Jamais de la vie... le monde des horreurs!.. ma Vénus ne peut être là.

BÉTASSON. — Mais ma femme y est peut-être allée.

BASCAILLOUX. — Elle serait la première!... Enfin, je commence à désespérer. Figurez-vous qu'en arrivant, j'avise un monsieur en train de regarder une jeune personne qui dégringole un escalier la tête en bas, manière de descendre pas commode du tout : «Pardon, monsieur, lui di-je, auriez-vous vu ma Vénus? » et je la lui dépeignis. Il resta rêveur, et comme j'allais entrer dans une salle, l'insolent me dit : « Pas par ici, ce sont les peintures étrangères, allez plutôt dans la section des peintures étranges. » Je commence à croire que ce gredin de Lantineau des Batignolles s'est moqué de moi, quand il me disait que ma Vénus avait

3.19 Draner, *La Caricature,* May 29, 1880, 2. Courtesy of Northwestern University Libraries.

3.20 Draner, *La Caricature,* May 29, 1880, 3. Courtesy of Northwestern University Libraries.

had feeble vision and that the electric light blinds him."[70] In another (fig. 3.20, right column), another variety of human wretchedness, vomiting, is activated by the intense lights. Standing on a balcony of the upper galleries, a woman seen from behind leans over the railing toward the nave of sculpture below. Her attentive husband has placed his hand in the small of her back to steady her while he responds sharply to a guard's accusation. "—But, monsieur, madame is damaging the sculpture. —It's your fault because the electric light indisposes my Émérance."[71] His most ambitious drawing (fig. 3.20, right column) imagines the dark. The square is all black, a negative space that contains only faint white outlines of female bodies. Draner's choices of female nudes for his thought experiment requires no gloss. The caption: "Weak glimpse (or impression) of what one would see if, after an accident with the machine, one could not see anymore."[72] The light may have been blinding, but if and when it failed—which it frequently did—close to nothing could be seen. The arc light is an intemperate two-edged sword: it yields surplus or nullity; blindness in either case. Arc light shining on art receives a guilty verdict every time in the realm of caricature.

The International Exhibition of Electricity (1881)

Nearly six months ago the French government resolved to hold an exhibition of all the arts and appliances connected with electricity. This branch of science, so new, so modern, so well destined to make the fortune of the twentieth century, as steam has been to make the triumph of our age, has never yet had an exhibition all to itself. . . .

By night the scene will be illuminated by thousands of electric lights, in which every system will be represented. The lamps of Jablochkoff, Jamin, Werdermann, Siemens, Gramme, Lontin, Noaillon, Merteas, Suisse, Wilde, Brush, Swan, Edison, Hiram Maxim, Arnaud, Crompton, Brockie, will shine in generous rivalry, some in simple globes, others aided by the reflectors of Baliestrieri's system. . . .

A very curious application has been made to the management of the exhibition. An optician has offered a high price for the exclusive right to sell blue spectacles in the building. The electric light, he argues, will be too dazzling for the naked eye.

"THE ELECTRICAL EXHIBITION AT PARIS," *HARPER'S WEEKLY*, MAY 28, 1881[73]

Trumpeting the modernity of electricity and its potential to generate prodigious profit in the new century showed foresight. *Harper's Weekly*, an American paper with overseas reporters, also accurately evoked variety as a defining trait of the lighting displays being readied for the mid-August opening of the International Exhibition of Electricity. The article fittingly envisaged a phantasmagoria of electric lights in the Palace of Industry; a wondrous display that would at the same time be very hard on the human eye. The writer also forecast, no doubt snidely, a brisk market for prosthetic spectacles—blue lenses to protect the eyes, a fad mined by Cham (fig. 3.21), "Blue glasses for both of you?"[74] As an enormous public display of curated goods under state-of-the-art lights, the Electricity Exhibition was clearly akin to the electric Salons convened in 1879 and 1880 in the same hall.[75] But commentators on the Electrical Exhibition were evidently not Salon goers, and the parallels went unacknowledged.

Because the exhibition combined scientific and industrial inventions with spectacular display—the elements pinpointed by the reporter for the *Harper's Weekly*—it was baptized "the Revolution of 1881" by its leading modern historians, Alain Beltran and Patrice A. Carré.[76] In 1883, Louis Figuier, the most important popular science writer of the late 1800s, responded likewise, averring that the exhibition was lived by its contemporaries as "one of the most important scientific events of the nineteenth century."[77] As written and pictorial responses to the event attest, Beltran and Carré did not exaggerate when they wrote that "close to 900,000 visitors crowded the Palace of Industry to discover technologies that would very shortly engender and characterize a *new material civilization*," and that "with electric light and

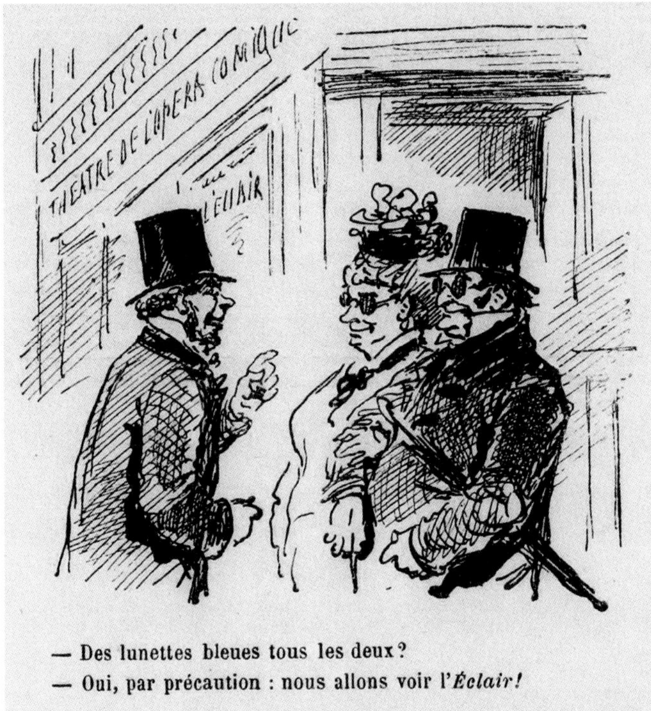

3.21 Cham, illustrations from "Théâtres," in *Les Folies parisiennes par Cham: Quinze années comiques, 1864–1879* (1883).

its slow diffusion there was a modification of a whole system of psychological attitudes."[78] The exhibition was thus judged consequential by contemporaries and historians alike in the spheres of the material, the psychological, and the scientific, as well as the realm of the spectacle (fig. 3.22).

Electric light caused the most excitement at the exhibition, as Beltran and Carré explain: "The telephone is among the electric innovations presented to the public one of those that seems to fascinate. But the major innovation, the one that attracted the attention of everyone (female and male alike) was without any doubt electric light."[79] American historian Ernest Freeberg has also stressed the importance of light at the event, but emphasized its conditionality, or rather that the exhibition was primarily viewed as a testing ground for *incandescent* light, and that the outcome ("the winner") of the contest was not foreseeable in advance.[80]

> The Paris exposition of 1881 . . . served as a most visible place for an excited public to see for themselves how well the incandescent light could deliver on its brilliant promises. Newspapers on both sides of the Atlantic followed each turn in the drama that pitted the gas companies against the underdog inventors. . . . The controversy had only grown more interesting as so many rival inventors staked their claims within months of each other. In this debate, few who pondered the future of the new technology were in any position to make a meaningful prediction.

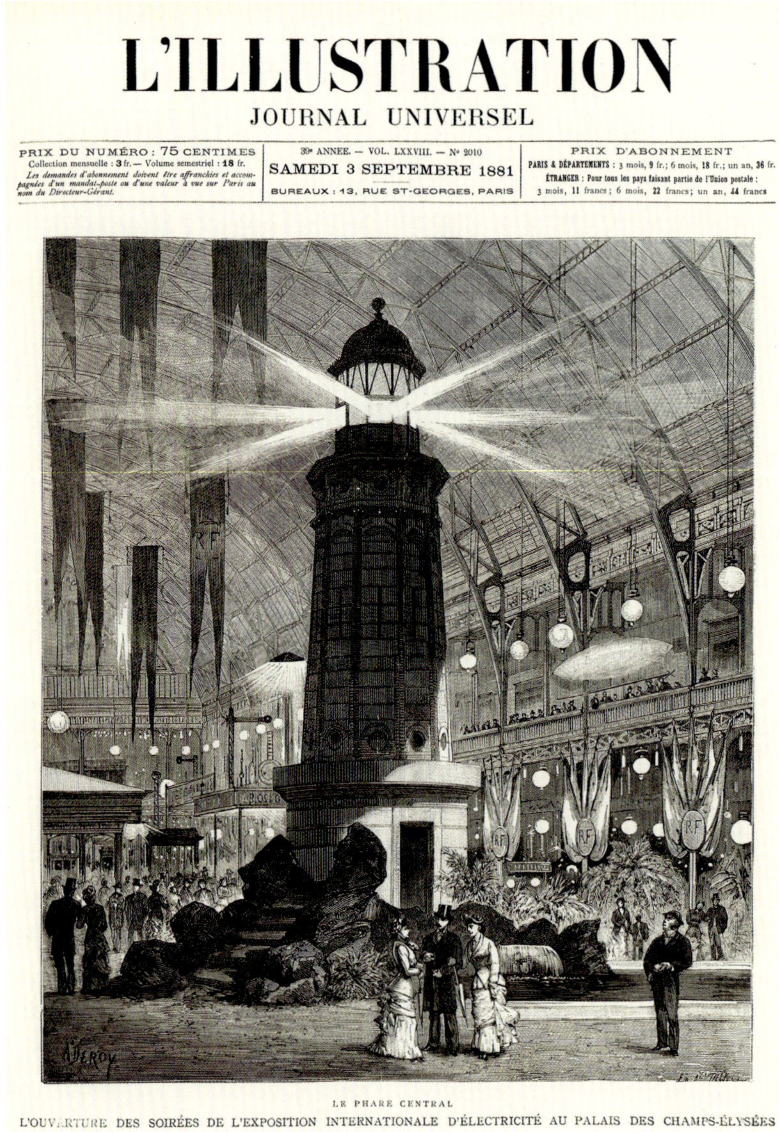
L'ILLUSTRATION

JOURNAL UNIVERSEL

PRIX DU NUMÉRO : 75 CENTIMES
Collection mensuelle : 3 fr. — Volume semestriel : 18 fr.
Les demandes d'abonnement doivent être affranchies et accompagnées d'un mandat-poste ou d'une valeur à vue sur Paris au nom du Directeur-Gérant.

39e ANNEE. — VOL. LXXVIII. — No 2010
SAMEDI 3 SEPTEMBRE 1881
BUREAUX : 13, RUE ST-GEORGES, PARIS

PRIX D'ABONNEMENT
PARIS & DÉPARTEMENTS : 3 mois, 9 fr.; 6 mois, 18 fr.; un an, 36 fr.
ÉTRANGER : Pour tous les pays faisant partie de l'Union postale : 3 mois, 11 francs; 6 mois, 22 francs; un an, 44 francs

LE PHARE CENTRAL
L'OUVERTURE DES SOIRÉES DE L'EXPOSITION INTERNATIONALE D'ÉLECTRICITÉ AU PALAIS DES CHAMPS-ÉLYSÉES

3.22 "L'Ouverture des soirées (The Opening of Evening Hours)," *L'illustration: Journal universel* 78, no. 2010 (September 3, 1881). Courtesy of Northwestern University Libraries.

Most narratives of the 1881 exhibition look back on it through the lens of the electric light–besotted twentieth and twenty-first centuries, and by way of the widespread if not universal assumption that, for his incandescent bulb, Thomas Alva Edison would deserve the unofficial title "Man of the Millennium."[81] Teleologists see the Paris event as the site of an inevitable triumph by two wizardly Americans, Alexander Graham Bell and Thomas Alva Edison, and their ingenious devices, the telephone and the incandescent electric lightbulb. The importance of Edison's bulb and the system that undergirded it is indisputable, well encapsulated by curator Bernard Finn: "Of all the products of electrical invention that made the great changes of the

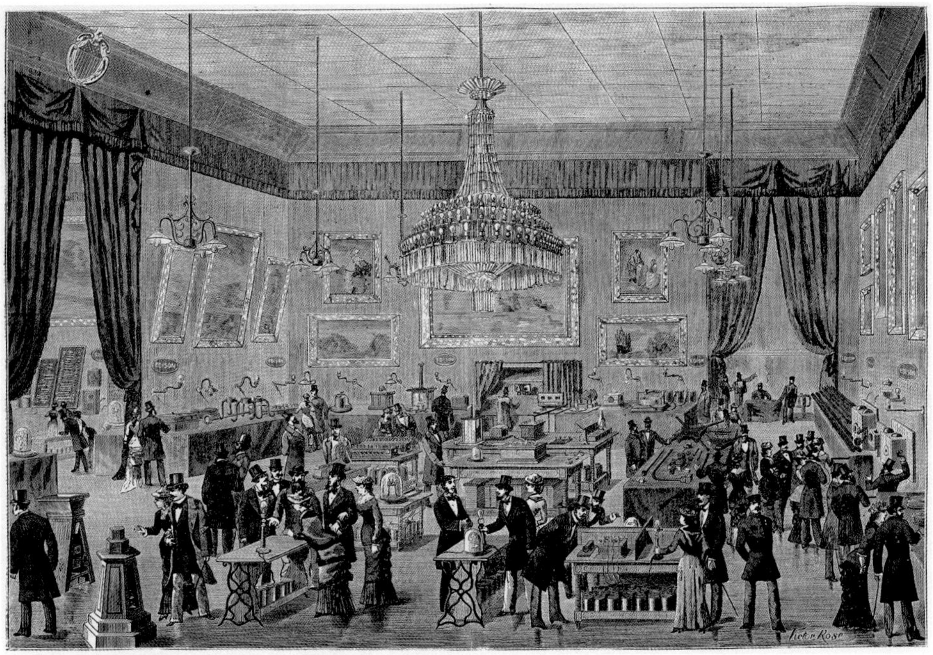

3.23 "L'Exposition de M. Edison," *L'Illustration: Journal universel* no. 2014 (October 1, 1881): 225. Courtesy of Northwestern University Libraries.

late nineteenth century possible, none had a greater impact than Edison's incandescent lamp of 1879."[82] But the winners are usually the authors of history, and this is especially true in the case of the history of the lights of Paris at the telltale moment of 1881, when Edison and his large display became centers of attention (fig.3.23) and Paris was baptized "la capitale électrique" for the first and the last time.[83] Historians of *éclairage* and electricity have studied the Exposition Internationale de l'Électricité exhaustively, but other implicated specialists—such as historians of France, Paris, and the nineteenth-century exhibitions—who might well have been expected to take an interest, have paid it scant attention. The 1881 exhibition is also not a site on the map of art history.

The 1881 International Electricity Exhibition was a radiant event, if only a transitory component of the famously kaleidoscopic visual environment of Paris. Though fleeting, its contribution to the city's visualities was crucial: the exhibition fielded a dazzling rivalry of electric lights; a twinkling assortment of diverse systems, shapes, intensities, and colors of lamps. The map published in *La Nature* (fig. 3.24) reveals the scope and international character of the event. Modern historian K. G. Beauchamp pinpoints the impact of the lights when he reports that the exhibition marked the first major use of electrical illumination in history, and that its use of 220 arc lamps and 2,220 incandescent lamps constituted an unprecedented visual spectacle.[84] Period observer William Henry Preece reported that "on the night of August 29th, there were in operation 277 arc lamps, 116 candles, 44 arc incandescent lamps, 1,500 incandescent lamps or a total of 1,837 electric lights in all. . . . I have little doubt

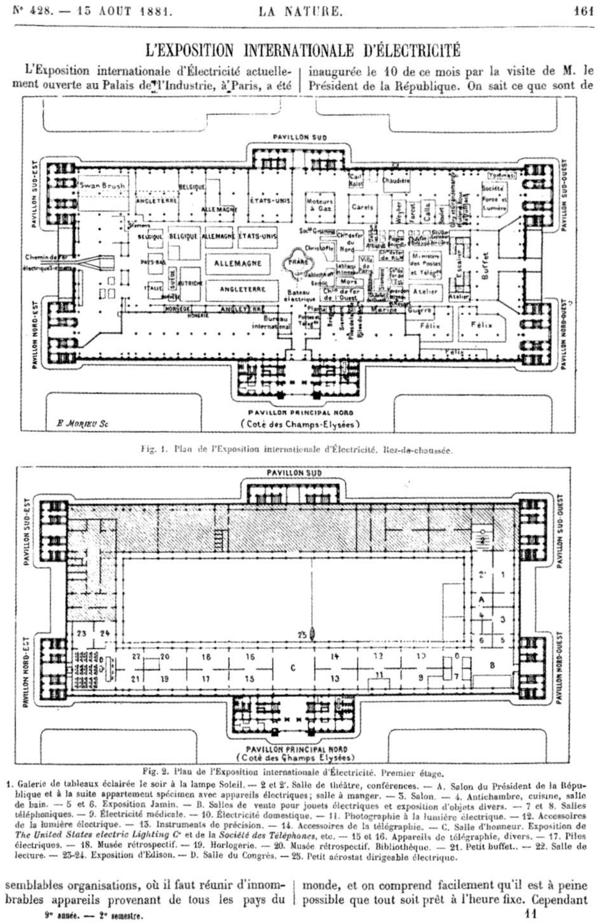
N° 428. — 15 AOUT 1881. LA NATURE. 161

L'EXPOSITION INTERNATIONALE D'ÉLECTRICITÉ

L'Exposition internationale d'Électricité actuellement ouverte au Palais de l'Industrie, à Paris, a été inaugurée le 10 de ce mois par la visite de M. le Président de la République. On sait ce que sont de

Fig. 1. Plan de l'Exposition internationale d'Électricité. Rez-de-chaussée.

Fig. 2. Plan de l'Exposition internationale d'Électricité. Premier étage.

1. Galerie de tableaux éclairée le soir à la lampe Soleil. — 2 et 2'. Salle de théâtre, conférences. — A. Salon du Président de la République et à la suite appartement spécimen avec appareils électriques; salle à manger. — 3. Salon. — 4. Antichambre, cuisine, salle de bain. — 5 et 6. Exposition Jamin. — B. Salles de vente pour jouets électriques et exposition d'objets divers. — 7 et 8. Salles téléphoniques. — 9. Électricité médicale. — 10. Électricité domestique. — 11. Photographie à la lumière électrique. — 12. Accessoires de la lumière électrique. — 13. Instruments de précision. — 14. Accessoires de la télégraphie. — C. Salle d'honneur. Exposition de *The United States electric Lighting C°* et de la *Société des Téléphones*, etc. — 15 et 16. Appareils de télégraphie, divers. — 17. Piles électriques. — 18. Musée rétrospectif. — 19. Horlogerie. — 20. Musée rétrospectif. Bibliothèque. — 21. Petit buffet.. — 22. Salle de lecture. — 23-24. Exposition d'Edison. — D. Salle du Congrès. — 25. Petit aérostat dirigeable électrique.

semblables organisations, où il faut réunir d'innombrables appareils provenant de tous les pays du monde, et on comprend facilement qu'il est à peine possible que tout soit prêt à l'heure fixe. Cependant

9e année. — 2e semestre. 11

3.24 Floorplans for the first International Exhibition of Electricity. *La Nature* 428 (August 15, 1881): 161.

that the number reached 2,500 in the beginning of November."[85] Gaston Tissandier, founder of the journal *La Nature*, asked rhetorically: "What remains to be said? That in the evening the Exhibition sparkles of fire the likes of which has never been seen before. . . . Never before has one been summoned to admire such beautiful illumination produced by the powerful machines."[86]

Henri de Parville, the science journalist, measured the total luminosity of the electric light in the palace by way of comparison with gaslight in order to stress its intensity: "In the Palace, the electricity made it possible to produce, with 1,800 horsepower, the equivalent of 55,000 gaslights or 6,000 more gaslights than ever existed in all the streets and promenades of Paris. . . . We are at the dawn of a new epoch."[87] William Henry Preece, Welsh electrical engineer, remarked on the shortcomings of the display layout: "The terrific mélange of lights that flooded the interior of the Palais de l'Industrie with great brilliancy [was] an impracticable and impossible means of comparing and judging the relative merits of different systems."[88] Or rather the lights commingled and could not be evaluated, let alone seen, separately. The radiance of the display foreclosed on fine-grained assessment, except by some of the *électricien* insiders. Because the nave of the Palace was divided into multiple rooms and there were galleries on the upper floor, different spaces were illuminated by various means: a theater lit by Werdermann lamps, a gallery of paintings lit by a Lampe Soleil, and so on, though Edison's display was well defined with rooms walled off (fig. 3.23).

Some period observers held that the main luminosity event was outside the building. Henri de Parville, for example:

> It was preferable to enter the Exposition for the first time in the evening. If you hadn't known in advance the location of the palace, you would have figured it out quickly by the glow it projected from afar over the city. You would have said that the fiery lighting [*le feu*] was on the Champs-Élysées or that a magnificent Aurora Borealis was shining in the west. Two powerful electric fires [lamps; *foyers*] equipped with reflectors and installed at the summit of the portal of the door of honor cast their sparkling rays onto the Arc de Triomphe and the place de la Concorde.[89]

Henri Valette agreed:

> The trails of this white light, characteristic of electricity, escaped from the windows of the Palace of Industry, plunging afar, and seemed to want to disperse (*fouiller*) the shadows of the somber avenues of the Champs-Elysées.[90]

Cham's humorous drawing of a blazing avenue de l'Opéra in 1879 (fig. 3.11) imagined a street dissolved by electric light, but no artist apparently left an impression of the radiant exterior of the Palais de l'Industrie in 1881 and its surroundings transformed by the glow. The September 3, 1881, cover of *L'Illustration* (fig. 3.22) did, however, a creditable job of showing the myriad beams in the vast palace on the first evening the lights were switched on.

Though the exhibition took place in the French capital, it was neither especially French nor Parisian. It was international inasmuch as it was associated with and had the imprimatur of an International Congress of Electricians, making it "the first comprehensive international gathering of electrical technologists."[91] That Paris was the venue was, however, a French coup. France was also abundantly represented: of the 1768 exhibitors, more than half (55 percent) were French. The other major contributors, in descending order (fig. 3.24), were Belgium, the German Empire, the United Kingdom, and the United States.[92] It was the American exhibits that drew the most attention, especially Edison's copious display (fig. 3.23).

The exhibition took place in 1881 for three primary reasons: the rate of innovation in the "field" of electricity had rapidly accelerated (especially since 1875);

a milieu of interest (comprising experts and amateurs) had developed; and public officials in France (most important, Adolphe Cochery, the minister of the post and telegraph) decided to act in response to the interest generated by the new conquests of science in this area.[93] The rise of a lobby of *électriciens* and a proliferating specialized press played important roles as well in the lead-up to the organization of the meeting. It was thus made possible by cooperation between government and private initiative.[94] Historians François Caron and Fabienne Cardot stress the auspiciousness of the conditions. The 1881 exhibition was the "resounding but also symbolic manifestation of the formation of a composite milieu of *électriciens* ready to mobilize its energy to carry the new technology to the highest levels of industrial and domestic applications, to conquer the world, in a word."[95] In the eyes of a partisan like Henri de Parville, the 1881 exhibition unquestionably exceeded expectations.[96]

> This Exposition of electricity, born in a milieu of nearly general indifference, will have proven to be fertile in more than one category: it gave much more than it had promised; it surprised by its brilliant success, even those who had been its most ardent partisans right from the start. It will leave perhaps more durable traces than its predecessor, the great Universal Exposition of 1878. Even to the least clairvoyant it made completely obvious the role of special expositions henceforth.

The headline news bruited ubiquitously following the conclusion of the fair (Edison, Edison, Edison) quickly erased the memory of the importance of French professional and political leadership, and the differences between the diverse wares on display. It was Edison's carbon filament bulb (to whose invention the English inventor Joseph Swan had an equal claim) and his integrated system that emerged triumphant.[97] Recall that there were four different kinds of incandescent lighting on show: Edison, Swan, Maxim, and Lane-Fox. That Edison did not show just lamps, but also exhibited an integrated twenty-ton generating machine (called "Jumbo") gave his scheme a great advantage.[98] The specific quality of Edison's light was also admired in many quarters (see below).

Regardless of the specific display or lamp targeted for comment, it was clear, in Beltran and Carré's words, that "for the majority of Exposition visitors, electric light represented the promise of a rupture with an old method of lighting."[99] They have stressed the change at the perceptual level: "The advent of electric lighting signifies—for numerous witnesses—a profound change in the modes of perception. Electricity effectively modifies ways of seeing, ways of looking. Very quickly it was understood that henceforth looking at things will not be the same. A new regime thus, one that touches first of all the organ of sight."[100]

The unprecedented congregation of electric lights at the 1881 exhibition and the excited discussion they generated inspired *La Caricature* to fashion a targeted crop of humorous drawings in which, among other scenarios and outcomes, various electric contraptions justified erotic adventure on the part of bedazzled Parisian men, and fostered duplicity and vulgarity on the part of Parisian women. According to Christoph Asendorf, "a part of the fascination with electricity might have arisen out of the closely related analogy to erotic attraction (or repulsion), so nicely characterized as tension between the sexes."[101] As in 1880, caricaturists acted upon this analogy, and intertwined electricity and sex again and again.

The first reaction to the exhibition was front-page coverage by Draner, "À l'Exposition d'Électricité," published on October 22, 1881, well into the second full month of the exhibition (fig. 3.25). The vertical page format is used to advantage to showcase a lighthouse that echoes the imposing light tower dominating the ground floor of the Palais de l'Industrie (fig. 3.22), a replica of the lighthouses that guarded the coasts of France.[102] Draner disposes seven vignettes around the tower, making good use of artifice, that of tinted light beaming through glass storm panes (or rather the illusion of light altered by tinted panes) enclosing the lantern room and its lens. Draner draws only three of the panes, but colored light also pours through unseen panes on the back of the lantern. The lighthouse in the nave of the palace did not beam colored light (nor did real lighthouses), but the variation of tint provided a means to demonstrate an awareness of the diverse colors and intensities of the lights in the exhibition hall. In Draner's clever scenarios, the color of light is used to transform the political, medical, and sexual status of the visitors upon whom it falls.

Verifying that Draner's variegated palette was in tune with the news of the day, the varied tones and intensities of the lights on display were leitmotifs of commentaries. Parville's description of the kaleidoscope of lights was itself *féerique* (fairytale-like).

> When one crosses the main entrance portal, the spectacle becomes magic. It is best to climb immediately to the first floor; your gaze will take in the full ensemble of the great nave and its innumerable lights; the illumination was incomparable in splendor. Here and there, everywhere, without order or symmetry, in the middle of machines, instruments in motion, electric lamps in their opalescent globes shone like sparkling venetian lanterns; the nave was as if punctuated by huge white diamonds that marked the location of nations and defined the place of the exhibitions: here, lamps of intense light; there, lamps that cast a soft and warm illumination; everywhere, chandeliers, candelabras, lampposts, candles sending their disparate clarities onto banners and multicolored flags. From the first floor one had the impression of seeing an immense mosaic of gold, silver, and precious stones. . . . The palace of the Champs-Élysées had truly become the veritable palace of Light.[103]

The varying colors of the beams of light trigger Draner's little fables (fig. 3.25). The top two scenes at left and right, green and red—the panels themselves are triangular and truncated to emulate the shape of light rays—are outlandish readings of the changed tint of visitors' otherwise uniformly white faces. At left, Ernest is green, and the panicked woman concludes, crying for help, that he may have cholera. At right,

Numéro 95 PRIX : 35 CENTIMES 22 Octobre 1881.

A. ROBIDA
RÉDACTEUR EN CHEF

La Caricature

PUBLICATION
DE LA
LIBRAIRIE ILLUSTRÉE

Abonnements d'un an, Paris et Départements : 16 francs. — Six mois : 9 francs. — Union postale : 18 francs. — Bureaux, 7, rue du Croissant.

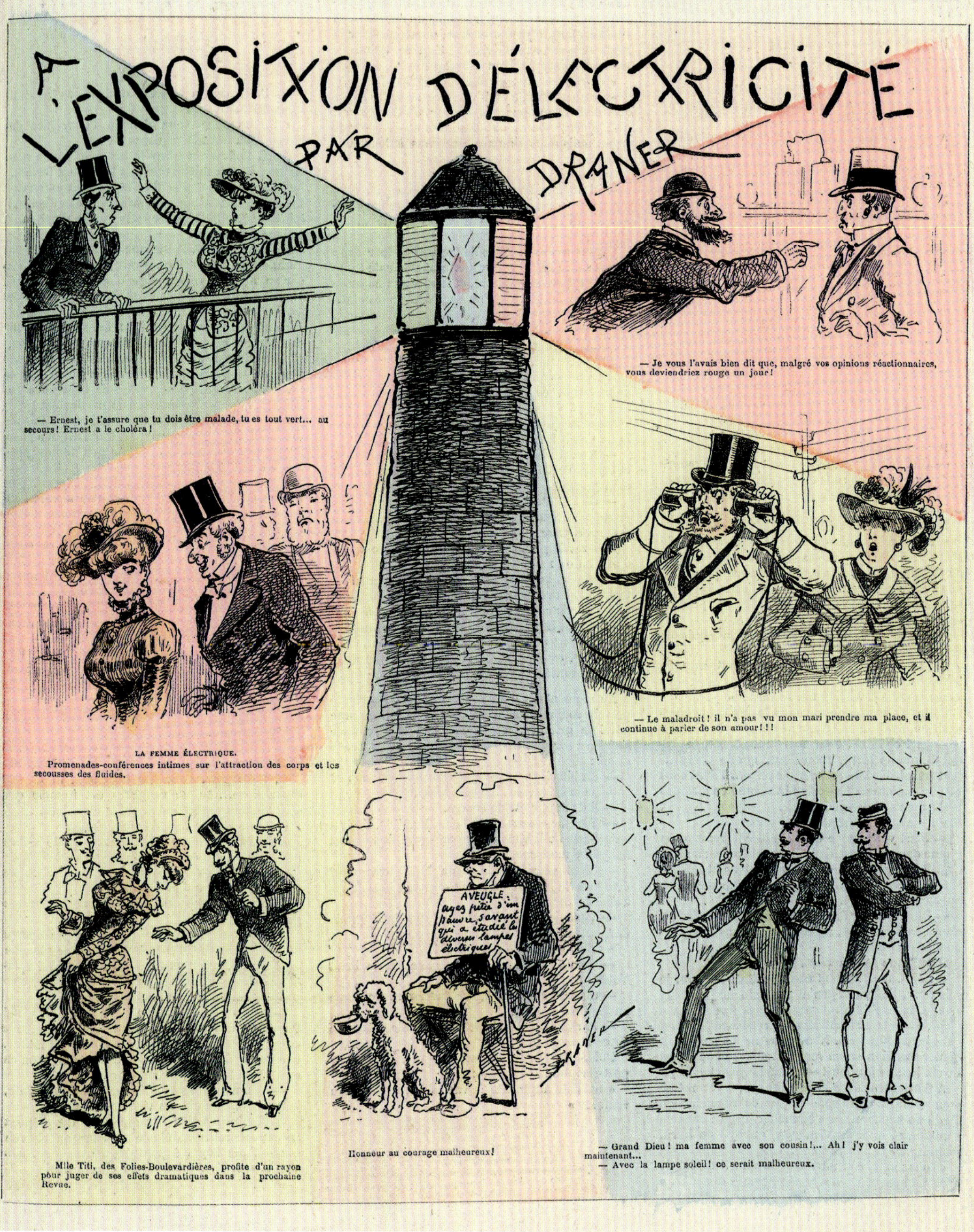

— Ernest, je t'assure que tu dois être malade, tu es tout vert... au secours! Ernest a le choléra!

— Je vous l'avais bien dit que, malgré vos opinions réactionnaires, vous deviendriez rouge un jour!

LA FEMME ÉLECTRIQUE.
Promenades-conférences intimes sur l'attraction des corps et les secousses des fluides.

— Le maladroit! il n'a pas vu mon mari prendre ma place, et il continue à parler de son amour!!!

Mlle Titi, des Folies-Boulevardières, profite d'un rayon pour juger de ses effets dramatiques dans la prochaine Revue.

Honneur au courage malheureux!

— Grand Dieu! ma femme avec son cousin!... Ah! j'y vois clair maintenant...
— Avec la lampe soleil! ce serait malheureux.

a politically reactionary fellow has finally become red, just as his scruffy bearded companion predicted. "I told you that despite your reactionary opinions, you would become red [a Red] one day!"[104]

Sexual innuendo drives the central scene on the left, "La Femme Électrique" (The Electric Woman), bathed in red. Its thematics deliver the program of humor fully onto Asendorf's territory.[105] Thanks to a fantasy dose of electricity, the young woman shows flirtatious interest in the older man alongside her, easily drawing his active attention. Electricity thus proves its power by enabling a mature, top-hatted man to secure the interest of a curvaceous Parisienne. The reader eavesdrops, as a result, upon "intimate promenade-discussions concerning the attraction of bodies and the jolts of fluids."[106]

One square in the series speaks to the conjunction of the utilitarian and spectacular benefits of bright illumination, building on period reactions to electric light like this one by Henri Valette:[107] "Of luminous intensity, electricity gives as much as you want. What our eyes ought to see is not the luminous point, but the illumined object; the sun was not created so that one could look at its brilliant disk, but instead the objects that surround us that the sun is obliged to illuminate." In the yellow square (lower left), Mlle. Titi of the fictional "Folies-Boulevardières" makes strategic use of a ray of light to show off her legs, or rather "she profits from a beam of light to evaluate its dramatic effects in the next Revue."[108] A familiar theme from Second Empire and earlier Third Republic boulevard humor is enhanced when harnessed to the properties of electric light. The bromide is this: an attractive young showgirl does anything to expose her body to public view, and the result is enriched by brighter illumination. But she knowingly yokes the enhanced visibility of her shapely legs to the prospect of augmenting the theatrical allure of a commercial attraction. Electric light serves commerce and female self-commodification all at once.

3.25 Draner, "L'Exposition d'électricité," *La Caricature*, October 22, 1881. Courtesy of Northwestern University Libraries.

The yellow and blue panels at right star a character familiar from Edison's laboratory as satirized in *La Caricature*, "La donna è mobile," but in new contexts. The yellow panel plays on the confusions wrought by the telephone, a favorite device at the exhibition. The adulterous wife is hard on her lover, the unseen fellow at the other end of the phone conversation with her husband. *Le maladroit!* she calls him. "The blunderer! He didn't notice that my husband took my place [on the phone], and continued to talk about his love!!!"[109] It updates the story of a cuckold learning the truth about his wife entrenched by Rococo genre paintings in which purloined letters were the medium of disclosure. The early telephone may have increased the technological complexity of clandestine communication, and even the possibility of being found out, but it also eliminated the material trace of an exchange of adulterous romantic sentiments. Here the yellow color makes a blunder legible, but it carries no particular coding or valence. Yellow-tinged light after all, pre-electricity, was standard, and well suited to the conventional narrative.

The blue panel showcases a clever woman who benefits from the bedazzlement of the fair. Draner cunningly mobilizes a central theme of illumination discourse:

the alignment of "seeing clearly" with "painful seeing." "Good God! My wife with her cousin! Ah, it's plain (clear) to me now."[110] Bluish light illuminated her treachery, but "[w]ith the *lampe soleil*! It would have been painful."[111] One electric light system reveals unsuspected treachery, but a brighter beam would bring ocular misery. Baron Haussmann, the influential partisan and implementer of the city's gaslight network, held the same opinion. Writing his memoirs in the 1880s, he objected to electricity's "unpleasant light" that in its extreme brightness hurts or tries the eyes.[112] Draner's central panel reinforces the danger of exposure to electric lights. The scene that emerges uncannily from the lighthouse's otherwise opaque stone cylinder features the modern illumination scholar's alter ego: a sightless beggar blinded by the lights accompanied by his seeing-eye dog. His handwritten placard reads: "Blind Man. Take pity upon the poor scholar who studied all the diverse electric lamps."[113] Draner's slightly condescending caption: "Hats off to misplaced courage!"[114]

A last double-page spread of cartoons on the exhibition's contraptions and personalities appeared in *La Caricature* on November 26, 1881, again by Draner.[115] The humor ran largely true to course, but the semantic intricacy of "À LA SORTIE" (fig. 3.26), set in twilit shadow, deserves a gloss. The image-caption unit instances the impressive level of complexity that ambitious, apolitical graphic satire could achieve. A snuggling bourgeois couple heads away from the shadowed palace, the glow of whose bright exhibits (including schematic arc globes) is framed by the open door. They saunter into the gathering darkness outdoors. An apparently straightforward scenario of retreat to intimate space is inflected by a complex caption whose highly contingent animal metaphor tests the present-day reader-viewer. Surely spoken by the man (witness his right hand's speaking gesture): "Now that we no longer see so clearly [i.e., see with the help of all that prior light], this would be the moment to light up my cat."[116] Light up his cat, as if it were a portable light source, like a flashlight or a candle?[117] Is it also a sexual double entendre?

Consultation of the *Grand dictionnaire universel du XIXe siècle* by Larousse (1865–90) reveals the prominence of cat metaphors in later nineteenth-century French proverbial speech. Where in English we would refer to a frog or pig, for example, in French it is a cat. The cat comes into focus as a linguistically indispensable term, which carries a great deal of semiotic weight and responsibility while being exceptionally flexible, even labile at the same time. It could even be likened to the dichotomous multiplicities of electric light itself: the agent of enlightenment par excellence that clears away the murk and darkness, but at the same time is overdetermined in the opposite direction as a circumstance—whether present or absent!—that causes optical disorientation that frequently in turn abets mischief.

The expression "wake up the sleeping cat" ("éveiller le chat qui dort"), for example, depends on a cat metaphor that resembles the case at hand. According to the Larousse, it means "arouse a slumbering issue" ("réveiller une affaire assoupie"); "stir up a danger that it would have been prudent to avoid" ("se susciter un danger que l'on aurait dû éviter prudemment.")[118] A sleeping cat is roughly equivalent,

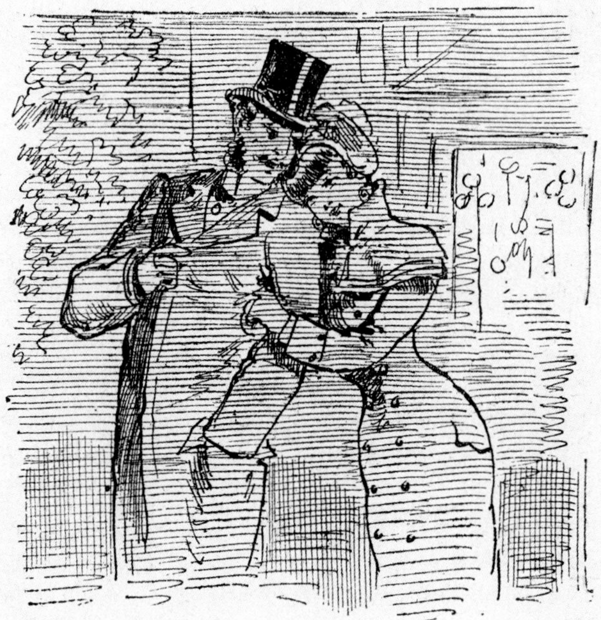

3.26 Draner, "À la sortie," *La Caricature,* November 26, 1881, 383. Courtesy of Northwestern University Libraries.

then, to a thorny issue or a danger that would best remain undisturbed. Might lighting up his cat then suggest firing or stirring up something (dicey or not) that would vividly compensate for the low-stimulus environment of the post-exhibition gloom? As in "throw a cat into someone's legs" ("jeter le chat aux jambes à quelqu'un ou de quelqu'un"), which corresponds to "cause someone trouble" ("lui susciter les embarras.")[119] The cat, in the latter instance, is again a bothersome thing or a hindrance, and presumably lighting up the cat could serve to aggravate the "embarras" (the trouble). But since it is not the cat itself that is vexing, but rather the thrusting of it into someone's path—a problematic and offending move—lighting it up would not eliminate the difficulty, though it might to some degree, borrowing a British maxim, "put the cat among the pigeons," in the strict sense of animating the status quo. The stated intention of the cartoon's top-hatted man reminds us that the practice of *éclairage* in the context of 1881 Paris was always a move toward enhanced vibrancy—whether visual, social, or sexual.

On New Year's Day, 1882, *La Caricature* carried an elaborate fold-out supplement, an oversized bonus picture having been customary in the journal back in the 1830s. At its upper right edge, editor Albert Robida presented a novel Edisonian contraption, the Sentimentalifère Électrique, clever enough to rival even the ingenious Fidèlimètre introduced in 1880 (fig. 3.27). In paradigmatic Robida hyperbole, the caption declares that this new Edison invention would eclipse all of his greatest devices. "Alongside the marvel given birth by Edison, the telephone, the phonograph, and the lamp are mere bagatelles."[120] It returns us to the topos of sex and

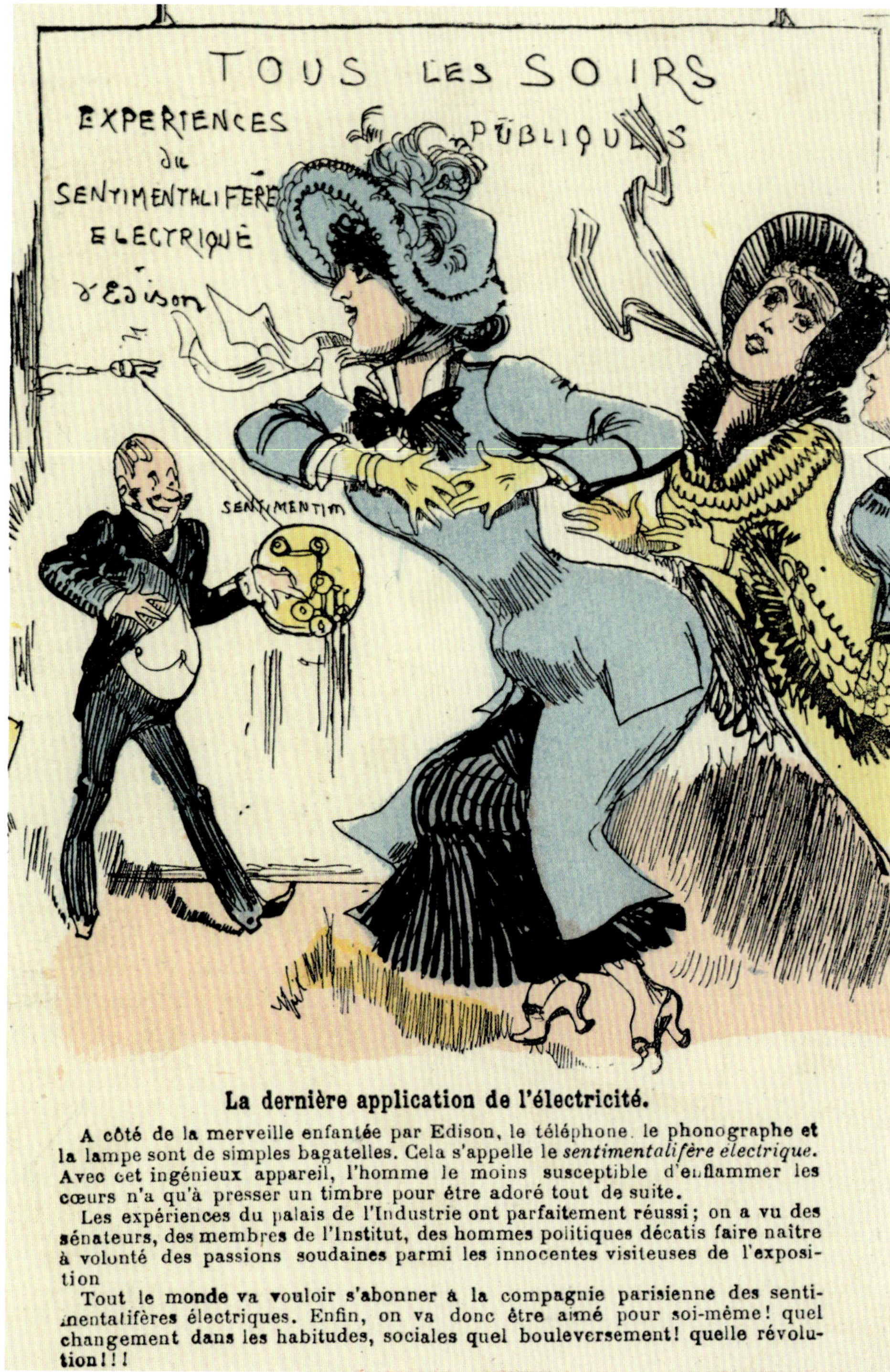

La dernière application de l'électricité.

A côté de la merveille enfantée par Edison, le téléphone, le phonographe et la lampe sont de simples bagatelles. Cela s'appelle le *sentimentalifère électrique.* Avec cet ingénieux appareil, l'homme le moins susceptible d'enflammer les cœurs n'a qu'à presser un timbre pour être adoré tout de suite.

Les expériences du palais de l'Industrie ont parfaitement réussi ; on a vu des sénateurs, des membres de l'Institut, des hommes politiques décatis faire naître à volonté des passions soudaines parmi les innocentes visiteuses de l'exposition

Tout le monde va vouloir s'abonner à la compagnie parisienne des sentimentalifères électriques. Enfin, on va donc être aimé pour soi-même! quel changement dans les habitudes, sociales quel bouleversement! quelle révolution!!!

3.27 Albert Robida, "La Dernière application de l'électricité," *supplément to La Caricature,* January 1, 1882. Courtesy of Northwestern University Libraries.

electricity, in this case electricity's singular value for the libidinous man who would otherwise lack sex appeal.[121] The text explicitly invokes the sexual lessons learned in the space of the prior year's exposition. By simply touching an electric bell, the least likely heartthrobs in France—senators, members of the Institute, and wrinkled politicians—gave rise at will to sudden passions among innocent female visitors to the exhibition.[122] A paunchy man of distinction activates the device, and beautiful young women swoon on cue.

Such satirical maneuvers are credibly interpreted by literature scholar Mike Goode: "The kinds of stereotype-based comedy in which the caricature form engages may hardly be conducive to imagining communities that tolerate difference."[123] In the graphic world of Robida (and Draner), it was gender difference that was most firmly delineated and enforced. The distinctions were cunningly and hilariously exaggerated at every turn, but the humor often entailed duping clever young women. Even electricity and its lights were recruited to secure Belle Époque patriarchy, we might conclude, but in other examples the Parisiennes outfox their male contemporaries, bringing the daring of *La Caricature* into a countervailing orbit, that of the imagination of cultural resistance, Richard Terdiman's remit.[124]

Another important thread to follow is the 1882 program's deep skepticism about electricity. Robida's text is shot through with poisonous irony but also shows profound doubt about electricity's future. As early as the beginning of 1882, the knowledgeable belief that subscribing to any organized Parisian electric company was nothing but a pipe dream has gone into wide circulation. "Everyone will want to subscribe to the Parisian Company of Electric Sentimentalifères. Finally, one will be loved for oneself! What a change in habits, what an overturning of social customs! What a revolution!!!"[125] The cunning doubt that undergirds Robida's last sentence gives onto a topic of central importance: What was the legacy of the revolution of 1881? Did lighting change as a consequence?

Electric Light in the Interior after 1881?

The belated electrification of street light in the French capital was discussed in the introduction. It appears that the lights on display in the Palace of Industry in 1881 convinced many observers that incandescent light was extremely well suited for domestic *éclairage*. This opinion was among the most striking to emerge from the exhibition, a genuine sea change in illumination discourse. Louis Figuier was an early enthusiast: "The veritable and great novelty, the revolution, one can say, that was revealed at the international exhibition of electricity, is the lighting of apartments, which is to say the production of a small volume luminosity, not with the strength of the voltaic arc, but finding itself reduced to the proportions of domestic lighting."[126] To conclude his technical discussion of incandescent light, he argues: "One thus obtains an illumination of a small power, but which is precisely what one has been looking for, for the lighting of the interior of homes."[127] William Henry Preece concurred: "The chief lesson of the Paris exhibition is this, that the arc light is specially suited for external illumination, and that incandescent lamps are eminently adapted for internal and domestic illumination." He moved seamlessly to praise of Edison: "There were several incandescent lamps shown at Paris besides those of Mr. Swan, notably those of Maxim and Lane-Fox; but that which possessed the greatest novelty, and was decidedly the most efficient, was that of Mr. Edison. The distinctive character of the Edison lamp is the remarkable uniformity of its texture and light-giving power."[128]

Did this enthusiasm change lighting practices, and did it leave its mark in the visual arts? In chapter 4, I address Mary Cassatt's begrudging and late adoption of electric light (in 1925) for her posh apartment in the eighth arrondissement in Paris. It turns out that many women of her class opposed the electrification of home lighting because of its deleterious visual effects, or rather because it was unflattering.[129] English sources are particularly rich on later nineteenth-century efforts to oppose or conceal electric lamps. Alice M. Gordon's influential *Decorative Electricity* (1891), for example, advised her readers to "domesticate the electric light by embedding it in traditional and contemporary aesthetic forms that disguised the electrical form of the lighting, thus maintaining the continuity of ocular experience."[130] The specific qualities of the electric light were to be erased rather than celebrated, in what Graeme Gooday called "a move that reveals the traditionalism of progressive electric lighting in the early 1890s, rather than any suggestion of thrusting modernity."[131] The effect of an overhead light was especially unsettling. In Wolfgang Schivelbusch's words: "Traditionally the illuminated space had been structured by the range of the light cast by individual lamps or candles, now . . . the whole room was evenly lit up."[132]

At the same time, there were fierce advocates of electrically lit apartments in the French capital on the grounds of salubriousness: gas was filthy; electricity was clean. The outlook of one incandescent light advocate, "chemist-*électricien*" Georges Fournier, is prototypical: "The better a house is lit [with gaslight], the more insalubrious for the persons who live there."[133] The perfect light should not consume oxygen, should not add breathable matter to the air, should have no element of danger, should produce an agreeable clarity, should be entirely controlled by the person who uses it, and should be available at a price that is not an obstacle to its daily use. "Does this light exist? Yes, it exists. It exists in the use of incandescent electric lamps."[134] Final sentence: "The nineteenth century developed rapid transport of people and ideas by steam, the telegraph, and the telephone; with electric light, it will have inaugurated the light of the future."[135]

The disinclination to employ electric lamps in elite domestic interiors on both sides of the Channel in the final decades of the nineteenth century helps to explain the paucity of imaged homes thus illuminated. Exceptions are the electric interi-

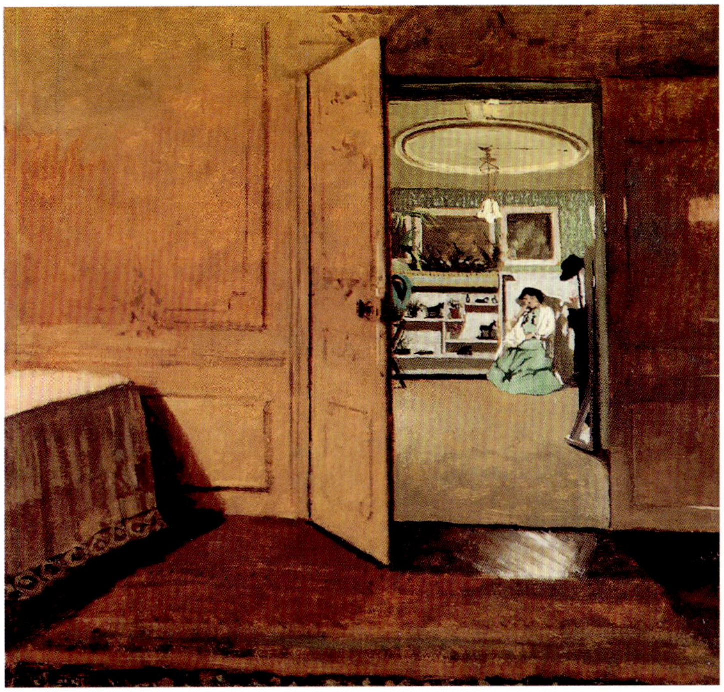

3.28 Felix Vallotton, *Intérieur, soir* (*The Green Room. Interior, Vestibule by Lamplight*), 1904. Oil on board, unframed: 61.6 × 63.5 cm (24¼ × 25 in.); framed: 73 × 74.6 cm (28¾ × 29⅜ in.), Virginia Museum of Fine Arts, Richmond, Virginia.

ors of Swiss-French artist Félix Vallotton (1865–1925).[136] In 1903, the Vallottons moved to a posh apartment on the rue des Belles-Feuilles in the sixteenth arrondissement, which clearly had electrical lighting. Vallotton dove into the rare genre of the painted electric interior shortly thereafter, with such works as *Intérieur, soir* (*The Green Room. Interior, Vestibule by Lamplight* [1904; fig. 3.28]). Looking from the space of the foreground vestibule into the adjoining living or sitting room, the picture boldly addresses the bright whiteness of a glaring pendant tulip lamp that, while minimally decorative, does not conceal the entirety of its incandescent bulb, which bulges into view and emits rays that reflect glaringly onto the shiny parquet floor.[137] He only painted electric interiors in 1904; none thereafter, which drives home the rarity of the small corpus he completed.

Last question: after the mixed response to the illuminations used in the electric Salons, did the 1881 exhibition demonstrate whether electric light was efficacious for the lighting of artworks? Art critic Ernest Chesneau's probing review of the exhibition of paintings electrically illumined on the first floor of the Palace of Industry can serve as a guide to informed opinion on the matter. Chesneau, like Jules Claretie, found antipathy rather than harmony in the relationship between paintings and electric light. Could the realms of the mechanical be reconciled with the expressions of human genius? he asks. The answer: "No, the problem of lighting painting by electric light is not yet resolved."[138] However, as he compared the two systems available, he found incandescent light superior or at least promising in contrast to arc light. His critique of the deleterious impact of arc light on the paintings displayed in the 1881 exhibition is the most detailed I have found. He called the light brutal, cold, abounding in violet rays, and particularly destructive of yellow and green colors; especially hard on landscape painting.[139] He argued that incandescent light, whether Swan's or Edison's, offered advantages in the spheres of both intensity and color.[140] However, paintings illuminated by Edison lamps appear charred (*charbonnées*). Under the circumstances, his advice to painters and architect/decorators who wanted to benefit from the best lighting outcomes was to produce and select paintings on the smallest scale possible, and to prefer matte finishes like wax. The bottom line for Chesneau in 1881: electric light will be with us from here on out, and will produce inconveniences that will need to be remedied as well as possible. This attitude of resignation exemplifies the mainstream, Robida's snide suspicion notwithstanding, in the years following the "Revolution of 1881." Indeed, the adoption of electric lighting in art spaces advanced more rapidly than in the city's apartments.

·4·

Night Lights on Paper

ILLUMINATION IN THE PRINTS OF
MARY CASSATT AND EDGAR DEGAS

A print can sometimes tell us more about the history of art than can a painting. . . . A print is much more revealing of the human network involved in its production than is a painting.

MICHEL MELOT, 1996

Mary Cassatt and Edgar Degas made radically inventive intaglio prints side by side for several years starting in 1879. Their exceptional works on paper (including Degas's monotypes and earlier lithographs) deserve full membership in the orbit of the "Painting of Modern Life." When the designation "painting" is not applied literally, a more inclusive category results: a grouping that delineates a multimedia modernist enterprise without a hierarchy of medium. There are aesthetic, material, ideological, and historical reasons for revamping the reach of the classification, first used capaciously by Charles Baudelaire (as the "Painter of Modern Life") in 1860 to denote a set of attitudes, visual and otherwise, toward urban modernity. He cited an unexpected exemplar, the graphic artist Constantin Guys. The term "Painting of Modern Life" is customarily used in contemporary art history as a placeholder for the Impressionist modernist camp of nineteenth-century painters based in Paris or, as T. J. Clark put it memorably in 1985, the "Art of Manet and His Followers."

Cassatt's and Degas's achievements in the graphic arts demand inclusion.[1] My discussion accedes fully to that desideratum because remapping the borders of nineteenth-century French modernism is logical, overdue, and, somewhat unexpectedly, in synch with debates underway in print circles that uncannily echo tendencies that surfaced among Impressionist print scholars in the 1980s and 1990s. In 1989, Jean-Paul Bouillon, for example, persuasively criticized the tendency to sequester prints from paintings in many histories of Impressionism, including major accounts of the eight Impressionist exhibitions held between 1874 and 1886 where they appeared intermingled with paintings. Bouillon noted acerbically that critics who responded to those Parisian shows at the time were very far from considering the prints on display a minor and negligible annex.[2]

Richard Brettell, Nicole Minder, and Michel Melot entered the discussion in the 1990s. Brettell, in line with Bouillon, insisted that the œuvres of the long list of late nineteenth-century *peintres-graveurs* simply cannot be understood without reference to their prints.[3] Minder and Melot addressed specific traits of period prints, especially etchings. Minder aligned the prejudice against prints in regard to paintings with their comparative severity: "The exclusion or indifference regarding the print was, and still is, essentially due to the qualities of the medium, the austerity of black and white, and the lack of prestige attached to works on paper."[4] Michel Melot's brief on behalf of the printmaking practices of Cassatt and Degas (and their colleague, Camille Pissarro), was also medium-centered. Building on his transformative scholarship of the mid-1990s, he argued in 2005 that their work reversed the connotations of an arsenal of techniques, still widely devoted at that time to reproduction (and hence the popular status of printmaking as frivolous), by placing them in the service of aesthetic innovation.[5]

Peter Parshall, eminent historian of early modern European art including prints, has very recently extended the polemic by challenging curators to rethink the routine sequestration of prints and their staffs in museums, and urging all historians of postmedieval art to make the history of prints an essential part of that history.[6] He aligns still current practice with various theoretical predecessors, but most powerfully for the purposes of this discussion with the c. 1939 interventions of Clement Greenberg.

> He [Greenberg] encouraged a retreat into self-containment, and in so doing he also provided a theoretical undergirding for the separation of curatorial responsibilities. . . . I doubt anyone turned to his criticism as a justification for something that was more or less set by the time he was writing. Nonetheless, one can legitimately claim Greenberg's formalism as theoretical support for this division of art history even though, as far as I am aware, he never wrote a word about a print. . . . The practice of art history should ultimately be one of reintegration, not parceling out.[7]

My embrace and elevation of Cassatt's and Degas's prints is thus located at a point of convergence of multiple historiographic threads.

•

The extremely fertile period of Mary Cassatt's first sustained exertions in printmaking began in the spring of 1879, at the close of the fourth Impressionist exhibition.[8] She began to collaborate with Edgar Degas, Camille Pissarro, and Félix Bracquemond, and she obtained results that stunned both Degas and Bracquemond, the two print veteran experts. Her dive into the world of etching was occasioned by wanting to prepare work for *Le Jour et la nuit*, the projected (failed) collaborative journal of original prints.[9] The first references to the magazine project occur in May 1879 in Degas's correspondence and in the journal of his friend, Daniel Halévy.[10]

The artists believed that they could somehow have their cake and eat it, too, even though the techniques the Impressionist printmakers preferred were those least favorable to reproduction: drypoint, aquatint, and soft-ground etching. As Melot put it starkly: "*Le Jour et la Nuit* could not survive such a contradiction. . . . The failure of *Le Jour et la Nuit* signaled the death of a dream in which art, faithfully served by industry, would survive reproduction without being definitively altered."[11] Indeed, according to every account, we shall never be quite clear what printing process they hoped to employ. The prints made no concessions to publishing in terms of mechanical reproducibility; they were instead in the same aesthetic category as pastels and wash drawings. According to Peter Parshall, by the second half of the nineteenth century, etchings were "the defining art of privacy."[12] This status furthers our awareness of the contradictory nature of the projected journal, and dovetails suggestively with Cassatt's frequent selection of the spaces and themes of the home in her graphic work, resulting in her uneasy valentines to an old-fashioned interior. Indeed, in Degas's discussion of printmaking technologies and techniques, according to Melot, "the close link between the original work and the reproduction, on which the art of printmaking is based, is stretched to [the] breaking point: the original print becomes a unique work of art, a monotype or a 'special edition,' and like a drawing, awaits the invention of a miraculous new reproductive process."[13] The same can be said for Cassatt's intaglio black-and-white prints.

Why did they choose the title, *Le Jour et la nuit*? According to Jean-Paul Bouillon, "the name that they chose for the publication of avant-garde prints, *Day and Night*, admirably summarizes a concentration upon the atmospheric values of black and white, but in the sense of pure form."[14] But there are other theories. Judith Barter, for example, hypothesized that the artists took to heart the ideas of reversal and opposition suggested by *Le Jour et la nuit*, and used them to structure oppositions in their graphic work. She proposes that Degas's c. 1879 monotype *Brothel Scene*, showing a naked standing woman seen from behind leaning her right hand on the back of a curve-topped chair, would have been *la nuit*, while Cassatt's 1881 etching (which exists in many states), *The Visitor*, which reverses the scenario—a fully dressed *bourgeoise* in a domestic interior leans her arm on the back of a similar chair facing the viewer—would have been her opposite, *le jour*.[15] Douglas Druick and Peter Zegers evoke the "Le Jour et la nuit" column that ran in *Le Moniteur universel*, written by Georges Lafenestre, a digest of the trivial doings of everyday life (a kind of "faits-divers"), as a possible inspiration for a printmaking enterprise concerned to focus on similar quotidian topics.[16]

The myriad theories are difficult to appraise, but Bouillon's view that tonal values were the aesthetic foundation of the projected joint venture is unassailable. Visual evidence allows for an amplification of his outlook. The imagery that Bracquemond used in his 1879 sketch for the cover of the journal (fig. 4.1)—especially when looked at alongside one of his earlier etched plate designs, *Notre-Dame*, from 1869–70 (fig. 4.2)—hints tantalizingly that the artist was thinking about the contemporary Paris night and its street furniture as appropriate vehicles for the formal categories of darkness and light. The radiating lines around the two calligraphically swirling instances of the letter L in the sketch for the journal cover seem to be protofuturist rays of light emanating from something like an upright streetlight, albeit stylized and fantastic, shining down and against the inky black night identified and punctuated by a bright crescent moon. The earlier plate design, centered on the Île de la Cité, includes a sequence of *réverbères* suggesting that Bracquemond recognized streetlights as integral and unavoidable elements of the city's appearance, even when the focus of the plate design fell upon Notre Dame.

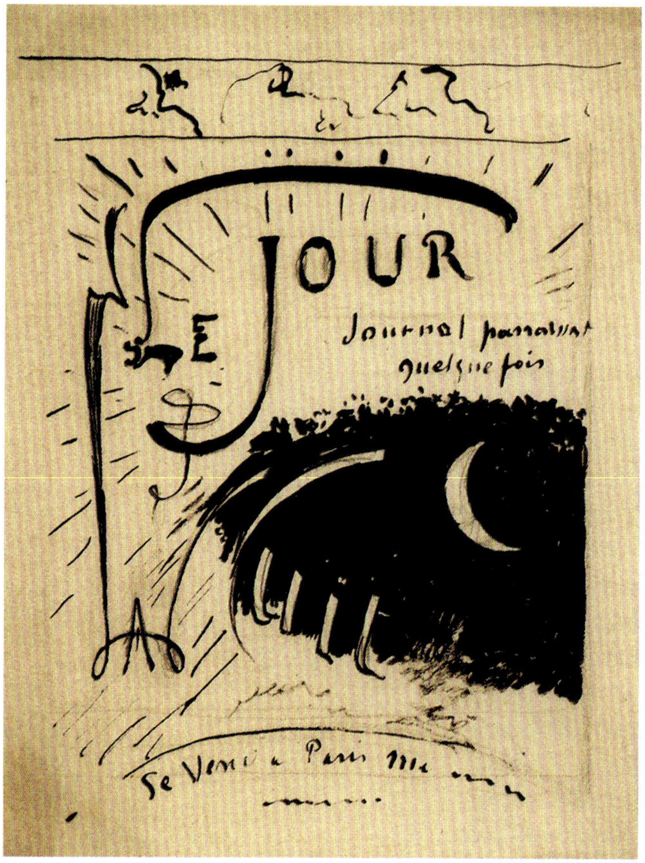

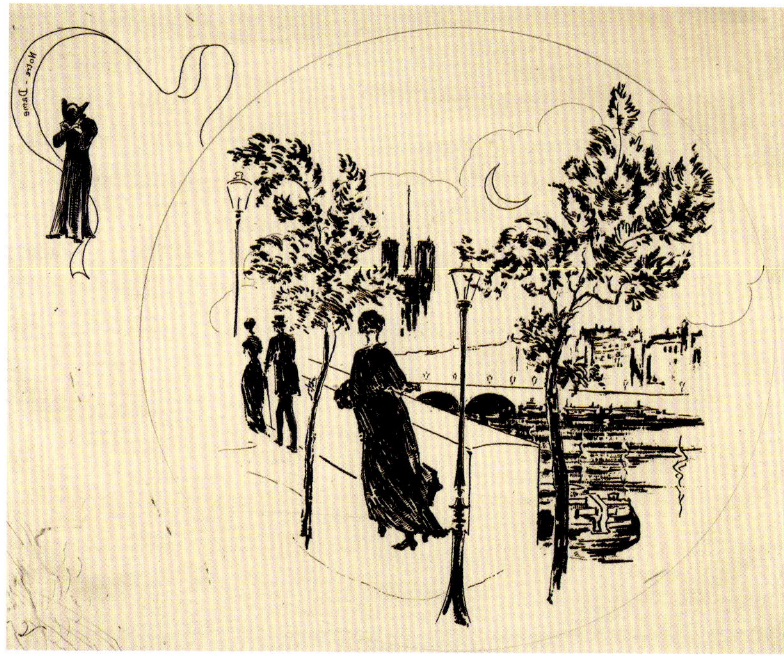

4.1 Félix Bracquemond, cover for *Le Jour et la nuit*, 1879. Pencil and Indian ink wash drawing, 36.5 × 26.5 cm. Private collection.

4.2 Félix Bracquemond, *Notre-Dame* (Decoration for a Plate), 1869–70. Etching on laid paper. Sheet: 30.3 × 47.6 cm ($11\frac{15}{16}$ × $18\frac{3}{4}$ in.). Metropolitan Museum of Art.

I argue that Cassatt and Degas thought that way, too. Another of Druick's and Zeger's insights supports that point of view. They note that Degas's contemporaneous images of the café-concert also "reflect his current fascination with the theme of evening illuminated by artificial light, one of the ideas included in his list for his projected journal."[17] This supports my contention that the artists' encounter with artificial illumination was decisive in fostering their signature artistic originality in their prints replete with brash and quirky tonal contrasts. This perspective on a richly inventive interlude might, even should, evoke the acuity of Clement Greenberg's assessment of the bond between perspicuous description and modernist formal innovation. "The paradox in the evolution of French painting [*sic*] from Courbet to Cézanne is how it was brought to the verge of abstraction in and by its very effort to transcribe visual experience with ever greater fidelity."[18] In the prints under discussion here, the struggle to capture the visual properties of urban illumination played a decisive role.

The magazine failed, but Cassatt's, Degas's and Pissarro's prints intended for the journal were seen in public after all by being shown at the fifth Independent exhibition in 1880. That showing turned out to be particularly notable because no prints appeared in the 1881 exhibition, and, in 1882, Degas and his friends, including Mary

Cassatt, were absent from the show. Barbara Stern Shapiro reminds us that the inclusion of multiple states of the same print in the 1880 show—some of which were signed and annotated—possibly marked the first time that printmakers exhibited trial impressions.[19] They are certainly splendid vestiges of an unsuccessful enterprise. And they raise essential questions about what kind of art printmaking was for these devotees of immediacy and light. Printmaking actually enabled these artists—whose practices were not oriented to the flux of sunlight—to make a radical light-based art. Throughout their collaborative pas de deux, to borrow George Shackelford's apt term,[20] Cassatt and Degas made heavy use of aquatint, as noted above. What matters here is that in such works, original prints—the great oxymoron of the later nineteenth century—became unique works of art, rather like drawings. As Baudelaire put it: something both personal and aristocratic.[21]

What printmaking must also have offered the artists linked by the *Day and Night* project was the benefit and risk of surprise.[22] Intaglio printmakers know well that you do not know exactly what you will get when you peel the damp paper away from the inked copper on the press bed; the effects transferred from plate to paper via ink under pressure are not completely predictable. Rather like turning on a light in an unfamiliar space, intaglio work adds uncertainty and risk to an aesthetic arsenal oriented toward recording aspects of the visible world. Its materiality was well suited to the techno-visuality of lighting—what Emily Dickinson called the "very lunacy of light."[23] The intaglio print was a medium in which optics and aesthetic translation converged to generate unprecedented tonal surprise and complexity.

The unpredictability of printmaking was itself the trait as well as the presenting condition of a kind of Impressionism in an expanded sense. In painting, every additional stroke erased what lay beneath; the newest impression always destroyed the prior painted marks. Prints work in a cognate albeit reversed way. Plates can be altered to define different states that are often accretions, and furthermore each inking (and printing) creates a different impression. The practice of this kind of printmaking is as far from an art of replication as can be imagined. The multiplicity of intaglio printmaking in particular allowed for diversity and variety that actually outstripped the parallel structural capacities of painting. Whereas prints can have multiple states and almost always produce changed impressions, in order for paintings to inscribe multiple responses or aesthetic ideas, they have to be done in a series. There has to be more than one canvas. The multiplicity of the print evoked here points in the direction of the print qualifying as the *ur*-Impressionist object.[24]

Germane to the clockwork, black-and-white pictures of Parisian life under consideration here is the fact that the dates of the pictures (mostly between 1877 and 1880) fall within a particularly active and exciting era in the history of the industrialization of light in Paris. I argue here that two developments that helped to create modern Paris, new urban lighting and innovative printmaking, were entangled—a paradigmatic if quirky instance of a link between technological modernity and pictorial modernism.[25] Specialists in the graphic arts have observed, though without linking the events, that the inauguration of a new light environment in Paris streets also marked an era of tremendous consequence for printmaking. Nicole Minder, for example, while not positing a causal connection between the phenomena, observed that "the years 1879–80 constituted . . . a pivotal date for the development of the graphic arts."[26] I argue here that the new *éclairage* in the street and the innovative prints coming out of the modernist's studio did not merely coexist, but were so closely coupled that the former was the ideational and visual matrix of crucial examples of the latter. In the process I fearlessly counter the judgment of two leading historians of light: "The graphic arts were probably the least affected by developments in lighting technology."[27] I do so by pressing "lamps" and "light" (or rather "lighting" and "light") into intimate association though my wager that Cassatt and Degas often thought *éclairage* and *lumière* concurrently.

I believe that the era's nonstop march of brighter and higher tech light systems into both private and public spaces informed Cassatt's and Degas's prints at an obvious and fundamental level insofar as the syntax of their innovative art literally was darkness and light. But the new light in Paris further informed their art because light as a constituent of Parisian modernity had both an economics and a poetics, and for progressive artists the new silent, glaring light was a figure for the contrary coexistence in modernity of sheltered privacy and freedom, but also of mechanization and control.[28]

The challenge of managing the dazzle, the often blinding brilliance, was a constant concern during the early years of the history of electric light out of doors. Indeed, illumination and bedazzlement were routinely construed as antagonistic components of the new light.[29] One of the striking features of Edison's 1881 display, according to his devotees, was his incandescent light's domestication of the blazing electric beast. According to Henri de Parville, an astute observer of the exhibition:

> We are accustomed to representing electric light to ourselves in the form of dazzling, scintillating fires, hard on the eye, burning, constantly changing in intensity, in variable and wan tones. Here, on the contrary, one has a light that has been in some sense civilized, accommodated to our customs, put at our disposal (or level); each lamp illuminates like gas, but like a gas that had to be invented, a gas providing light of a perfect fixity, gay and brilliant without bothering the retina.[30]

In a social world where the new *éclairage*—its benefits, its drawbacks—was on everyone's lips (and in everyone's eyes), Cassatt and Degas plunged into etching.[31]

Edgar Degas

When Degas began his collaboration with Cassatt, he was already a graphic artist with an interest in artificial illumination, its visualities and machinery, often as the lights struggled to forestall and withstand the surrounding murk. J.-K. Huysmans singled out a pastel that manifested that interest, *Étude de loge* (fig. 4.3), in his review

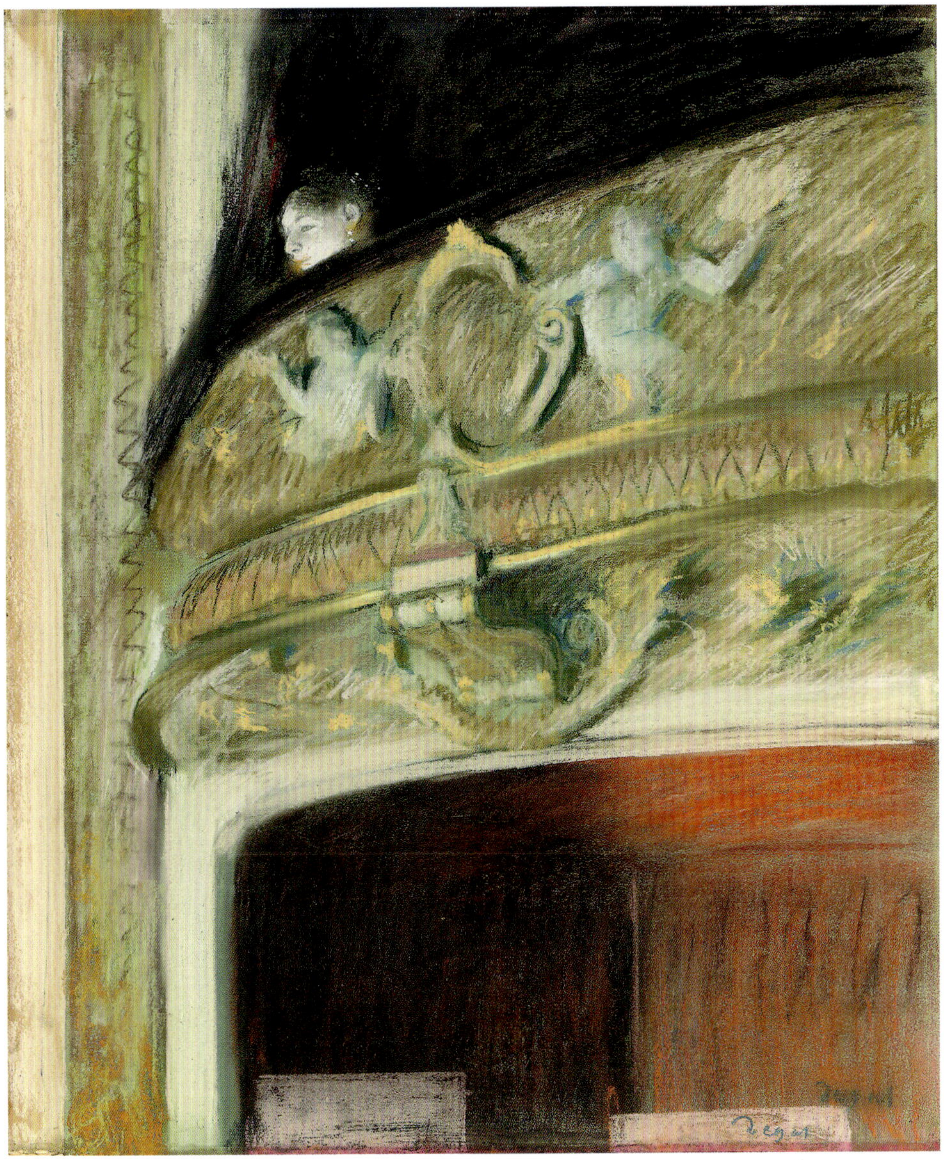

4.3 Edgar Degas, *La Loge*, 1880, detail. Pastel on paper laid down on board. 66 × 53 cm (26 × 20⅞ in.). Private Collection. Photo © Christie's Images/Bridgeman Images.

of the 1880 Impressionist exhibition. The critic admired its record of the effects of light falling on the face of a solitary woman at the edge of a theater box. "What a study of the effects of light!"[32] Elsewhere in his review, Huysmans again praised Degas's ability to capture the look and impact of contemporary *éclairage*. He is thrilled by a ballet picture that he does not identify: "The time of apprenticeship is finished and there they are appearing in public, on point, in the rays of gaslight, in the jets of electric light."[33]

Huysmans was right to identify Degas as an artist tuned in to lighting effects. We know that Druick and Zegers proposed that the interest in artificial illumination

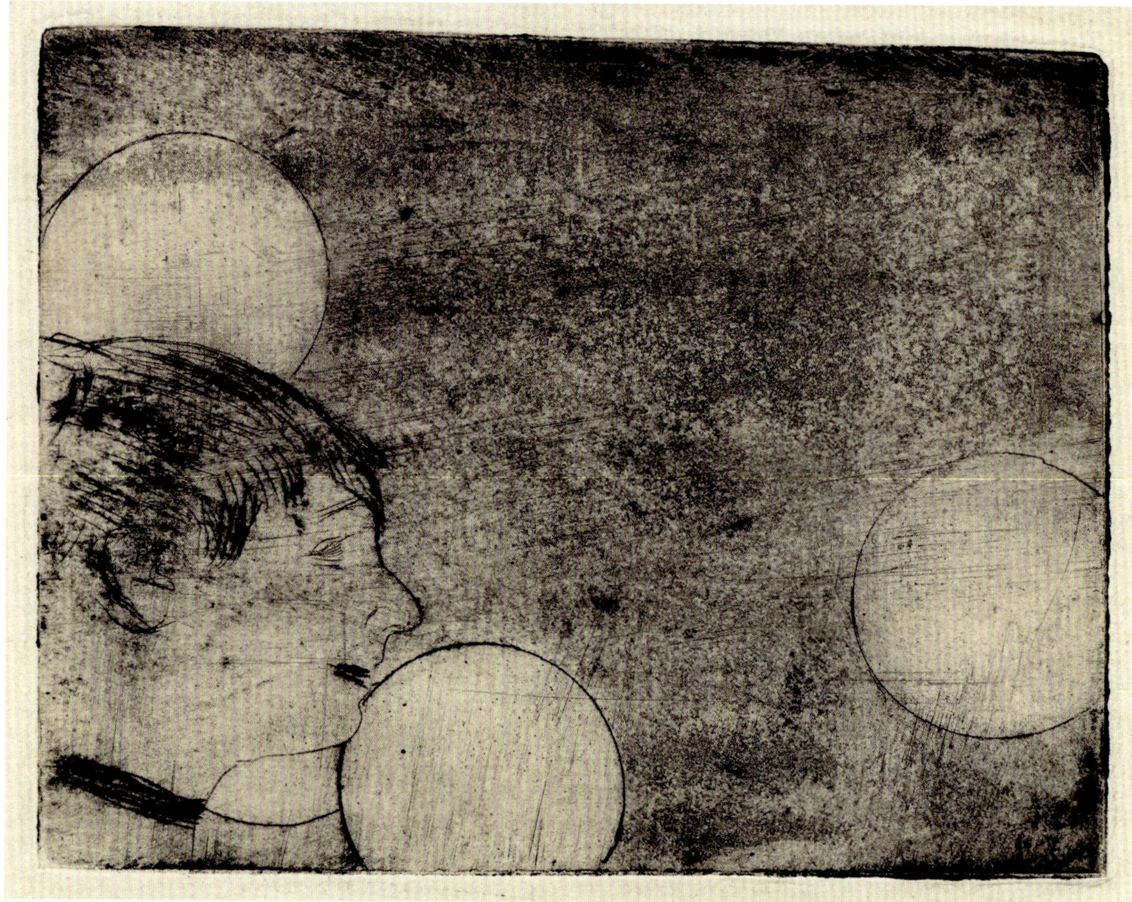

4.4 Edgar Degas, *At the Café des Ambassadeurs*, c. 1879–80. Etching, softground, drypoint, and aquatint on laid paper. Sheet: 47.6 × 31.6 cm (18¾ × 12⁷⁄₁₆ in.). Sterling and Francine Clark Art Institute, Williamstown, MA, 1962.31. Image © Sterling and Francine Clark Art Institute, Williamstown, MA (photo by Michael Agee).

may have been behind the journal project, *Le Jour et la nuit*. That the artist wrote about the thematics of light is further proof. He recorded this idea, for example, in a notebook in about 1876: "On evening—infinite variety of subjects in cafés—different tones of the glass globes reflected in the mirrors."[34]

Degas incorporated commercial lighting contraptions expertly and often in his intaglio prints of the mid- to late 1870s. In them—preponderantly indoor (or threshold) pictures—he often used industrialized light as a marker of mechanized urban modernity, as well as the gaudiness of working women, via the frequent juxtaposition and rhyme of light fixtures with their heads. *Singer's Profile* (1875–78; fig. 4.5), a whimsical, tiny etching (only about 7 × 9 cm [2¾ × 3½ in.]), is the *locus classicus* of Degas's ingrained tendency to align lamp globes with women's heads, suggesting an analogy between vulgarity and brashness. The adjacency of a female performer's profile to four light globes in this work sets up a kinship between and among the spheres. The young woman's winsome expression distinguishes her from the machinic object world, but the quartet of circles reads as a family of cognate forms nonetheless.

The shape and meaning of the spheres in Degas's etching echo a metaphor used in 1882 by Guy de Maupassant in his "Claire de lune," later archived by Walter Benjamin.

> I reached the Champs-Elysées, where the cafés concerts seemed like blazing hearths among the leaves. The chestnut trees, brushed with yellow light, had the look of painted objects, the look of phosphorescent trees. And the electric globes—like shimmering, pale moons, like moon eggs fallen from the sky, like monstrous, living pearls—dimmed, with their nacreous glow, mysterious and regal, the flaring jets of gas, of ugly, dirty gas, and the garlands of colored glass.[35]

The shared attentiveness to the glow of "moon eggs" is striking; gas globes in Degas, electric orbs in Maupassant. *At the Café des Ambassadeurs* (c. 1879–80; fig. 4.4) endows a gas candelabrum with considerable power. The stage performer appears involuntarily but inescapably drawn to the grapelike cluster of moon eggs, gleaming but remotely, ineffective against the nightclub's murk, floating like a specter, but luring the female entertainer toward it; she leans into its orbit; toward its sphere of influence.

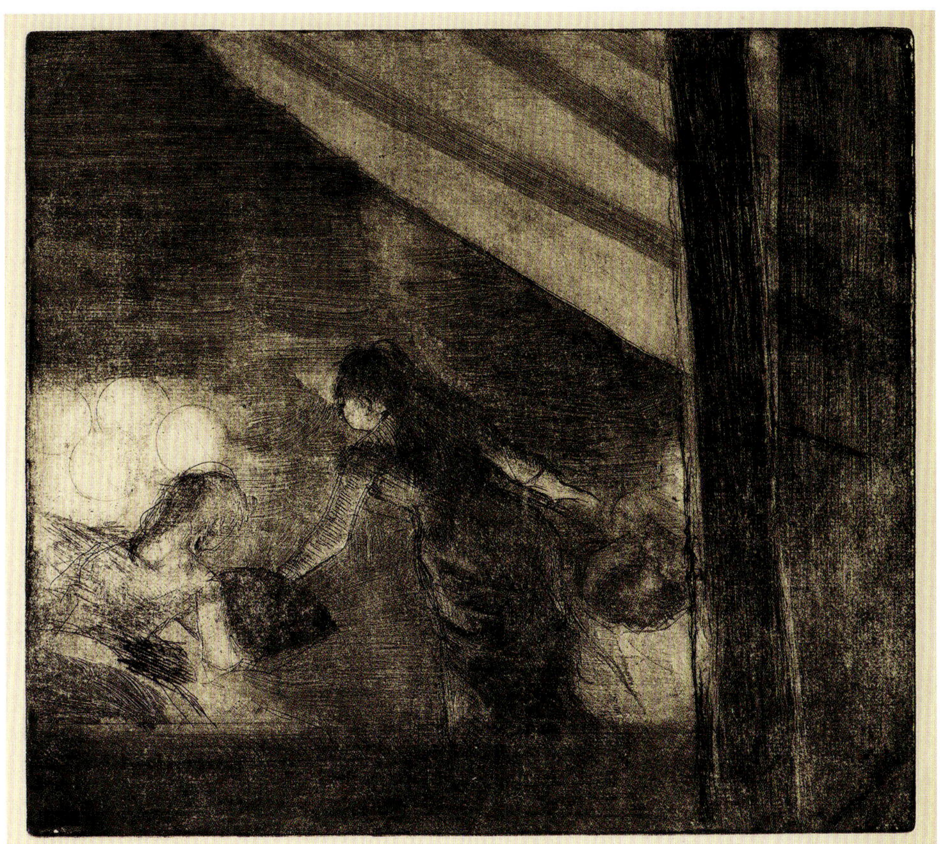

4.5 Edgar Degas, *Singer's Profile* (*Profile de chanteuse*), c. 1875–78. Etching, drypoint, and aquatint on paper. Plate: 6.8 × 8.6 cm (2 11/16 × 3 3/8 in.). Sterling and Francine Clark Art Institute, Williamstown, MA, 1967.21. Image © Sterling and Francine Clark Art Institute, Williamstown, MA (photo by Michael Agee).

A postscript to an 1879 letter to Bracquemond about plans for the next Impressionist show displays Degas's familiarity with a specific technology then appearing amid great ballyhoo in his immediate environment: "The Company Jablockof [*sic*] proposes to do the lighting with electric light."[36] An irresistible anecdote concerning Degas's ad-hoc art supply choice strengthens the case for identifying him as a lighting maven. Alexis Rouart reported that when Degas was detained by bad weather one day around 1879 without his tools, he just picked up what he found lying around the Rouart household that struck him as adequate. He only needed, after all, to find something harder than the copper plate he would be working. He reached for a carbon rod, no doubt discarded from an electric arc lamp, which he took to using because it made an agreeable gray mark upon the copper. He baptized it his "electric pencil."[37] This takes the posited connection between technologized illumination and printmaking onto the terrain of the material, quite literally, while it confirms that in these years Degas's and Cassatt's practice occurred literally in the shadow (or the light) of electric technologies.

While Degas's attention to the tonalities and social valences of lighting frequently inflected his intaglio prints, the monotype was his defining (*ur*-) medium. It was also his most important pre-intaglio graphic medium. Marc Rosen and Susan Pinsky observe: "By the time Degas had invited Cassatt to join in the etching project for *Le Jour et La Nuit*, he had made over one hundred monotypes."[38] I argue that his monotypes allegorize the complexity of his intermedial work and imagination. Michel Melot observes that in the artist's discussion of printmaking technologies and techniques, "the close link between the original work and the reproduction, on which the art of printmaking is based, is stretched to [the] breaking point: the original print becomes a unique work of art, a monotype or a 'special edition,' and like a drawing, awaits the invention of a miraculous new reproductive process."[39] Following this logic, all his prints can be considered monotypes. The monotypes concentrate the originality of Degas's graphic art, and many of them show an interest in the visualities of artificial illumination, as in the two cognate monotypes discussed in what follows.[40]

The setting of Degas's monotype *Café Singer* (c. 1877–78; fig. 4.6), is a largely indistinct café-concert equipped for stage entertainment. The shapes that bracket the main woman are essentially human-shaped blurs against which only one gloved hand holding an open fan is clearly defined. The dark-haired female performer, the sole legible figure, is, by contrast, a creature of theatrical illumination: beams of light starkly model her body and head from below. The glare of unseen footlights sculpts her arms in stark darks and lights. Her brightly lit face (on a head that tilts forward dreamily) contrasts sharply with her dark smudge of a mouth and barely defined gray eye sockets, which suggest eyes closed to avoid the glare. Degas frequently whitened female entertainers' faces with white pastel strokes to enhance the effect of glaring lights—for example, in *Singer with a Glove* (c. 1878; pastel on canvas), in Harvard's Fogg Museum.

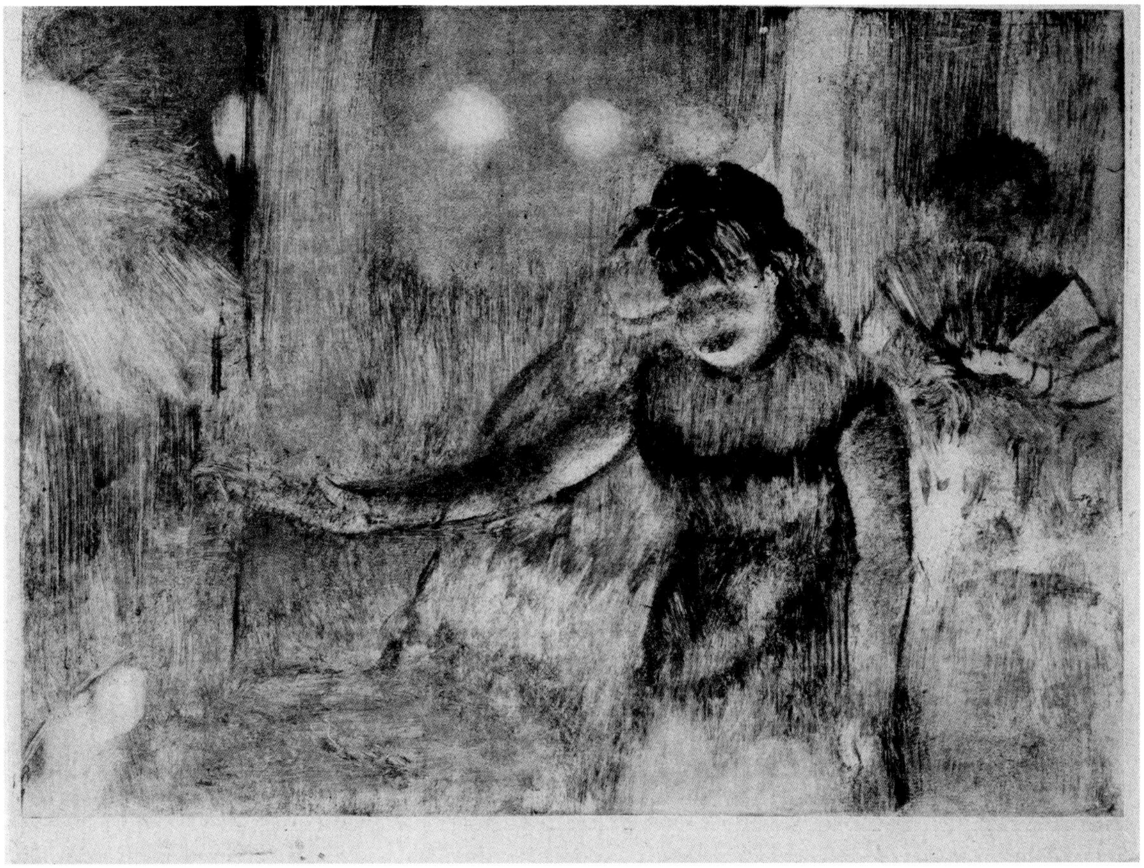

4.6 Edgar Degas, *Café Singer (Chanteuse du café-concert)*, c. 1877–78. Monotype on paper. Private collection.

The print is chockablock with lamps that shine into view. A bright white, round globe blazes at left, emitting rays that form a corona to denote the light's piercing brilliance. It must be a Jablochkoff candle (an electric arc light), in use in some clubs in the later 1870s, like the ones Degas evoked in his letter to Bracquemond and that served as Maupassant's points of departure for his 1882 "moon eggs." It unambiguously defines an outdoor space. The three white balls just above the singer's head are surely gas globes, defined here as elsewhere, as unmodeled spheres. Following the connotative logic of rhyming things and people, the trio of light fixtures echoes and verifies the presence of three women performers. The identity of the fifth lamp (second from the left) is less certain, but, despite the slight irrationality of its placement, its oval spot of light and dark body suggests a gas streetlight (*un réverbère*). This sheet deserves to be counted among Degas's most virtuosic monochromatic orchestrations of multiple kinds of night light. Two others are well-known lithographs: *Mlle Bécat at the Café des Ambassadeurs: Three Motifs* (c. 1875), based upon monotypes, and *Mlle Bécat at the Café des Ambassadeurs* (1877–78; fig. 4.7). The latter is a symphony of illumination; it includes a large gas lamppost (it may be an arc light) and a cluster of gas globes above a string of gas fixtures, all at the right, which are

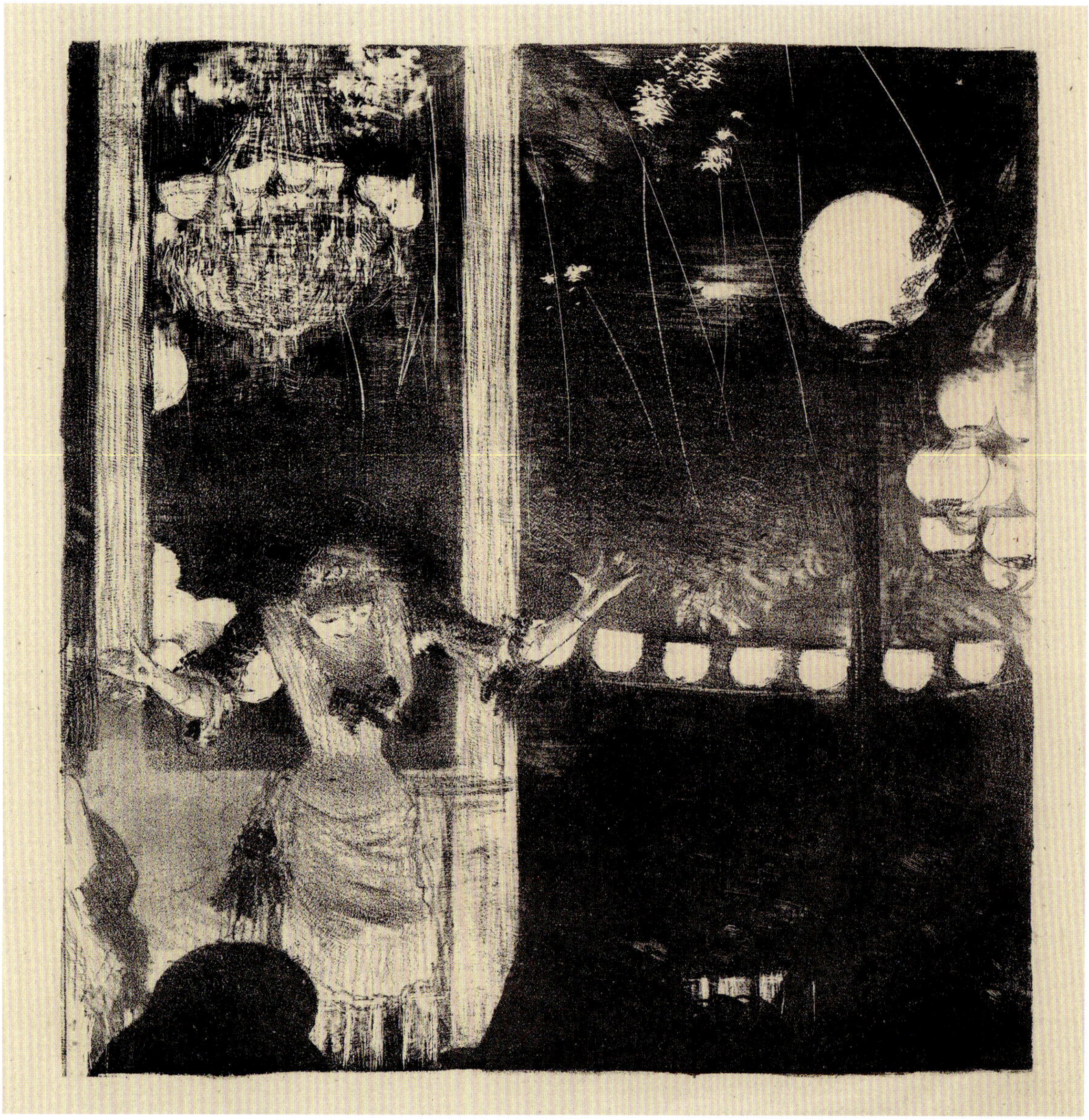

4.7 Edgar Degas, *Mlle Bécat at the Café des Ambassadeurs, Paris*, 1877–78. Lithograph on wove paper. Metropolitan Museum of Art.

reflected in the large mirror at left along with a spectacular chandelier. The sky is shared by a clouded moon and the tendrils of falling fireworks. Final touch: the performer's face, like that of the star of *Café Singer*, is brightly lit by unseen footlights.

If complex *éclairage* is one of the distinguishing features of the monotype *Café Singer*, the other is its solitary and striking performance gesture: the woman's expansively extended right arm, a cylindrical pneumatic sausage stretched horizontally across the page. Her limb appears too long, even distended, and as if buoyed up by and floating upon the light that blazes onto its lower surface.

The pastel *Singers on the Stage* (1877–79; fig. 4.8) significantly modifies the logic of the tonal foundation common to it and the monotype *Café Singer*. The makeover of the figures and their spatial setting pushes to the breaking point our belief in the sibling relationship between the two works on paper. The transformed *mise-en-scène* alters both the axis and temporality of the routine pursued by the main figure: she is no longer facing the audience, now obviously off to the left, so she either has finished her turn (now heading toward the wings) or has not yet begun singing.[41] The expression of the pink-frocked lead performer shifts from a smile accompanying a gentle melting into the light, ostensibly directed toward limelight and customers, to a pinched hesitancy before the now mitigated, even erased onslaught of stage lighting. (And the woman at right is jarringly caricatured.) Not just the subtly altered axis of the central woman's head but her newly cautious or even worried facial expression, coupled with her swankier hairstyle, redefine her comportment.

4.8 Edgar Degas, *Singers on the Stage*, 1877–79. Pastel, over monotype, on ivory wove paper, laid down on board. Art Institute of Chicago.

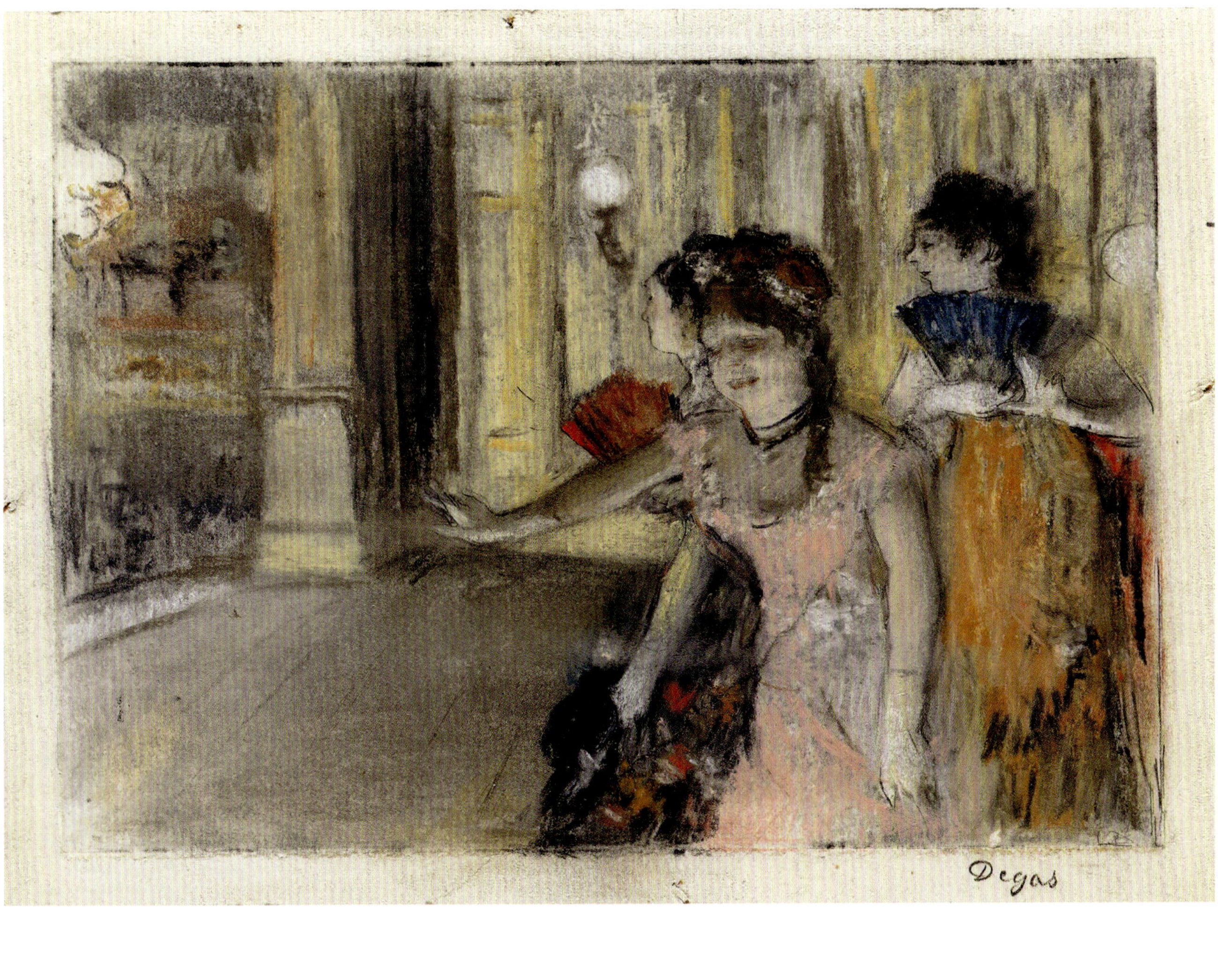

That a tense demeanor defines an entertainer far away from the space of performance is puzzling. Her right arm still extends to the right, now toward the audience, but it is shortened, and its distortion and stark modeling are gone. Its ramrod straightness, however, augments the strain, not to mention the odd note struck by the eye-catching acute angle between her arm and the downward pointing arm of the sharply drawn performer with a red fan. This conjunction of two lean arms, which resemble the hands of a clock sitting almost at the center of the sheet, is disruptive, and muddies the definition of the space occupied by the two women with the spiky limbs.

Another telling modification is the transformation of the lighting. The lamps in the pastel are less numerous, motley, and ferocious; all the light is gaslight. An upmarket gas sconce (*un applique à gaz*) replaces the naked gas "moon eggs." Most significant, the dazzling arc light at the far left of the monotype, as well as the ostensible gas streetlight, both markers of outdoor space, are gone; they have been replaced by one elegant, multiglobe gas chandelier (*un lustre à gaz*) suspended over the audience. It secures the establishment's identity as a theater. Theaters in Paris were not lit by electricity in the later 1870s.[42]

Two extraordinary further details illustrate Degas's abiding interest in the visualities of artificial light. At the top left, to the right of the chandelier, is a sequence of sawtooth, gray lines that imply both the rounded shape of an apparitional lamp and the jaggedness of its rays. These strokes are surrogates for the dazzle seen in the monotype *Café Singer* and concealed by the pastel surface of *Singers on Stage*. The artist could not resist experimenting with indications of the brilliance of artificial light. Final observation: on the lip of the stage (the diagonal at far left), the sequence of bright white marks undeniably denotes limelight (a type of gaslight).[43] Might this glare explain the pained expression, angular gesture, and averted gaze of the woman in pink?

Mary Cassatt

Mary Cassatt was introduced to multiple innovative printmaking methods in the milieu of intensive collaboration that she entered in the spring of 1879. This included the possibilities of mixing and superimposing intaglio techniques—etching, drypoint, softground, and aquatint (what Lucien Pissarro called "an unorthodox cuisine")—which she then applied in unconventional ways and combinations to copperplates.[44] She applied what Melot calls "the Degas method" of working over the plate until it was either completely filled up or worn out. She shortly became his equal. "She moved as easily as he did between paper and copper, and copper and canvas."[45]

Melot, Cassatt's most persuasive interpreter, has stressed the inseparability of work in the media of painting and printmaking in the case of a new *peintre-graveur* like Mary Cassatt. "She is a typical example of the painter-printmaker, a painter who practiced printmaking as if it were another branch of painting. She worked on her plates as an artist works on a canvas."[46] She was also industrious: between 1878 and

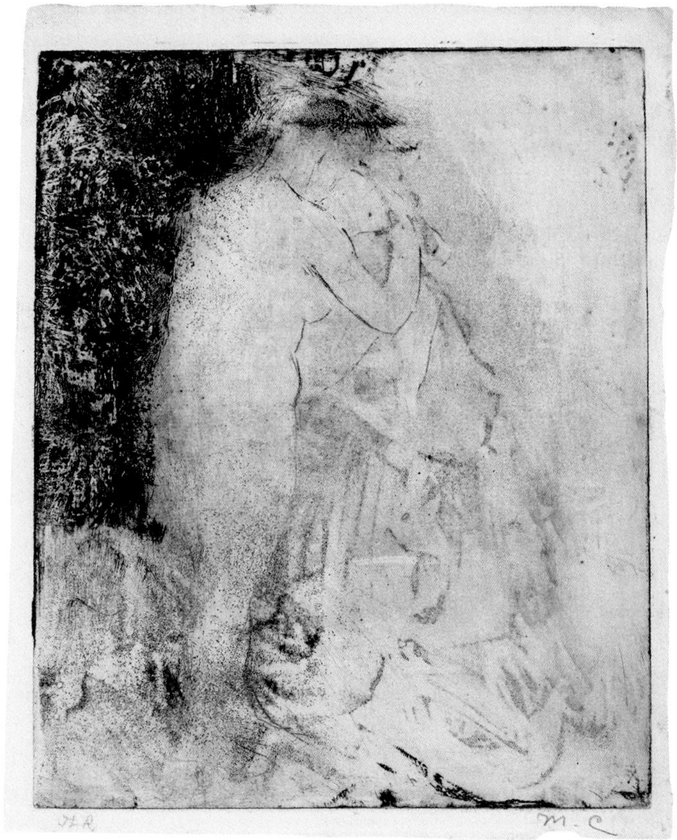

4.9 Mary Cassatt, *Standing Nude with a Towel*, c. 1879. Soft ground etching and aquatint. Courtesy of the Mary and Leigh Block Museum of Art, Northwestern University, 2009.1.

1883, she made approximately eighty prints, a very sizable number. She was painting in the same years. Melot's assessment of the similarity between her approaches to canvas and to copper pertains to the bravado and painterliness of both her soft ground and aquatint use. A highly experimental 1879 soft ground and aquatint, *Standing Nude with a Towel* (fig. 4.9), is an enthralling demonstration of her technical fearlessness.[47]

Melot has also pinpointed the contradictory and singular other side of Cassatt's intense aesthetic and technical marriage of painting and printing: "The very attributes that now appear to be transforming [Cassatt] into an emblematic figure are curiously those for which she was slighted at the outset of her career: she was a foreigner, a woman and she devoted time, talent, and energy she diverted from painting—or so the reproach must have gone—to the frivolous and popular art of printmaking."[48] He speculates credibly: "Perhaps it took an American woman to become seriously interested in the 'minor' art of printmaking and to struggle stubbornly and ingeniously with the technical issues of the medium."[49] He even proposes that "it was printmaking that kept Cassatt in France until her death in 1926 despite its nationalism and misogyny."[50]

Cassatt did not—could not—represent outdoor light, but she engaged the reputation and startling visuality of public *éclairage* in her theater interiors.[51] Among the first batch of pictures that Cassatt put on display in an Impressionist exhibition (four of them set in theater boxes, among nine pieces in the 1879 show) was a remarkable pastel, *At the Theater* (*Au théâtre*) (1878–79; fig. 4.10), that incorporates a reflected, twinkling gas chandelier, highlighted with metallic silver pigment.[52] The rhyme and interweave of the multiple rounded and curving shapes create an intricate interlaced design comprising the light fixture, the woman's volumetric profiled head and reflected hair, and her splayed, semicircular fan up against the curving balcony. The many strokes of acid yellow-green evoke the garish gaslight color spectrum. The tension that Cassatt engineered between the prim pink and aversive theatergoer, on one hand, and a lurid and glitzy public entertainment environment, on the

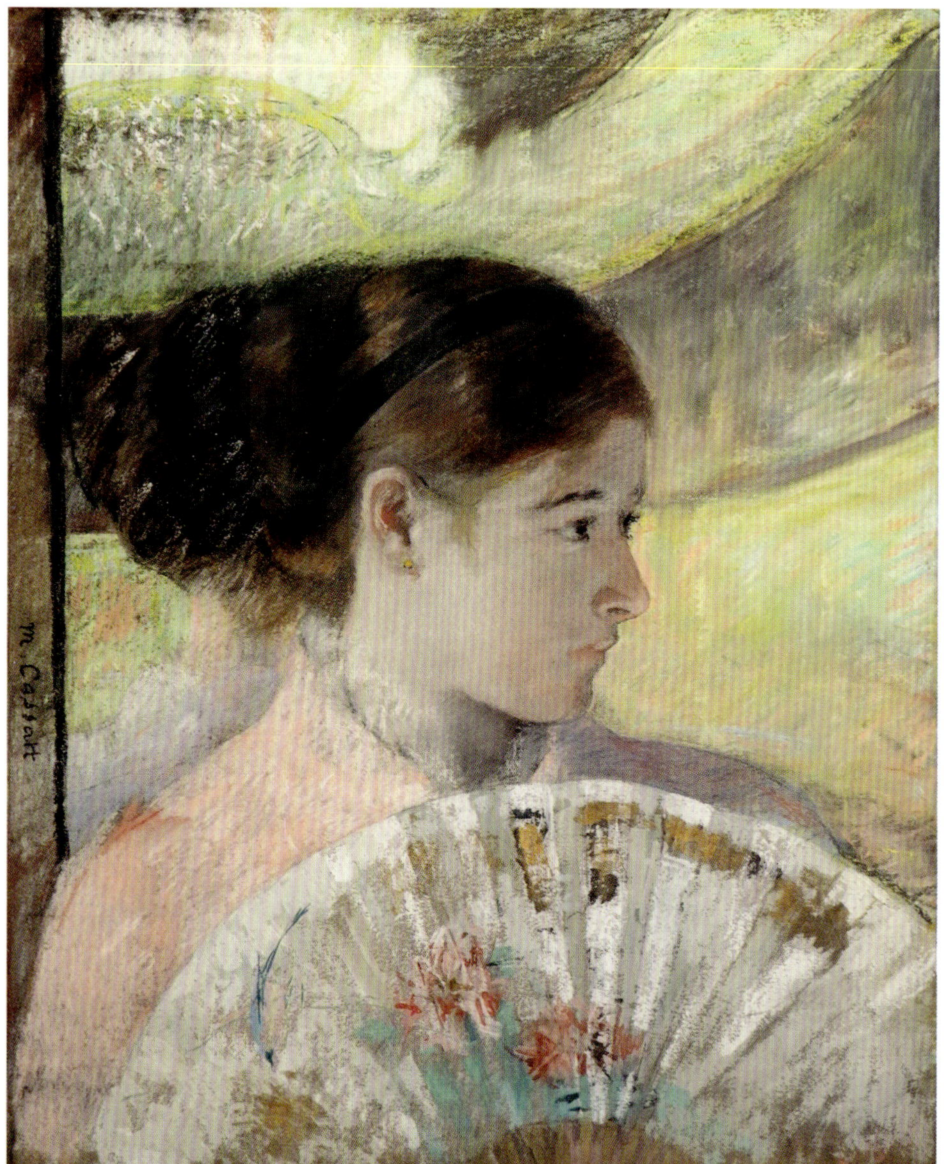

4.10 Mary Cassatt, *At the Theater*, c. 1880. Pastel and gouache on paper. Overall dimensions: 64.8 × 54.6 cm (25½ × 21½ in.). Photo: Colin McRae, August 27, 2006. Courtesy of the Ann and Gordon Getty Collection.

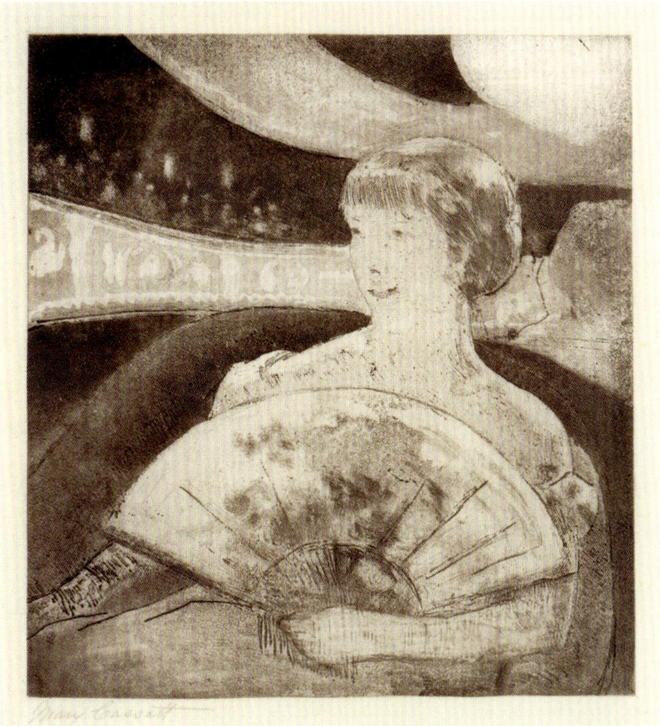

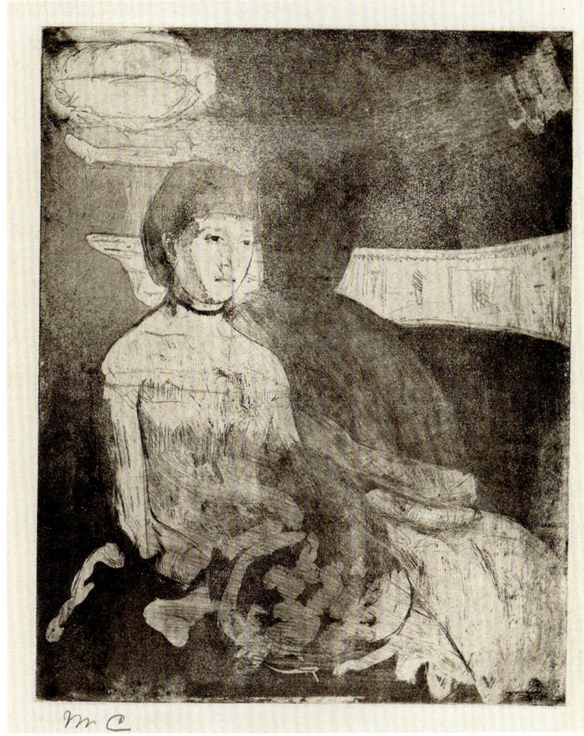

other, is a conspicuous ideological dimension of the image. While Degas assimilated the female *dramatis personae* of his brashly lit entertainment contexts to the technological world of the light apparatuses themselves, Cassatt tended to erect, or at least imply, a gulf, even a clash, between them.

In the Opera Box (*Femme au théâtre*) (1879–80; fig. 4.11), was to be her contribution to the first issue of *Le Jour et la nuit*; it was printed in a large edition of fifty.[53] It was obviously a print she considered significant and successful. Its subtle devising of variegated zones and spots of light is a defining trait of the picture. The irregular spots of light appear to pulse among unseen audience members. They are principal aesthetic events in the print, and they conjure more life, lively atmosphere, and organicity than does the woman's voided-out, flat face. Again Cassatt has built a disparity between a female attendee and a light-saturated public environment.

Two Young Ladies Seated in a Loge, Facing Right (1879–80; fig. 4.12) contains a bewildering phantom double.[54] A ghostly twin? A shadowy and indistinct reflection? Or are there really two women? The experts have not always been so sure. Here Cassatt pictures artificial light (the gas chandelier) as a Janus-faced harbinger of illumination as clarity and indistinctness simultaneously. She has again crafted the eerie appearance of a female subject destabilized, redefined, and, in this case, split by dazzling lights.

A contentious 1881 statement by J.-K. Huysmans about Cassatt's domestic scenes will introduce a close consideration of her library etchings:

4.11 Mary Cassatt, *In the Opera Box (No. 3)*, c. 1880. Aquatint and soft-ground etching with open bite, brush bite, and burnishing. 30.6 × 26.8 cm (12 1/16 × 10 9/16 in.). Photo: Rick Hall. Courtesy of Blanton Museum of Art, University of Texas at Austin, Gift of Susan Garwood in memory of Sue Ann Reagan, 2004.

4.12 Mary Cassatt, *Two Young Ladies Seated in a Loge, Facing Right*, 1879–80. Soft-ground and aquatint on paper. Plate: 27.6 × 21.6 cm (10 7/8 × 8 1/2 in.); sheet: 44.8 × 28 cm (17 5/8 × 11 in.). Initialed: M.C.; annotated: JLR. Courtesy of Adelson Galleries, New York.

> Mlle. Cassatt, who is American, I believe, paints French women for us; but, she endows her very Parisian abodes with the benevolent smile of the "at home": she delivers, to [us in] Paris, what none of our painters would know how to express—the joyous quietude, the tranquil good cheer of an interior.[55]

Huysmans's opinion of Cassatt's painted domestic interiors might apply to some of the exhibited canvases, but cannot be made to characterize her graphic works on the theme, insofar as the prints frequently shatter the tranquility of an interior, and do so speaking a language of troubling *éclairage*. In that regard, the graphic interiors repeat and echo some of the woman-versus-light-turbulence scenarios she fashioned in the theater prints. Royal Cortissoz, the irascible American antimodernist critic, understood the exceptional conjunction of the *mise-en-scène* of the domestic interior and ferocious formal innovation that characterized her work. To wit, shortly after Cassatt's death in 1926, he observed, "It is an amusing paradox in her history that her force, her penetrating vision, her technical clarity were wreaked largely upon the most fragile of themes."[56]

The prominence of a glowing lamp in almost a dozen of her etched domestic interiors, made between 1879 and 1882 and set in the library of her own apartment in the ninth arrondissement in Paris, is an aesthetic and thematic choice of great interest.[57] For example, *Mrs. Cassatt and Lydia in the Library* (fig. 4.13), an 1882 soft ground and aquatint, accords the spherical globe of the solitary visible light source, a colza oil–fueled moderator lamp, such tonal and design importance that the artist hands over some of the work of fashioning the image and its tonal harmonies to the givens of shape and luminosity of a household lamp, to the degree that a print like this one could be called an allegory of the "madeness" of illusion.[58] Also striking is that the blank and luminous disk of the lamp globe would be the only nonworked portion of the print were it exclusively an etching. Insofar as it is also an aquatint, however, the shining circle is, on the contrary, the zone accorded the most attention, knowingly blocked out before the resin-covered plate was plunged into the acid. Equally startling is the undeniable similarity between the unhatched orb and a luminous full moon, though Cassatt's moon globes stint on illuminating their environment. They glow, but without appearing to shed light. Cassatt's library etchings were conceived, drawn, bitten, inked, and printed during the years of debate about the efficacy and visual qualities of the electric lights newly shining in numerous, high-profile Paris streets. I contend that the new culturewide alertness to light—the rise of an "incipient light cult"[59]—helped to bring Mary Cassatt's innovative intaglio prints into being in and around 1880.

In a key linked pair, both called *Under the Lamp*—an 1879–80 conte crayon drawing (fig. 4.14) and the related soft-ground etching and aquatint (fig. 4.15)—the huge, spherically globed table lamp is a third being at the table, a full-fledged constituent of an intimate gathering. It is a towering sentinel, a beacon for a parlor. Its featureless frontal, facelike disk stands watch over the two adjoining human zones. It sutures the two spaces, but its strong mechanical identity and object stateliness work

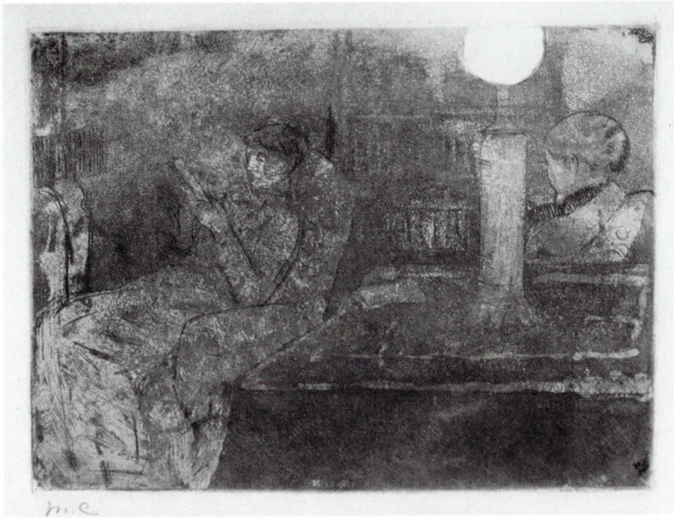

4.13 Mary Cassatt, *Mrs. Cassatt and Lydia in the Library*, 1882. Aquatint and etching on paper. 27.31 × 35.88 cm (10¾ × 14⅛ in.). Initialed lower left: M.C. Courtesy of Adelson Galleries, New York.

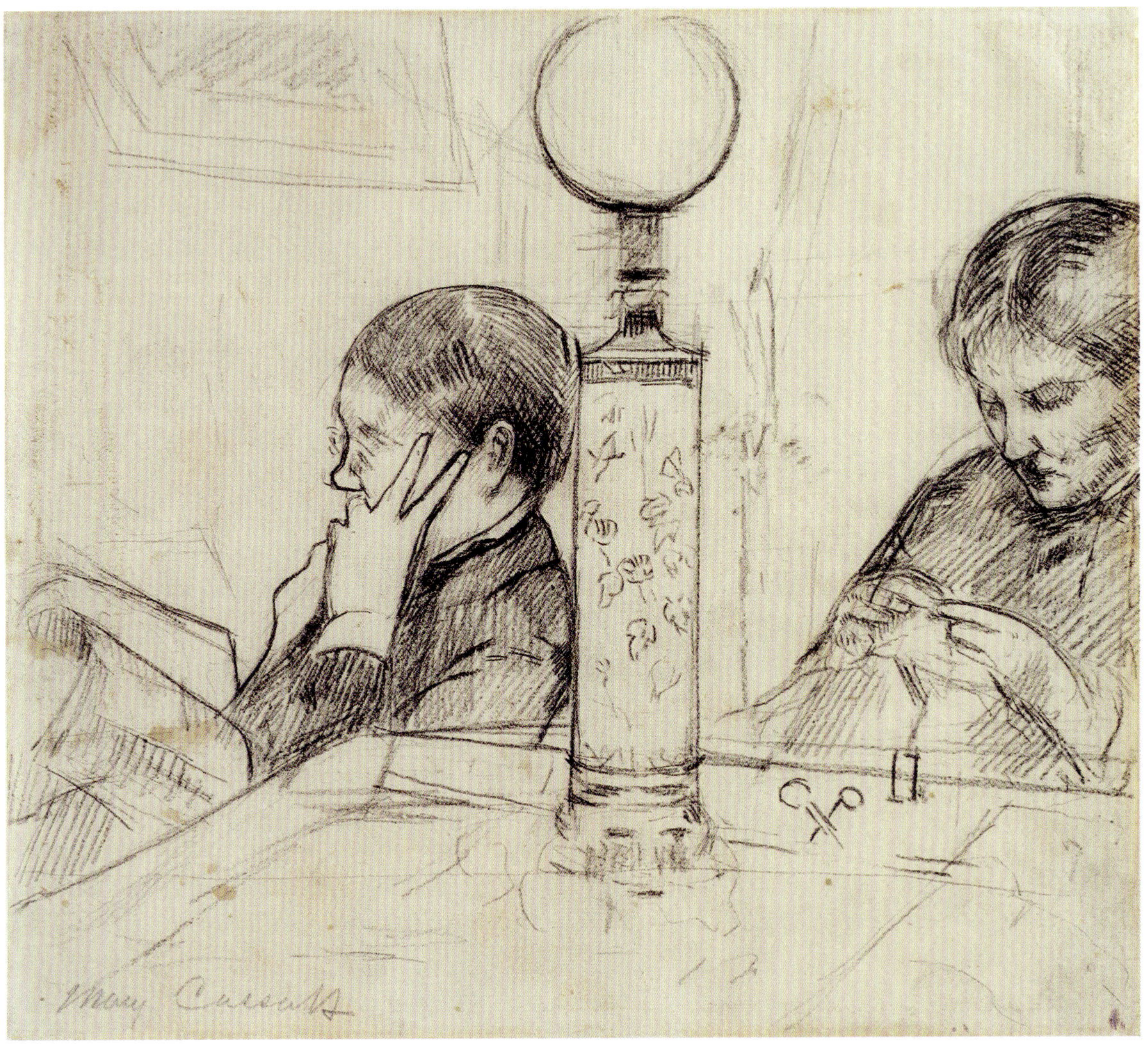

4.14 Mary Cassatt, Drawing for "Evening" (*Under the Lamp*; *Two Women at a Lamp*; Drawing for "Under the Lamp"), 1879–80. Conte crayon on thin, light wove paper. Hood Museum of Art, Dartmouth College.

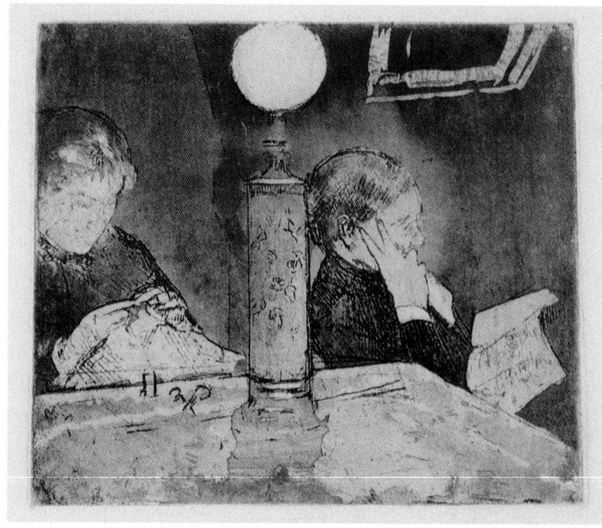

4.15 Mary Cassatt, *Under the Lamp*, c. 1882. Aquatint and etching on paper. 26.67 × 30.8 cm (10½ × 12⅛ in.). Courtesy of Adelson Galleries, New York.

to separate one busy woman from the other. In spite of their shared morphology, the three orbs do not coalesce into a network of cognate spheres. (The contrast to the planetary system often created by Degas, discussed above, is clear.) But despite its being a machine, the lamp's humanoid qualities push its presence and membership in a trio gingerly into the camp of the uncanny, while the small, round-headed creature nonetheless retains the alterity of an alien.

In both the drawn and etched versions of the subject, Cassatt certainly fussed over the form and placement of the lamp. The rim of the globe skims the upper sheet (and plate) edge exactly, and forces the void of the lamp globe into contact with the margin, the end of the image. And she went out of her way to establish sharply demarcated contours for the lamp. Emphatically parallel marks outline the leaf-enhanced cylindrical body, and repeated contours delineate the crisp spherical shape of the globe. The starkly stylized hand and erect pointing, head-cradling fingers of the older woman (at left in the drawing) strikingly contrast with the bulbous volumetric heads, which echo in turn the centralized towering globe (though the "species differentiation" has been noted). The pointed pattern of that dramatic hand seems itself to have been somehow assimilated into the machine world of the imposing lamp, though the floral motif on the cylindrical lamp body inflects its mechanical identity with organic *bona fides*. The barely sketched sewing implements (ostensibly scissors and a spool of thread) strewn on the table are minimal, but somewhat witty or even impish, additions to an otherwise unrelievedly stolid and imposing company. They lend a small note of understated levity to the sober or even slightly sinister effect of the vexing lamp. The women appear comfortable in their intimate coexistence and yet are distant from one another, thanks to the intimidating lamp, which is just too psychologically and environmentally daunting to function apotropaically.[60]

The quiet, nonverbal sociability at home instanced by the print is clearly enabled by the lamplight. Indeed, reading to oneself (silent reading) was a very modern practice. According to historian James Smith Allen, "Once a collective and public activity, reading became an individual and private concern. . . . By the end of the nineteenth century . . . reading within groups and in pairs gave way to men and women reading by themselves."[61] A resultant parallel is surprising but trenchant. There is, on the one hand, the modernity of the two figures' quiet coexistence, and, on the other hand, the blazing contemporaneity of Cassatt's rendering of a lamp globe that glows (like the moon), but fails to shed light on its surroundings. Its up-to-dateness resides in its emulation of the very features that Maxime du Camp assigned to undivided electric streetlights in the mid-1870s. Like the problematic arc lamps that he described so memorably, Cassatt's lamp dazzles and does not illuminate. It is a lighthouse for a parlor.[62]

Reading the Newspaper (fig. 4.16), dated speculatively to around 1883, another soft ground and aquatint, is very much a member of the same family of prints, although its scenario is altered slightly compared to *Under the Lamp* (fig. 4.15). A journal-reading woman coexists with an older bearded man, both facing in the same direction, but the authority and disjointing work of the lamp are familiar. This etching has

4.16 Mary Cassatt, *Reading the Newspaper, No.2*, c. 1883. Soft-ground etching and aquatint on cream laid paper, 13.5 × 16.5 cm (image/plate); 23.6 × 31.7 cm (sheet). Through prior gifts of Stanley Field, Katharine Kuh, Potter Palmer II, and Marjorie H. Watkins, 2000.417. The Art Institute of Chicago. Photo: Art Institute of Chicago/Art Resource, NY.

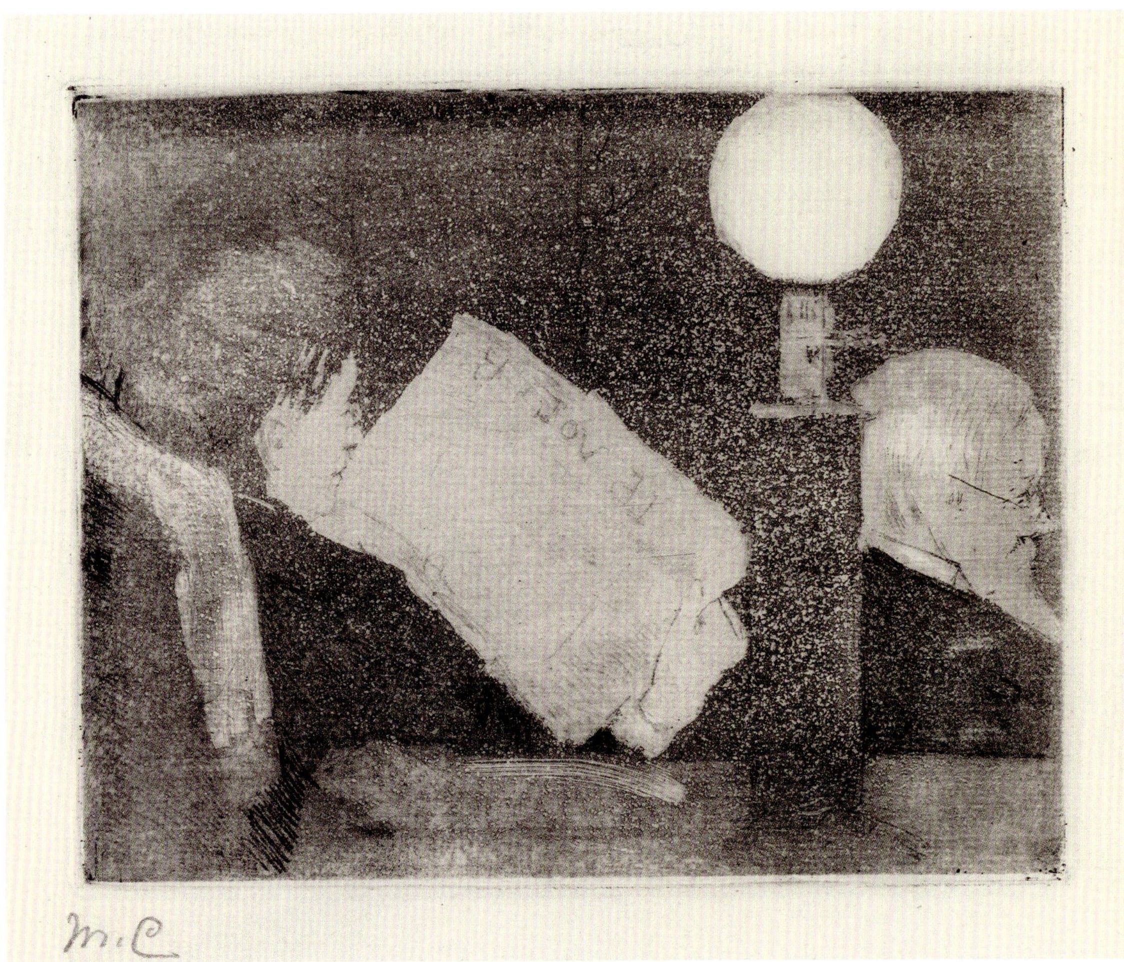

comparatively fewer middle tones; it is largely constituted of blacks and whites. It is both the latest chronologically and the most tonally contrasting in the small series. The shape of the lamp is also simplified compared with its earlier cognates, and the penetratingly white blaze of the globe's luminous void is exceptionally emphatic. But neither its dramatic tonal structure nor the boiled-down form of the lamp alters the precision with which the luminous globe's perimeter skims the sheet edge.

What power source could have fueled such an intense glow? Or rather what kind of indoor lamp served as the objective referent of Cassatt's investigation of the quality of illumination in the increasingly technologized City of Light? While it is very important to understand what kind of lamp she had, in her art it functioned beyond itself in order to signpost light outside in the street. In other words, her investigation of a domestic interior lighting scenario that she knew intimately was a metaphorical displacement of the news-making light environment of the Parisian outdoors.

By the 1890s, indoor electric lighting was becoming somewhat more common in Paris, but Cassatt did not have electric light, disliked by many elite women, put in her own Paris apartment until the fall of 1924.[63] Indeed, as Graeme Gooday explains, it was in the context of the first uses of electric arc light indoors (long before the existence of lamps equipped with incandescent bulbs) "that electric light was first judged 'unbecoming' by female observers; specifically that it was not merely unpleasant to look at, but also uncomfortable to be seen beneath."[64] Illumination discourse was an explicitly classed and gendered philosophy and debate.

Each of Cassatt's graphic interiors emphasizes a stately cylindrical lamp topped by a searingly bright globe, and each was set in the library of Cassatt's gas- and candle-lit Paris apartment (as it is usually described), but the "actual" lamp featured in the prints was not gas, nor was it an arc lamp. It was unquestionably an oil lamp. The French distinguish between two varieties of oil lamp unavailable to the English tongue: on the one hand, *lampes à l'huile* (vegetable oil), and, on the other, *lampes à pétrole* (petroleum-based oil). Mary Cassatt's lamp was emphatically the former; fueled with colza oil, it is a *lampe à modérateur* (moderator lamp) with a Japoniste, probably *cloisonné* or majolica base (fig. 4.17).[65]

While I am not the first to observe Degas's and Cassatt's incorporation of round light fixtures in their prints during the years of close collaboration, I am the first to identify Cassatt's table lamp, and to separate it from the gaslights she includes in works set in theater interiors.[66] As her adopted city was about to be baptized the Electric Capital, she accorded celebrity status to a rather old-fashioned and demanding, yet elegant lamp, invented in 1836. It had a spring-activated plunger that, when wound, forced the oil through a narrow tube up to the base of the burner, which enabled the colza oil (used on the Continent) to be burned efficiently.[67] The moderator lamp, as might be guessed from the intricacy of its internal mechanics, was a sensitive and demanding appliance, rather more like a person than a machine. Quoting an 1872 source, "These lamps are very apt to get out of order, and after being once

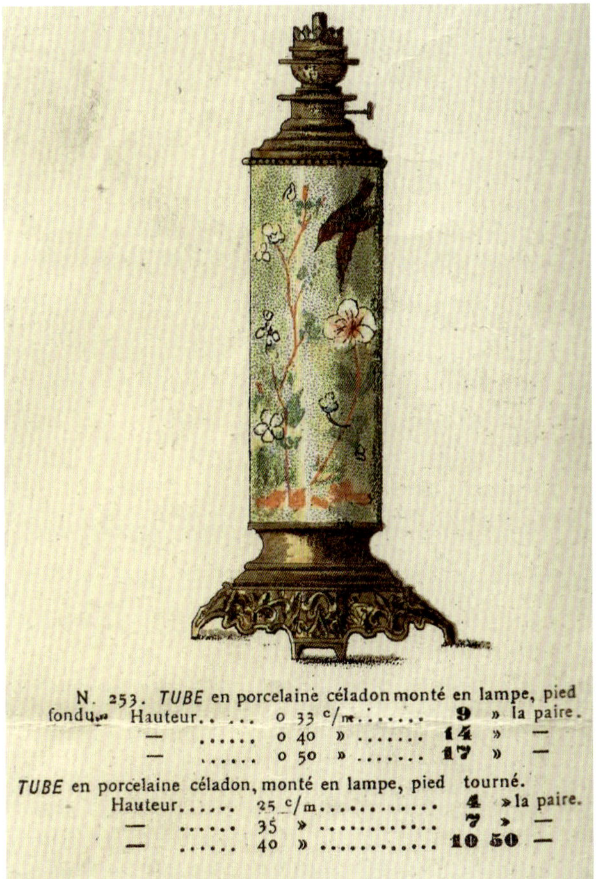

4.17 Porcelain lamp body, from *Chine and Japon* (Oppenheimer Frères catalog) catalog, Paris, ca. 1900.

repaired are still more liable to being so; the greatest care and cleanliness is required in the management of them. . . . Every day the lamp must be trimmed, the black edge of the cotton wick must be cut off very evenly with scissors made for the purpose, and fresh oil must be added."[68] And so on.

Fascinatingly, "moderator lamps were fashionable among wealthy Americans traveling in Europe (or wishing to give the impression of having traveled in Europe) in the 1870s and 1880s, who imported and used them as ornamental table lamps, even though technically inferior to gas and kerosene lamps."[69] Without doubt a well-to-do household was the usual destination, as confirmed by Ara Kebapçioğlu: "The initial cost [of such oil lamps] and of their upkeep being elevated, they were often reserved for bourgeois households."[70] This in turn echoes an older principle: in the bourgeois nest, there was usually a preference for the flame,[71] though Cassatt's etched representations certainly flatten out and abstract what may have been the variegation of a flickering light. Indeed, the "moon egg" disks of the etched lamp globes work in synch with the severe monochromatic tonalities of the prints to erase the flame and forestall the flicker.

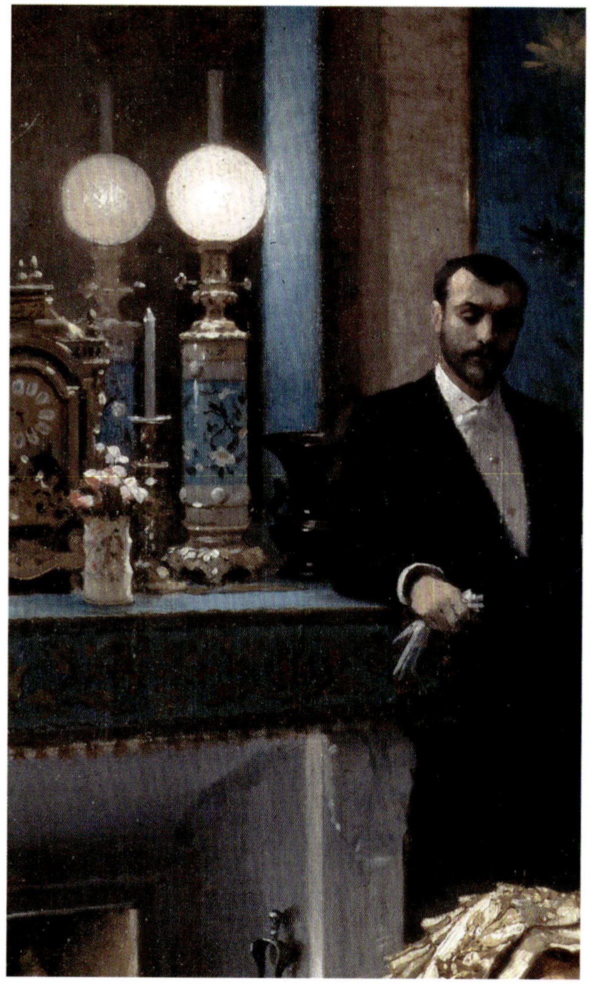

4.18 Édouard-Bernard Debat-Ponsan, *Avant le bal* (*Before the Ball*), 1867, detail. Museum of Tours. Image © Roger-Viollet/ The Image Works.

A virtually identical moderator lamp appears on the mantelpiece in Édouard Debat-Ponsan's highly detailed painting, *Avant le bal*, a posh domestic interior (fig. 4.18), notably brighter and paler than Cassatt's scenes.[72] Once a related painted lamp like Debat-Ponsan's is studied, several of Cassatt's elisions or erasures jar immediately into view. She saw no need to describe the two little devices that operate the lamp, right at the throat of the perforated burner. The thumb wheel (which operates the wick raiser) and the "key" that winds up the internal piston (which raises the oil up to the wick) are carefully described by Debat-Ponsan, but only alluded to by Cassatt's renderings. Neither did the particulars of the four-footed metal base interest her in the least. But by far the most significant eccentricity of Cassatt's depiction of her lamp is the consistent practice of eliminating the chimney. It was sawn off, so to speak, in each of the pictures, decisively and definitively, which makes her perspicuous rounding off of the globe exactly at the picture's edge in every instance a conscious calculation. Cassatt's globe is reduced to one hundred percent luminosity and pure aesthetic form, which indeed mimics the "moon eggs" that intrigued Degas. As already noted, they glow but do not cast much light upon the people close at hand. Hence, as representations of contemporary sociability, her versions are both tonally shadowy and psychically uneasy.

Cassatt's aesthetic move was a gesture against the denotation of the lamp's specific functionality; indeed, it inscribes the nonaffirmative nature of the prints' relationship with the material culture of their immediate ostensible surroundings. Her objective referents came from the public sphere—the world of street furniture. Her moderator lamp, which dazzles without illuminating, functions as if it were an electric arc streetlight rather than an oil parlor lamp. Her lamp globes resemble the bright white, flawlessly circular white fixtures of the Jablochkoff lamps installed along the avenue de l'Opéra in 1878, as stylized in the illustrational wood engraving published in *La Lumière électrique* in 1881 (fig. 4.19).[73]

Cassatt's oddly unlit, indeed gloomy, interiors overseen by glaring lamp globes should be understood as displaced evocations of the controversial blaze on the boulevards beyond her apartment walls in the ninth arrondissement. Her own library space provided a pretext or alibi for an examination of the signature visuality, and

4.19 "L'Avenue de l'Opéra éclairée par les lampes Jablochkoff" (Paris, 1878), *La Lumière électrique: Journal universel d'électricité* 38 (August 10, 1881): 186. University of Chicago, John Crerar Library.

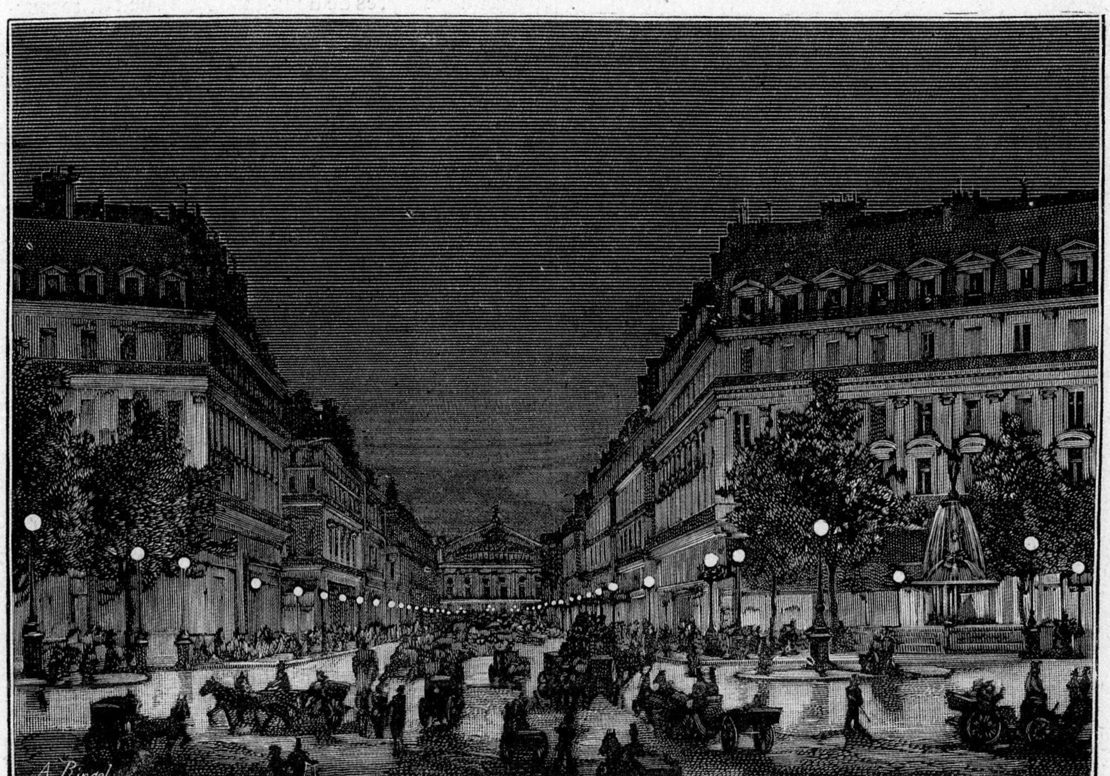

(FIG. 1. —L'AVENUE DE L'OPÉRA ÉCLAIRÉE PAR LES LAMPES JABLOCHKOFF.)

contrasting tonalities of the most topical urban illumination environment of her day. In this cluster of graphic works, the table oil lamp is a metaphorical and portmanteau type of equipment: its distinctive circle of luminosity is an artifact of recognizably upper-bourgeois female taste—not to mention a world of lighting convenience unthinkable without servants to keep a very complex lamp operating smoothly—which mirrors at the same time the notorious eye-burning lamps of the street.

Though the prints emerged from an intensely aesthetic project—radically revived etchings—they were steeped in the artist's awareness of the topicality and disputed valences of industrialized light in her adopted home city. When Cassatt tackled *lumière* via *éclairage* in her first intensive printmaking campaign (all black-and-white between 1879 and 1882), luminophilia was splintered by luminophobia. Her luminophilia was the foundational and enduring attraction to the pleasures and comforts of the flame-lit private bourgeois interior, while the counterthrusting luminophobia was the contingent upper-class female dislike of glaring lights. Other factors entered into this dialectic. Cassatt's light aversion was at the same time an anxiety, both ocular and social: a dislike of the noxious glare, but also of the inelegance and vulgarity of the new light, which crosscut the trendiness and modernity of the light cults of the moment. Her light enthusiasm was traditionally bourgeois but also, it can now be seen, a sign of her up-to-dateness, and a component of her desire (the aspiration of a voluntarily uprooted American lady) to be seen as an assimilated Parisian artist fully clued into the codes and signature trends of urban modernity.[74] The etchings' engagement with modernity and its signature visualities was an intimate union (with an old-fashioned and elegant apartment interior conjoined with the codes of modernist representation) that was also a hostile encounter (manifest in her mobilization of a surrogate "lamp for a nightmare").[75]

· 5 ·

Outsider Nocturnes

AMERICANS IN PARIS

It sounds like a paradox, but it is a very simple truth, that when today we look for "American art" we find it in Paris. When we find it out of Paris, we at least find a great deal of Paris in it.

HENRY JAMES, 1887[1]

The Franco-American Relationship

American artists painting in Paris during the Belle Époque were apt to enchant the French capital after dark.[2] My approach to comprehending their pictures of Paris at night is rooted in my belief that issues of identity and technology were entwined in their imagination of the French capital. A core interpretive proposition is that these American artists saw Parisian *éclairage* through the lens of their attitudes about the lighting in the rapidly electrifying cities of the northeast United States. This optic matters, inasmuch as American pictures of the nocturnal city were done for the most part during a singular transitional phase in the history of the illumination of the French capital: between the mid-1880s and the early 1900s, or rather, after the conclusion of the brashest experiments with arc streetlights, and during the time of haphazard hybrid lighting of the street and loud arguments about the electrification of the city's street lighting, and thus before the systematic positioning of incandescent electric streetlamps. I shall argue that national differences fostered later nineteenth-century metropolitan luminophilia *and* luminophobia. The second principle in play is twofold and also involves the artists' outlook on their erstwhile home base: the peripatetic painting Americans of the Belle Époque under scrutiny here were all urban- and modernity-oriented artists, and are theorized in what follows as sometimes brusquely treated yet perennially upbeat voluntary exiles in *La Ville Lumière*.

The evening pictures that I call "outsider nocturnes" were the sanguine inventions of American artist-visitors to Paris, including Maurice Prendergast (1858–1924), Childe Hassam (1859–1935), and Charles Courtney Curran (1861–1942).[3] My analysis of their artwork has two thematic (or iconographic) bases. First, the utopic nocturnal French capital coined by US painters (the outsiders) has no indigenous French modernist equivalent. William Sharpe, a scholar of the American nocturne in art and literature, has stressed the striking absence of equivalent Parisian examples: "The French do not seem to give special emphasis to the idea that the city itself is lovely at night."[4] Second, there is discord between the optimistic outsider nocturne and the frequently dystopic *Ville Lumière* fashioned by Parisian modernists in their portrayals of commercial entertainments in the same years, including, for example, well-known posters by Henri de Toulouse-Lautrec that promised scurrilous entertainment, phallic musical instruments, and piercing lights (*Moulin Rouge: La Goulue* [1891] and *Jane Avril* [1893]), and Louis Anquétin's painting of a venal woman, *Woman in a Veil (Femme à la voilette)* (1891; fig. 5.1), positioned against an acid corona of reflected orange-yellow chandelier light, endeavoring boldly to draw the interest of passersby.

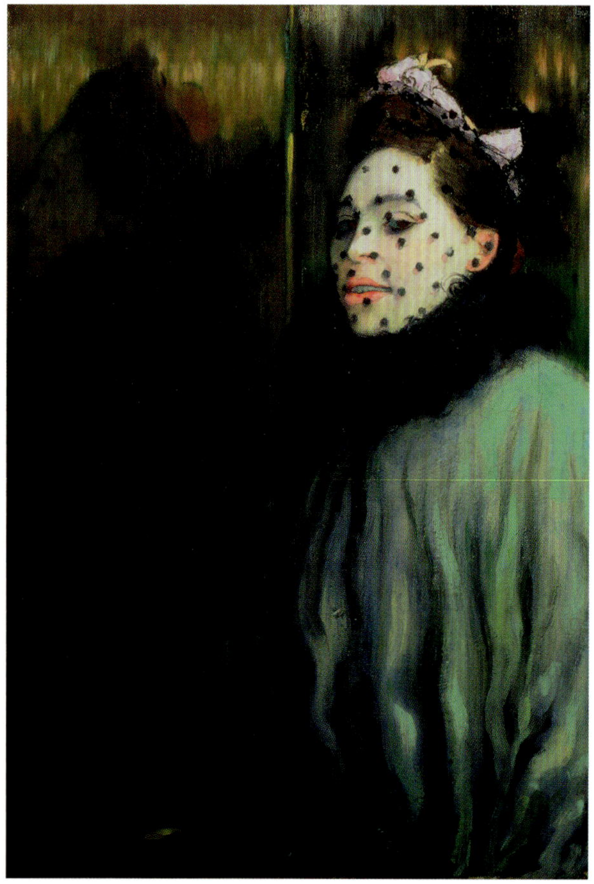

5.1 Louis Anquetin, *Woman in a Veil (Femme à la voilette)*, 1891. Oil on canvas. Private Collection/ Bridgeman Images.

A useful way to define the era when night paintings by Americans proliferated is to note that it coincided with the onset of mass transatlantic travel as well as the decisive acceleration of the dependence of American art upon Paris. The art world that Paris housed and nurtured in these years was cosmopolitan and emphatically transatlantic; it was a genuine "oceanic interculture," a pertinently material term coined by performance historian Joseph Roach.[5] For the thousands of visual artists from the United States enmeshed in this complex transatlantic axis of cultural exchange, the increasingly obligatory pilgrimage-cum-educational-destination was Paris (replacing or supplementing London, Rome, and Munich). In view of the vast numbers of artists moving back and forth across the Atlantic Ocean, Henry James could observe in 1887, as quoted in the epigraph, "that when today we look for 'American art' we find it in Paris. When we find it out of Paris, we at least find a great deal of Paris in it."

Whether residing in the French capital during the Belle Époque as students for a time or as expatriates—my focus here is the short-term visitor—the Americans went to gain a superlative art skill set in the world headquarters of contemporary art and prestigious art instruction, and to acquire cultural distinction in the most chic and international city on earth.[6] American sociocultural formation—unfinished and rough by definition—could only be completed by hewing to the Parisian model of sophistication and cosmopolitanism. That their experience of Paris as an erstwhile contact zone was frequently beset by conflict and resentment that sometimes fostered disavowal is at the center of my investigation of the voyagers' artwork.

American artists like Charles Courtney Curran, Childe Hassam, and Maurice Prendergast, who worked and studied for stints in Paris in the 1880s and 1890s, and returned to careers in major US cities thereafter, have been studied as travelers, art students, bohemians, cosmopolitans, tourists, and Impressionist epigones, as well as markers of "American national identity"—as if a birth certificate mattered but never a curriculum vitae.[7] But they and other artists converging on Paris from elsewhere are not typically studied structurally as strangers, foreigners, outsiders, or voluntary exiles engaged in painting as, to use Edward Saïd's formulation, a "construction of realities that served one or another purpose instrumentally."[8] I endeavor here to mount a fresh theorization of the *optique* of the voluntary alien in the big modern city. The most powerful theories of the optic and experience of the stranger or foreigner in the metropolis may help us to understand the outsider nocturne.

The signature of Georg Simmel's stranger, for example, is a distinctive objectivity. "Because he is not bound by roots to the particular constituents and partisan dispositions of the group, he confronts all of these with a distinctly 'objective' attitude, an attitude that does not signify mere detachment and nonparticipation, but is a distinct structure composed of remoteness and nearness, indifference and involvement."[9] Simmel's proposition holds out the promise of a gimlet-eyed perspective on the borrowed temporary home. Theodor Adorno's assessment of the lot of the émigré (written in exile in 1944) was altogether less sanguine. "Every intellectual in emigration, is without exception, mutilated, and does well to acknowledge it to himself, if he wishes to avoid being cruelly appraised of it behind the closed doors of his self-existence. He lives in an environment that must remain incomprehensible to him . . . he is always astray. . . . His language has been expropriated, and the historical dimension that nourished his knowledge sapped."[10] I am interested in the idea of the inevitability of incomprehensibility, a notion that may help to account for the anachronisms and idealizations in the US canvases.

Edward Saïd's figure of concern is the exile. "Once banished, the exile lives an anomalous and miserable life, with the stigma of being an outsider. . . . [but the exile carries with him compared to the refugee] a touch of solitude and spirituality."[11] "Much of the exile's life is taken up with compensating for disorienting loss by creating a new world to rule. . . . The exile's new world, logically enough, is unnatural and its unreality resembles fiction. . . . No matter how well they may do, exiles are always eccentrics who feel their difference (even as they frequently exploit it) as a kind of orphanhood."[12] The unreality of the new world that resembles fiction bears on the matter at hand. There is a resonance between Simmel's and Saïd's perspectives when the latter writes (cautiously) regarding positive things about the conditions of exile. "Seeing 'the entire world as a foreign land' makes possible originality of vision."[13] The internal quote is from Hugo of St. Victor, a twelfth-century monk from Saxony, but it could just as well have been Henry James speaking in 1877: "to be—to have become by force of circumstances—a cosmopolitan is by necessity to feel a good deal alone."[14]

The link between Simmel, Adorno, Saïd, and James is the concept of critical (sometimes melancholic, other times invigorating) disconnection from the (in this case freely chosen) new home that disallows place-specific complacency. This positionality makes it possible to see the new circumstance clearly in light of the discarded homeland, because the uprooting produces the denaturalization of home. In the cases under study here, the stay in Paris (about which much was known and believed in advance, producing the need and desire to go) fostered representation as dreamwork, therapeutic dreamwork. In order that the obligatory trip to Paris be worthwhile and that the object of desire, Paris, live up to expectations, it had to have encouraging environmental qualities as a city that differentiated it from America's brashly developing cities (New York, Boston, Chicago, and Philadelphia). Social attributes were also sought or devised that compensated for the affronts experienced to varying degrees by every member of the US passport-carrying and American English-speaking community. And following the logic of the French proverb, *la nuit tous les chats sont gris*, darkness provided a place to hide, and to erase difference.

My approach to the American pictures insists that they, unlike many French ruminations on the discordant and edgy qualities of modern Paris, were instead soothing images of a comparatively reassuring outdoor Paris that were personally comforting and professionally enabling in the sense that they helped to affirm that Paris was not a disappointment. They were depictive imaginings that originated in the painters' trenchant aesthetic goals, but were clearly inflected by their experience of the sometimes bruising particularities of transatlantic American social experience during the Belle Époque. I argue that the defining sensitivities of American artists in Paris, shaped by their spirited pursuit of pro-tem transnationality, shaped their aesthetic translation of the nighttime City of Light.[15]

The pictures are also distinctive moments in the history of the imagination of nighttime street life in Paris, which elucidate outsiders' reactions to the city's lighting technologies and its shifting purchase upon technological modernity—seen through a distinctively electrified American consciousness. The outsider nocturne repeatedly softens the night light environment, avoids institutions of commerce and organized leisure, disavows the lurid and the tawdry, and engineers a consistently interstitial *mise-en-scène*. It contains congenial or extremely minimally described sociability characterized by salubriousness or rather the absence of overt vice or even the absence of incident *tout court*. Paris is often Eros for the visitor, but not in these pictures. They instead figure the City of Light as a City of Virtue, and endow the French City of Vice with the equivalent of a happy ending. All of them represent the urban night as both visually striking and socially benign, with their telltale congenial, reputable forms of nighttime social life unfolding under and enabled by the accelerating industrialization of light. Yes, they were all motivated by Whistler's nocturnes, but surely they were also inspired to do nighttime Paris outdoor scenes to be *au courant*; to show they knew that streetlight, industrialized outdoor light, was the index par excellence of French metropolitan modernity.

Concerning the all-important matter of the European tradition in American art, Malcolm Bradbury, the leading analyst of the European connection in American literature, underscored the magnitude of the attachment by observing: "A good proportion of American works of fiction, poetry, criticism, aesthetics, and history have been written in Europe."[16] (Remember that *Huckleberry Finn*, that quintessentially American novel, was written in Italy.) As Bradbury explained, postcolonial America's reinvention of a mythic Europe demonstrated that "perhaps America more than any other nation needed Europe as an idea."[17] All across the nineteenth century up to the eve of World War I, the need for Europe remained pronounced, yet was a source of nonstop tension and anxiety in the United States, especially around the practice of expatriation for cultural reasons; it was part of the struggle between national particularity and cosmopolitanism. The desire to become cosmopolitan by shedding nationality was not and is not the exclusive aspiration of US artists in the years under discussion; it belonged and belongs to a broader syndrome: the "progressive disaffiliation of artists from the nationalistic model."[18]

After the US Civil War, the tension around Americans choosing Europe sharpened as Europe became identified "not only with cultural and traditional values absent in the United States but also with the American past that was being lost."[19] Hence Europe was often dehistoricized, its modernities overlooked for ideological reasons. But around 1900, for some artists—an expatriate like Mary Cassatt and a student-traveler like Childe Hassam—conceptions changed, and two understandings of Europe came into circulation: Old Europe was the realm of culture and tradition, while New Europe was the sphere of modernism and modernity, vividly exemplified by the glittering gaslit boulevards of the Second Empire and, for some, sustained by the arc-lit thoroughfares of the late 1870s and early 1880s, and the Eiffel Tower of 1889. The voluntary exile in Paris, the social actor of central importance here, is instructive insofar as he or she "demonstrates the tensions about location and nationality, obligation and independence that mark modern art."[20]

Another way of presenting the Franco-American dynamic would be to evoke an American archive called "France"—an archive that engrosses, arouses, and shapes many Americans. France became an American archive as a consequence of a singular (and signature) fixation with the French in the United States, to the degree that American speech about France—whether verbal or visual—evokes something already fixed or frozen. A recognizably Foucauldian kind of archive in many ways, the American France is a system that governs the appearance of statements as historical events.

Edith Wharton's 1919 book *French Ways and Their Meanings* is a crucial—I am tempted to say, pure—exemplification and cultural precipitate of the pervasive fascination with France among the East Coast American gentry and literati in the later nineteenth and early twentieth centuries. Her bracing small book (only 149 pages) is a *locus classicus* of US cosmopolitanism in the Gilded Age as insider knowledge and adulation of the French. *French Ways and Their Meanings* collected pieces that she wrote during the last two years of World War I for American soldiers and sailors bound for combat duty in the Great War on the Continent, to help them understand the people and customs of France—"the collection was in fact placed in all ship libraries, on the order of the US Department of the Navy."[21]

"I have singled out as typically 'French' in the best sense of that many-sided term," Wharton writes, "the qualities of *taste*, *reverence*, *continuity*, and *intellectual honesty*."[22] Some of her comments (many of them comparative) concerning "Intellectual Honesty" apply to the complex Franco-American relationship under scrutiny here. "The singular superiority of the French has always lain in their intellectual courage. . . . France is of all countries the most *grown up*."[23] "[France] is not afraid of anything that concerns mankind, neither of pleasure and mirth nor of exultations and agonies."[24] "The very significance—the note of ridicule and slight contempt—which attaches to the word 'culture' in America, would be quite unintelligible to the French of any class."[25]

Her analyses of French "Reverence" are also germane. "There is a reflex of negation, of rejection, at the very root of the French character: an instinctive recoil from the new, the untasted, the untested: and no one can hope to understand the French without bearing in mind that this unquestioning respect for rules of which the meaning is forgotten acts as a perpetual necessary check to the idol-breaking instinct of the freest minds in the world."[26] And concerning "Continuity," she justifies French harshness: "They are passionate and yet calculating, and simply uncalculated kindliness—the vague effusion of good-will toward unknown fellow-beings—does not enter into a plan of life which is as settled, ruled off and barricaded as their carefully measured and bounded acres. It savours too much of Adventure, and might lead one into the outer darkness of Risk."

Art history has not been forthcoming in acknowledging the Whartonian component of the move of American artists to Paris during the Gilded Age—neither her enthusiasms for various elements of French sophistication nor the masochism evidently necessary to understand and admire the French. Or rather, in most art-historical accounts of the Lure of Paris for Americans in the later nineteenth century, the force that drew increasing numbers of American artists to the French capital to study following the end of the Civil War, "French ways" in Wharton's sense receive scant attention as possible incentives. For example, Kirsten Swinth has written: "To come to Paris was to experience the simultaneous pleasures of the best contemporary art and the most vibrant art center. With such glamour and allure, it is not surprising that Paris became 'the Mecca of art students of both sexes.'"[27] Did the Lure consist solely of Art? Are the routinely unexamined but constantly invoked "glamour and allure" meant to be understood as the self-same admirable qualities of the French character and way of life trumpeted by Wharton? Did US artists uproot in droves only to improve themselves at the level of technique, or because they admired and hoped to assimilate the archival French virtues and to expose themselves, even if painfully, to a brusque people whose cosmopolitanism and sophistication were reputed to surpass their own by far? Or, to put the question even more reductively, did the Whartonian archive shape what US artists painted in France? If so, can we see it, for example, in their representations of the city of Paris?

I argue that the outsider nocturne pictured a Whartonian Paris. This affirmative response to my questions is enabled by looking at an assortment of the nocturnes painted between about 1880 and 1906 by displaced Americans in Paris—whom I like to call Makeshift Parisians or Outsider Panglosses. Their Parisian nocturnes bustle with after-hours activity (unlike Whistler's nocturnal world, which is characteristically silent and still). The congenial sociability of these pictures is categorically salubrious and usually, as already suggested, a bracketing of the erotic. Wharton quoted a statement she admired by William Dean Howells: "Yes, what the American public wants is *a tragedy with a happy ending*." Wharton comments: "What Mr. Howells said of the American theatre is true of the whole American attitude toward life."[28] In the outsider nocturne, American painters endow the French City of Vice with the equivalent of a happy ending.

Another perspective entirely upon the Franco-American "rapport" during the Belle Époque comes from theater critic and social observer Jules Claretie, a key voice in chapter 3. He gives us another way to understand the frictions in the transatlantic relationship. He discussed what he called *l'américanisme parisien* in the opening installments of his *La Vie à Paris* (1880–1910). His acerbic disquisition on so-called Americanisms in Paris provides a strong foundation for a discussion of the US "share" of Parisian energy equipment and infrastructure, or rather the self-protective constraints on the American ability to see Illuminated Paris in a time of changes in its lighting environment, albeit in a transitional hybrid condition. Claretie helps us to consider the very strong possibility that the chilly reception sensed by many US artists working and studying (whether long term or *en passant*) in the French capital was linked to Parisian qualms about the accelerating technologization of the city in the final decades of the nineteenth century. The frosty Parisian responses to American aesthetic actors thronging *fin-de-siècle* Paris museums and art studios were linked, I argue, to a mounting tendency to read new technologies in the city (including electric lights, telegraph wires, phonographs, and telephones) as metonymic signs of a rising tide of *Americanisms*.

In *La Vie à Paris 1882*, Claretie opens the passage dated February 10, 1882, with a quotation from Paul de Saint-Victor (1827–81) expressing his disgust, at the end of his life, with the status quo of literature by targeting its Americanization: "The gods are leaving us. We are decidedly becoming Americans."[29] Claretie continues in his own voice, taking the lionization of billiard stars as a baleful sign:

> It is perfectly clear that *americanism* is making progress recently, almost every day. The upsurge of curiosity, anxiety, fever directed toward Maurice Vignaux [famous billiardist who triumphed in the United States], from the Toulouse school, and George Slosson, master of American billiard table cushions [another billiards great] is a symptom of this change brought to our national temperament. It's by infiltrations that Yankee humor is penetrating us. But the *Yankee* is defeating us and, in every way imaginable, transforming us totally.[30]

He shifts to an indictment of the ostensible Americanization of the physical fabric of the French capital. And with these observations, he provides a glimpse of a likely material basis for the rising tide of anti-Americanism in France:

> Have you ever, when crossing the avenue de l'Opéra, which itself takes on aspects of a corner of London or New York with its department stores and offices, lifted your eyes up to the sky and the roofs? A network of telegraphic wires, a pile up of lines in a musical score in which some chimney makes the clef, crosses and recrosses themselves above our heads. The telegraph jumps from one block of houses to another, from pipe to pipe, and cuts, up there, its thin lines, straight like the lines in a schoolboy's notebook and which are like the spiderwebs of civilization. This telegraph, which cuts the horizon and the sky the way a wire cuts butter, will end up giving to Paris the aspect of San Francisco, it's the advance guard of an uninterrupted suite of examples (or types) of material progress destined to singularly modify the Parisian landscape. We already have a steam tramway; we will have an aerial railway, omnibus-locomotives, tunnels under Les Variétés, bridges suspended in front of the Gymnasium: all of Paris heated by gas, lit by electricity, woken up by the telephone, and taken for a ride by steam. . . . Yes, look closely again at this Paris surrendering to every American *utilitarianism*.[31]

Nocturnes by American Painters in Paris

Paris at Night (1889; fig. 5.2) and *A Paris Nocturne* (1889–90; fig. 5.5), painted by Curran and Hassam, respectively, when they were young American art student-visitors to Paris, are the works featured in this discussion. My analysis leans hard on these two portrayals of the illuminated nighttime French capital as a twinkling idyll in order to put my hypotheses to the test. Close scrutiny of these paintings will decide the degree to which such artworks exemplify a Panglossian approach to the representation of central Paris under sundry illuminations.

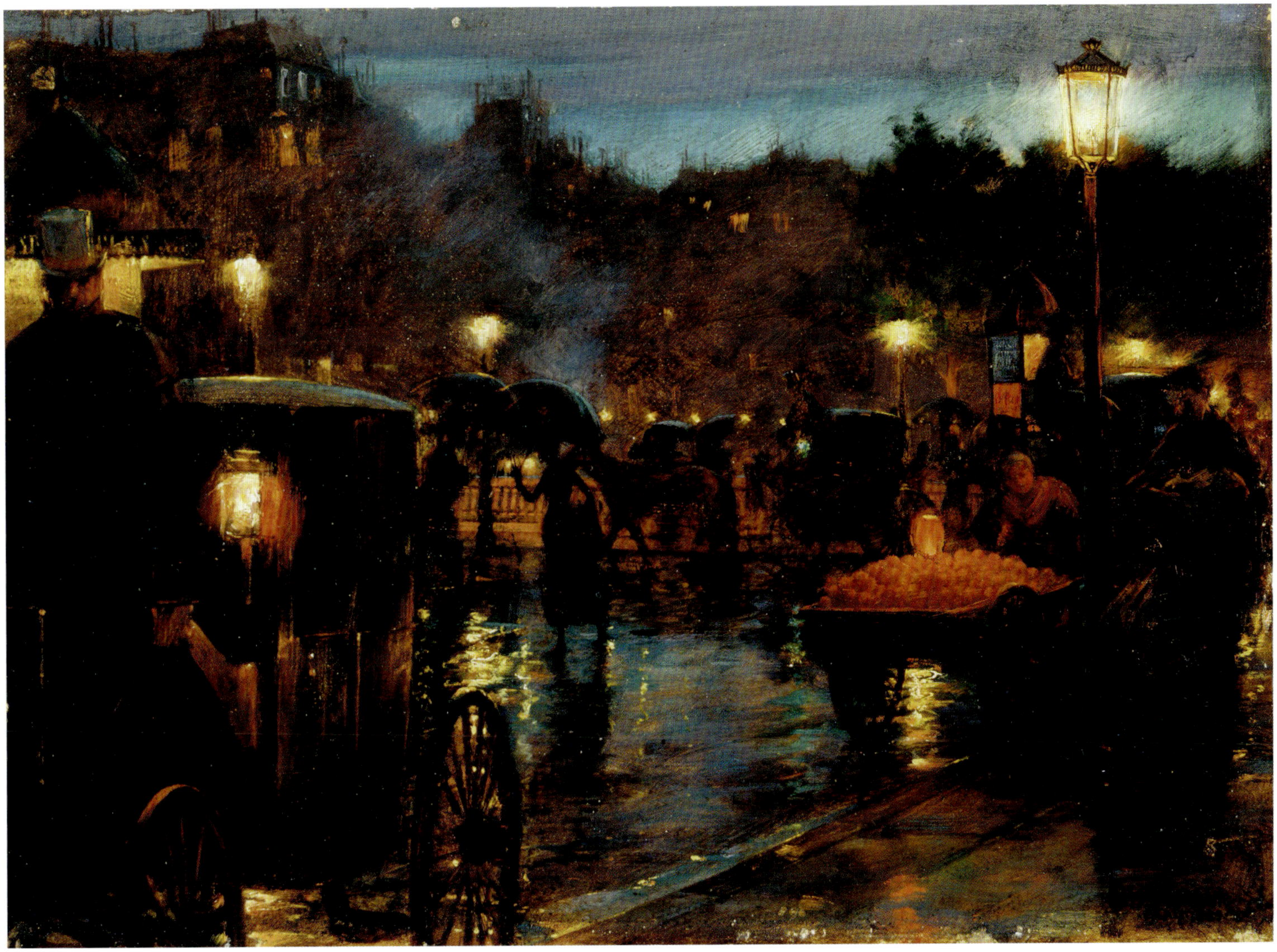

CHARLES COURTNEY CURRAN IN PARIS

5.2 Charles Courtney Curran, *Paris at Night*, 1889. Oil on panel. Unframed: 23.0 × 31.1 cm (9 1/16 × 12 1/4 in.); framed: 34.6 × 42.9 cm (13 5/8 × 16 7/8 in.). Terra Foundation for American Art, Daniel J. Terra Collection, 1989.12. Terra Foundation for American Art, Chicago. Photo: Terra Foundation for American Art, Chicago/Art Resource, NY.

Curran was a figure painter who also painted landscapes. Many of his most vivid genre paintings showcase figures set in brightly colored and botanically detailed landscape settings. More precisely, he was considered a painter of the ideal figure, a Romantic landscapist, and a prolific portraitist.[32] The New York dealer Richard York described him as "one of the most popular artists of his time—among his peers, with exhibition juries, and with the public."[33] Curran was born in Hartford, Kentucky, and spent his formative years in Sandusky, Ohio, the town to which his family moved at the outbreak of the Civil War to depart from the slaveholding South.[34]

In 1880 Charles enrolled in the Cincinnati Fine Arts Academy, but shortly thereafter (1882) he moved to New York City where he enrolled in the National Academy of Design, one of the most prestigious art schools in the United States at the time. To strike a balance, he also took courses at the more innovative Art Students League. His success was almost immediate: he had his first exhibit at age twenty-three at the National Academy of Design, and one of his paintings won a prize there in 1889.

Curran and his wife, Grace Wickham Curran, spent two and a half years in Paris (1888–1891), where he studied at the Académie Julian with Jules-Joseph Lefebvre (1836–1912), and showed work in the Paris Salon three years running (1889, 1890, 1891); he won an honorable mention in 1890. Three of his paintings were shown in the first exhibition in Paris devoted exclusively to American art and artists, at the Durand-Ruel Gallery.[35] After the stint in the French capital, he spent the reminder of his life in New York: in New York City, where he played an important role in various art organizations, and at his home and studio in the Cragsmoor region of New York state, an artists' colony, where he spent forty summers.[36]

Curran's small painting *Paris at Night* (fig. 5.2; 9¹⁄₁₆ × 12¼ in.), betrays a strong interest in the imagery of urban life, but that concern in his art was confined to his time in Paris.[37] Debra Force argues that Paris altered the tenor of his work: "In Paris, Curran veered away from his attractive scenes of elegant ladies that had achieved great acclaim in New York. Inspired by the city's industrialization and innovation, he immersed himself in the Paris landscape and explored the modern urban environment."[38] At the same time, a contemporary critic observed: "The scenes which he paints are at once Ohio and Paris combined."[39] Or rather his figure types were considered American, while his technical virtuosity was found French.

Paris at Night is a rich amalgamation of divergent human-made luminositics heightened by multiple wiggling trails of yellow streetlight reflected in the rain-soaked wet pavement. (The lemon-yellow color denotes gaslight.) Natural light is included as well. The twilit sky is a radiant blue, punctuated subtly by a tiny rising moon at the picture's upper edge. Streaks of the sky's blue are boldly reflected on the surface of the watery street. The paths of reflected yellow light resemble Van Gogh's orchestration of riverine night light in *Starry Night over the Rhône* (1888; fig. 6.9), but minus the stars, and, in spite of Curran's reflected evening sky and little moon, absent the highlighted contest between natural and artificial lights, won by gaslight, that is at the very center of Van Gogh's conception, discussed in chapter 6. Curran has also added spots and zones of artificial lighting to enliven the hulking dark expanse of silhouetted apartment buildings. Those orangey red patches of illumination seen through apartment windows are the glow of oil lamps or candles, sure signs that residents are awake and stirring.

A horizontal stretch of sidewalk and stone balustrade in the middle ground, along with the diagonal border between street and sidewalk at right, helps to articulate the structure of the painting's depicted swath of compressed urban space, which seems to be a square of some sort. Three horse-drawn carriages bearing oil lantern lamps crisscross the busy space, each heading in a different direction. Curran's square houses five sizable gas streetlamps (*réverbères*); many more on a smaller scale are visible farther back. The most conspicuous of the lamps is the tall one at right. Its size and prominence take us back to our discussion in chapter 1 of the frequent privileging of the street furniture of illumination, especially isolated lamps, in the painting and photography of modern life, with respect to works by Marville, Caillebotte, and Van Gogh. The key difference is that Curran's commanding lamp is illuminated, and glows gently.

Enhancing the modernity and variety of Curran's busy evening tableau are two advertising pillars, glowing Morris Columns (named for their midcentury inventor, the printer Gabriel Morris) of contrasting designs that bracket the scene at left and right.[40] The placement of Morris Columns, although they served commerce, was, like the humanizing scale of municipally installed streetlights, benches, and Wallace fountains, part of a "quest for a certain coziness in the urban space," according to François Loyer, which "corresponded to the city's explosion in scale."[41] The umbrella-toting pedestrians in the painting, whose silhouetted bodies reflect from the wet surface of the pavement, add another animating note to Curran's elaborately choreographed episode of nocturnal social life outdoors. But in spite of all the lights and activities we have inventoried so far, Curran's painting is implacably, unrelievedly, and unrealistically dark. Each of the many glass-enclosed lamps twinkles brightly, contrasting with the surrounding murk, but without actually illuminating its surroundings. The figures are all dark silhouettes indistinct in the gloom, with the exception of the fruit woman at right (and the *cocher*, another worker, high up in the driver's seat at far left).

Street commerce—fruit selling—unfolds in the far right middle ground. The presence of labor in the square enlivens and diversifies the scene socially and atmospherically. It is either an instance of preparation for early morning sales, or the closing down of business for the "day." The fullness of the cart strongly suggests the former. In Curran's slice of Paris, labor thus coexists unproblematically and cheek by jowl with the transiting of bourgeois pedestrians and carriages; workers share the very dark urban night with strollers and carriages, but they are among the few illuminated beings in the painting. The kinship between the illuminated vendor and the glowing coachman is fostered by a clever compositional gambit. The rear carriage wheel (left foreground) and the vendor's left cartwheel, virtually identical in form, succeed one another in space and skim the same diagonal curb.

The similarity and alignment of the wheels construct a continuity between two otherwise contrasting conveyances, albeit both instrumental, and enhance the cordiality of their coexistence. The disposition of the dark painting's lights both knits the two cohorts (passersby and laborers) together, and gingerly distinguishes between them at the same time. The woman fruit vendor's oil lamp creates a pocket of warm orange, bathing her red-orange wares (apples or oranges) in a warm glow that also lights her face and upper body, as well as those of several apparent child helpers at her side. The adult man hefting large bags at the right edge, near the cart but on the other side of the streetlight, may be a participant in the inchoate episode of fruit selling on the streets, or, in view of the vendor's anxious demeanor, he may be involved with another form of demanding nocturnal work. In either case, he exists in complete darkness.

Summing up the large number and range of incidents that constitute Curran's small picture, it is certainly a virtuosic assembly of highly diverse ingredients, fitted together like jigsaw puzzle pieces to make up one very dark painting of night. The darkness of the picture surely made pulling together its diverse ingredients easier. Its very dark tonalities preclude a detailed reflection on the likelihood or even plausibility of the coexistence of all its social actors, temporalities, lights, and activities. That said, the dark atmosphere does a lot of valuable aesthetic and ideological work in the painting. It paradoxically makes the bright blue tonalities of twilight (twilight in the dark?) shine luminously, and shows the many points of yellow *éclairage* to advantage. It is an unlikely scene, in many ways, but its manifold contrasts of tone and hue make for a vivid painting: an arresting and congenial image of the shadowy hustle and bustle on a nonetheless alluring Paris night, gently punctuated by points of seven different kinds of light.

Reference to its multiple points of light may spark reflections on the five sources of light used by Sargent in his Luxembourg Garden paintings, discussed in chapter 2. The contrast between the darkness of Curran's picture of night and the gray-mauve middle tones of Sargent's twilight (not to mention its prominent full moon) makes all the difference. Another very dark Paris night painting underscores the singularities of Curran's small painting. Henri Evenepoel (1872–99) was born in Belgium but spent the final nine years of his very short life in Paris. He was trained in Gustave Moreau's studio, a celebrated crucible of artistic excellence and diversity. *La Nuit à Paris* (c. 1895; fig. 5.3) has many striking but superficial similarities to Curran's picture: its overall darkness pierced by multiple points of light and the hulking presence of carriages. Two differences are the most consequential and account for the contrast in atmosphere between the paintings, both social and psychological. First of all, Evenepoel includes four tall, white electric streetlights. Their intensity and coronas of rays signal the artist's knowledge and interest in capturing their defining glare. They contrast unmistakably with the gentler array of orange gaslights elsewhere in the shops lining the sprawling square. Two more spots of saturated orange are two oil coach lights. Curran's range of lights appears symphonic in its harmony by comparison. The foreground woman in the Evenepoel, who manifests stress, perhaps alarm, strikes a note of human emergency and distress that creates a stunning climate of danger, completely unlike the congeniality of Curran's bustling picture.

Curran's only other night picture set in the French capital, *Illuminations at the Exposition* (fig. 5.4), parallels *Paris at Night* aesthetically in many ways with its cognate orchestration of zones of saturated color set against an otherwise very dark

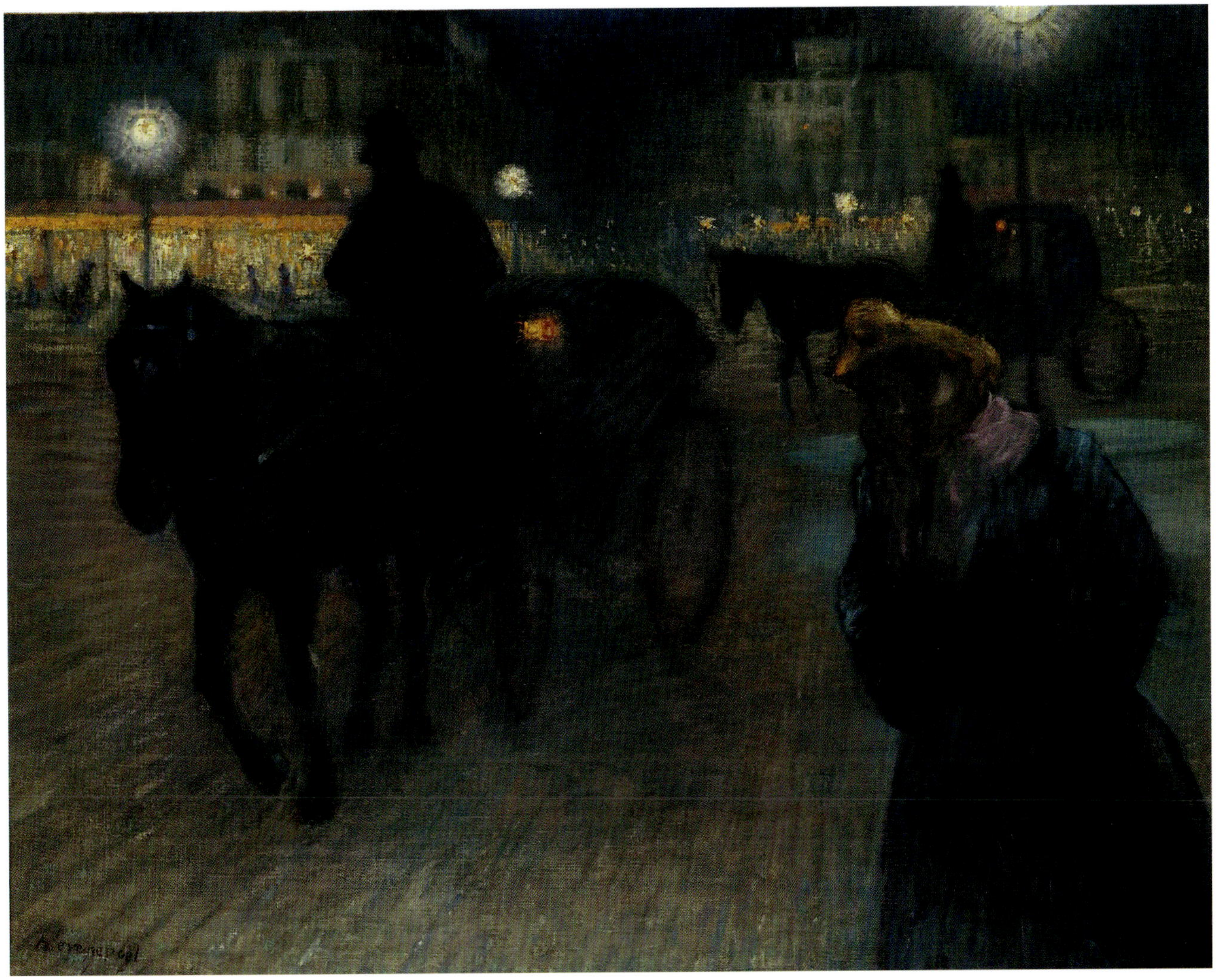

atmosphere, but its subject matter is extraordinary in the annals of Paris nocturnes. Dated "89," the tiny oil on panel (4½ × 12⅜ in.) depicts the second night of ceremonies at the Exposition Universelle of 1889.[42] It is the only painted representation by an American artist of the lights cast by the multicolored, glowing Eiffel Tower (centerpiece of the 1889 event, not depicted here) during the world's fair. A very dark picture as well, the small horizontal panel inventories a diverse crowd, shown largely in silhouette, gathered across the river from the blazing Tour Eiffel. The lights that shine across the horizon line defined by the line of diverse onlookers transitions from the yellow of gaslight at left (familiar from *Paris at Night*) to the white of the electric beams shone by the tower and the rainbow colors of its gaslights (in opalescent globes) and Bengal lights.[43] It splendidly demonstrates Curran's ability to orchestrate deepest black darkness in counterpoint with contrasting bright zones of colored light, and also showcases his talent for staging a complex gathering of diverse social types who congregate tranquilly in close proximity at night.

5.3 Henri Evenepoel, *La Nuit à Paris*, c. 1895. Oil on canvas. 65 × 81 cm. (25.6 × 31.9 in.). Clemens-Sels-Museum, Neuss. Image © Clemens Sels Museum Neuss. Photo: Jörg Schanze, Düsseldorf.

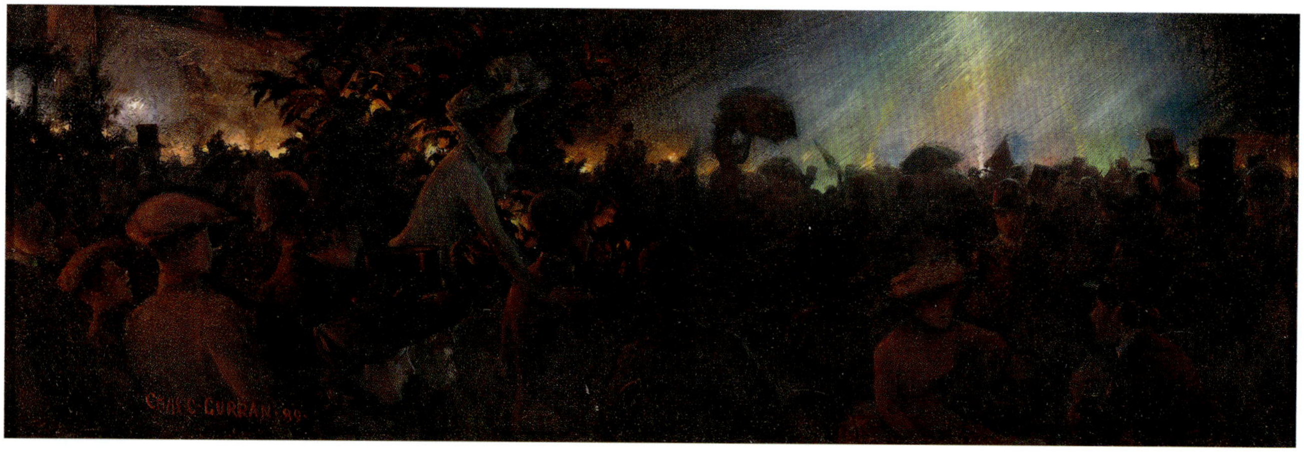

5.4 Charles Courtney Curran. *Illumination at the Exposition*, 1889. Oil on board. 11.43 × 31.4 cm. (4½ × 12⅜ in.). Signed and dated at lower left: Chas C Curran 89. Courtesy of Adelson Galleries, New York.

CHILDE HASSAM IN PARIS

Hassam's work is rich terrain for this line of analysis, because as William H. Gerdts observed: "Hassam was the initial American artist to gain renown as a painter of urban views."[44] Hassam, considered a key American Impressionist, was a Bostonian (from Dorchester) by birth. Largely self-taught and a high school dropout, he worked as an illustrator from 1882 and began exhibiting his watercolors at a gallery in 1883. A friend persuaded him of the need to get to Europe to look at art there. He took a five-month trip to six countries in the summer of 1883. A sojourn in Paris was a necessary and logical next step. Hassam and his wife, Maude Hassam (née Kathleen Maude [or Maud] Doane), remained in Paris and the nearby countryside (at Villiers-le-Bel) for most of their three years abroad, from 1886 to 1889.[45] Like Curran, he studied at the Académie Julian and exhibited in the Paris Salon; his work was also seen in the Exposition Universelle of 1889.[46] He turned thirty on shipboard on the way home.

Upon their return to the United States in 1889, the Hassams settled in New York City. Childe Hassam made common cause thereafter with other Impressionists (Theodore Robinson, John Henry Twachtman, and J. Alden Weir) and the poet Celia Thaxter, and painted frequently in leading East Coast artists' colonies, including Cos Cob, Appledore Island, Gloucester, and Old Lyme. While frequently painting the American countryside, he continued to hone his reputation as a painter of cities.

For the most part, while he was in Paris, Hassam was offhand or cheerful about the Franco-American relationship. He did make caustic remarks about contemporary art, both academic and vanguard, but judging from his letters he estimated that personally he got on all right. "We lived among French people, spoke French," he recalled. "It was a French house and we wanted to speak French."[47] Aggravation with the French did surface in some of his letters. In an April 8, 1889, letter to William Howe Downes, a critic for the *Boston Evening Transcript*, he wrote acerbically: "If you could live over here a couple of years to see how things are run here in art matters it would make you sick. Americans ought to know this. It ought to be shown up. The

'French disinterestedness' does not exist for any *stranger*."[48] And Hassam certainly had a low regard for the official and unofficial wings of the much-admired French art education system. In another letter to Downes (May 28, 1889), he observed: "I believe too much of this foreign study for Americans is going to do what the Prix de Rome system has done and is doing for France. The Julian Academy is the personification of routine, and with the Beaux Arts has spoilt more than one artist who goes there too long."[49]

Shoring up Hassam's complaints are other indications of tension between French and foreign students. A warning sign, posted at the entry to the Académie Humbert et Gervex around 1888, speaks to the xenophobia then current in the Paris art world. "No-one may enter the studio who cannot speak French."[50] In one of J. Alden Weir's letters home (October 1873), he reported, "There is no honor or anything else among the French students."[51] Tense relations in the studios were also caused by professional competition for the limited spaces in the system.[52] In line with my opening remarks, I argue that the experience of these tensions helped to shape the upbeat tenor of Hassam's pictures of evening Paris.

Compared to Curran's busy, dark pictures, Hassam's *A Paris Nocturne* (1889–90; fig. 5.5) is statelier, more luminous, and contains a cast of considerably more visible (if not emotionally legible), homogeneous, and elegant *dramatis personae*. Hassam actually completed it very shortly after his return to the United States (in New York City), thereby giving the picture a somewhat retrospective cast.[53] Hassam's luminosities and brighter palette contrast with the more saturated hues and darker darks employed by Curran, but constructing an atmosphere of geniality was part of Hassam's vision as well. By the time Hassam painted *A Paris Nocturne*, he was something of an old hand at the genre, having done previous nocturnes, set in both Boston and Paris, all of which show a fundamental commonality. Deborah Chotner has rightly observed: "Hassam early showed an interest in depicting the more attractive aspects of city life under varying conditions."[54]

The suave demeanor and prominence of the elegant and respectable *bourgeoise* in black at the center of Hassam's *A Paris Nocturne* are notable, and unlike the women in public seen in French Impressionist painting. Think, for example, of the dependency of the chaperoned woman in black in Gustave Caillebotte's *Paris Street: Rainy Day* (1877; fig. 1.9), or the downcast dissolution of the unaccompanied female drinker in Edgar Degas's well-known *Absinthe* (1876; also called *At the Café*). Hassam's Parisienne is, however, akin to the women pedestrians in the work of the popular French *pompier* painter, Jean Béraud, though she is comparatively more independent and prepossessing. While there are four carriages in Hassam's crowded street scene (numerable via the *cochers* whose heads surmount the traffic flow), there is only one male pedestrian. By endowing an unaccompanied woman on the street in the evening with supreme confidence and self-possession, Hassam disavows or at least does not mobilize the indigenous modernist iconography of the venal or hyperanxious solitary woman of nighttime Paris. There is, however, a slight hint of strain between the two prominent female figures that nuances the principal woman's

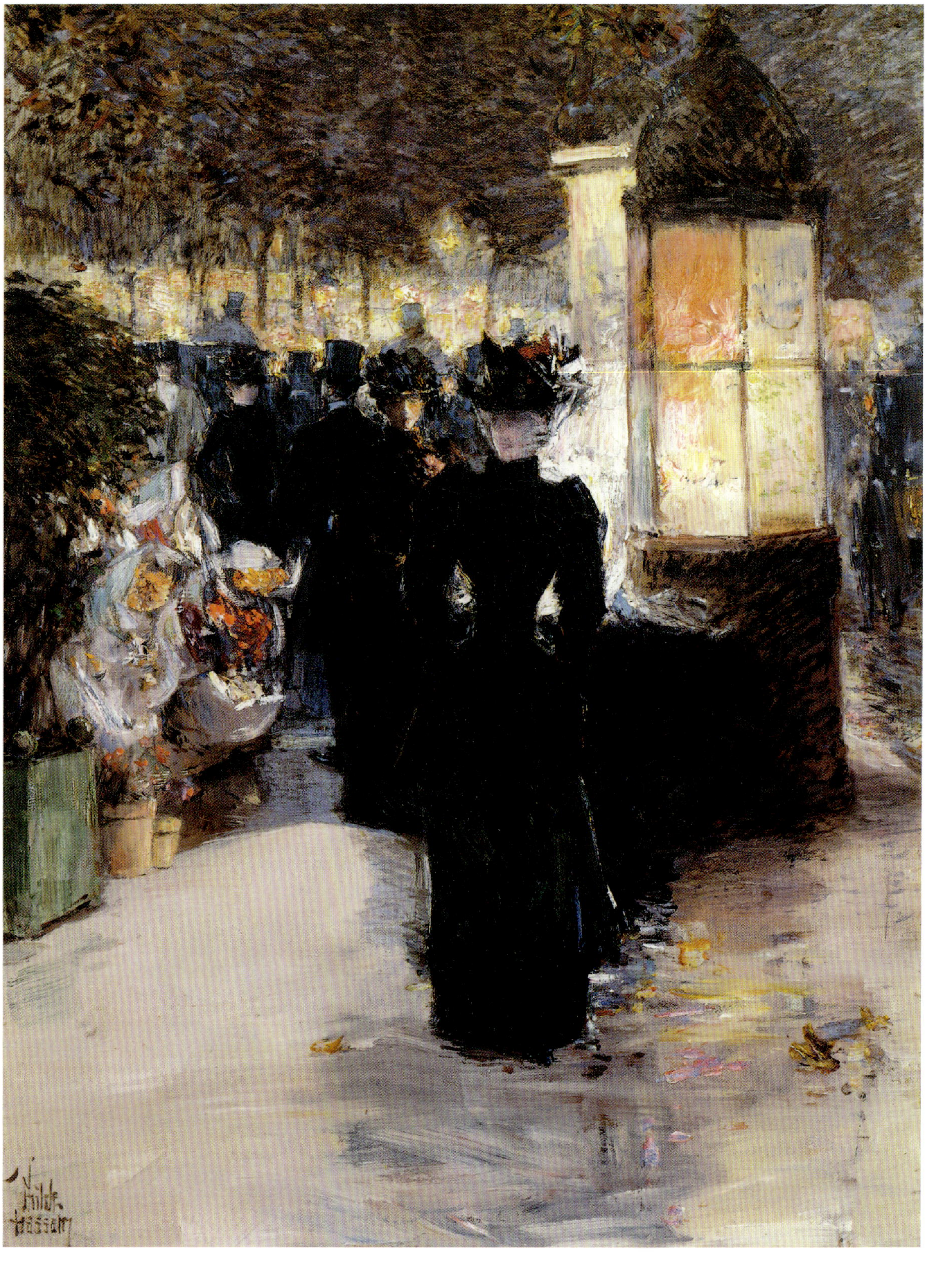
Childe
Hassam

otherwise implacable or even narcissistic poise: the woman behind her in the somewhat less fashionable hat appears to be scrutinizing the imposing foreground woman who shares the broad sidewalk. An understated suggestion is thus made of an investigation under way, woman to woman, of the prominent female social actor. An investigation of her character? We cannot say. The implied scrutiny of one woman out in public by another adds narrative interest to an otherwise stagy, even frozen pageant of a painting, but the subtle episode does not impugn the prominent female stroller's social standing or moral character.[55] This interaction unfolds *sotto voce*, and does not puncture the painting's beguiling overall mood of evening concord.

A prominent Morris Column links the light environment of Hassam's *A Paris Nocturne* to that of Curran's *Paris at Night*. (There is a second one farther away off to the right in Hassam's picture.) Hassam makes effective use of the multicolored glow emanating from the rectangular panels of the huge structure. These hexagonal or octagonal columns were reserved for posters, and in the evening they were lighted on the interior, producing a colored glow through the posters that were mounted behind glass.[56] While Curran's Morris Column panels produced a stained-glass kind of luminosity, Hassam's illuminated rectangles are much brighter and more variegated; they emit a pastel-hued and whiter glow. The light seems to shed paradigmatically Impressionist blue and mauve tones upon the sidewalk, creating a foreground pool of pale light that contrasts sharply with the trees in brown leaf flanking the boulevard (sometimes identified as the Champs-Élysées, but not decisively).[57] We must understand the sidewalk to be moist to make sense of the colored accents that appear there.

5.5 Childe Hassam, *A Paris Nocturne*, 1889–90. Oil on canvas. 69.2 × 51.4 cm (24¼ × 20¼ in.). Formerly Manoogian Collection.

The intense, pale yellow blaze of street light along the boulevard—one *réverbère* stands out, but not baldly—blends with the lights of illuminated shops and other establishments, unnamed, in a mottled and painterly but continuous horizontal zone of light above the central woman's head. The effect is that of an atmospheric gaslight glow rather than a descriptive capturing of the appearance of specific street furniture. There may have been electric streetlamps on a prominent boulevard like this one in Paris by 1889–90, but the painting's pale yellow tonalities guarantee that Hassam's referent is gaslight.

Inasmuch as Hassam painted a considerable number of Paris nocturnes, we will inventory a cluster of them, if only briefly, to round out our understanding of his *modus operandi* and vision. *Along the Seine, Winter* (1887; fig. 5.6) is a gentle tonalist record of a *fiacre* on a roadway near the Seine on a snowy day. Hassam's attraction to the animation brought to a canvas by moderate spots of artificial illumination, white and yellow, is seen again. The small oil's gray palette, points of *éclairage*, and minimal social incident show Hassam at his most Whistlerian, albeit including a volumetric stone wall and approaching carriage, which are not the kind of thrusting sculptural forms to be found in Whistler's nocturnes. At the same time, there is nothing saliently Parisian about its ingredients.[58] *Street Scene with Hansom Cab* (1887; fig. 5.7) is a comparatively lively scene that includes a variety of pedestrian activity on a broad boulevard. A horse-drawn carriage—a conveyance that was apparently

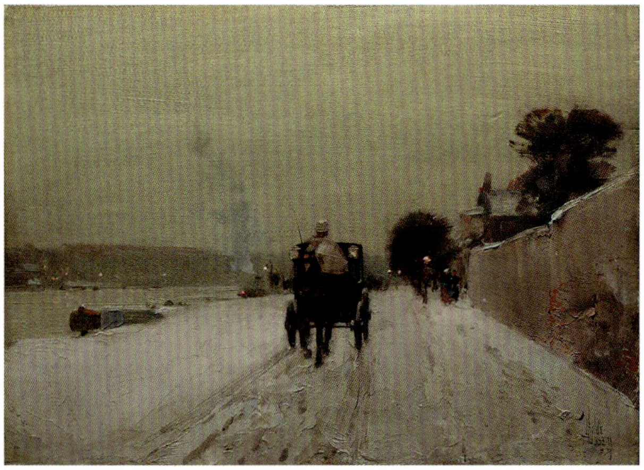

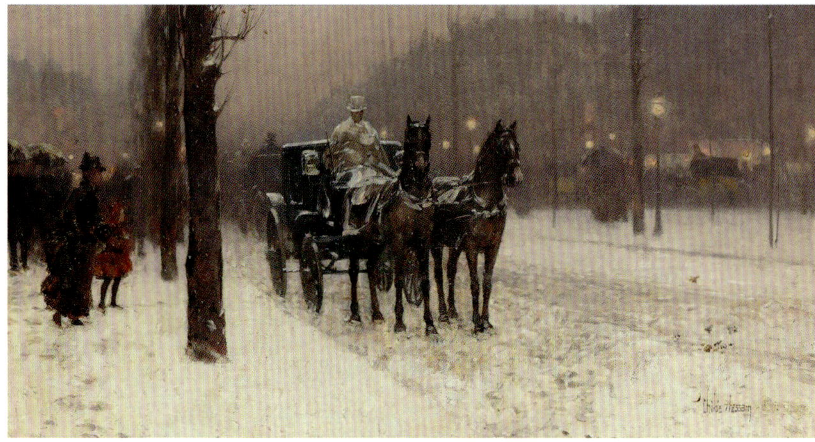

5.6 Childe Hassam, *Along the Seine, Winter*, 1887. Oil on wood. 20.32 × 27.94 cm (8 × 11 in.). Dallas Museum of Art, bequest of Joel T. Howard, 1951.40. Courtesy of Dallas Museum of Art.

5.7 Childe Hassam, *Street Scene with Hansom Cab* (Paris), 1887. Oil on canvas. 43.2 × 78.7 cm (17 × 31 in.). Private collection.

indispensable to Hassam's imagination of Paris public life—is front and center in a very subdued record of a city street and its black-clad pedestrians in the snow. The terms "muted" and "restrained," with respect to atmosphere and narrative, apply well to these paintings.

An even quieter aesthetic disquisition on urban dusk emerges in Hassam's large picture *Twilight* (*Le Crépuscule*) (c. 1888; fig. 5.8).[59] In this work, the moment of sundown that coincides with the igniting of the city's streetlights becomes the occasion for a quiet moment of remote contemplation by two rooftop women in white, whose appearance and quiet demeanor are worthy of Hassam's American contemporary, Thomas Dewing, another student at the Académie Julian. The city in this case is not conceived as a space of involvement, but rather as an object of rapt attention. Another American in Paris (albeit an expatriate), Frederick Frieske, painted a very similar situation in 1904, *The Balcony* (fig. 5.9). His female figures lean more actively into the act of viewing, though they are similarly mesmerized by what is before them. The objects of their concentration include the rising full moon as well as the lights in the park below. It is a restrained, quiet, and passive image in a key consistent with that established by the earlier Hassam.

Hassam's slightly later evening scenes accelerated the pace and density of Paris street life. But both *Paris Street Scene, Autumn* (1889; fig. 5.10) and *Walk at Sunset in Paris* (c. 1888–89; fig. 5.11) maintain an exclusive focus upon bourgeois passersby in a smart location. The former imagines a single-sex urban scene. All of its pedestrians are women: a combination of bourgeois shoppers (note the hatbox) with their servants or interspersed with other working women in white caps. In both these canvases, done shortly before Hassam's departure for the United States, the principal narrative trope of his bellwether *A Paris Nocturne* (fig. 5.5) is in full flower: well-turned-out and unimpeachably respectable solitary women on the sidewalk in a *beau quartier*. This "device" was a constant feature of another American traveler's Paris nocturnes as well, those of Maurice Prendergast, as seen in *Lady on the Boulevard* (1892–94; fig. 5.12).[60] Prendergast, also a Bostonian, was likewise a student in Paris at the Académie Julian.

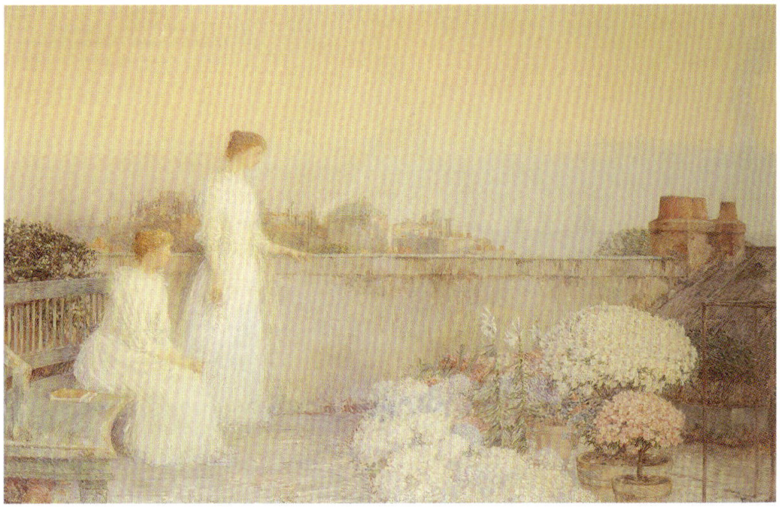

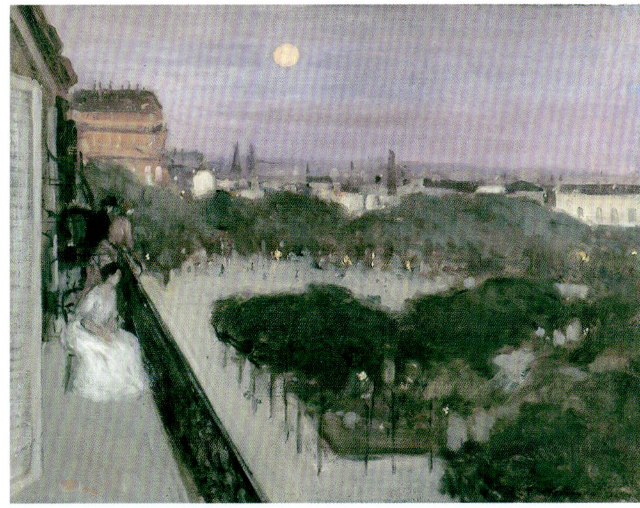

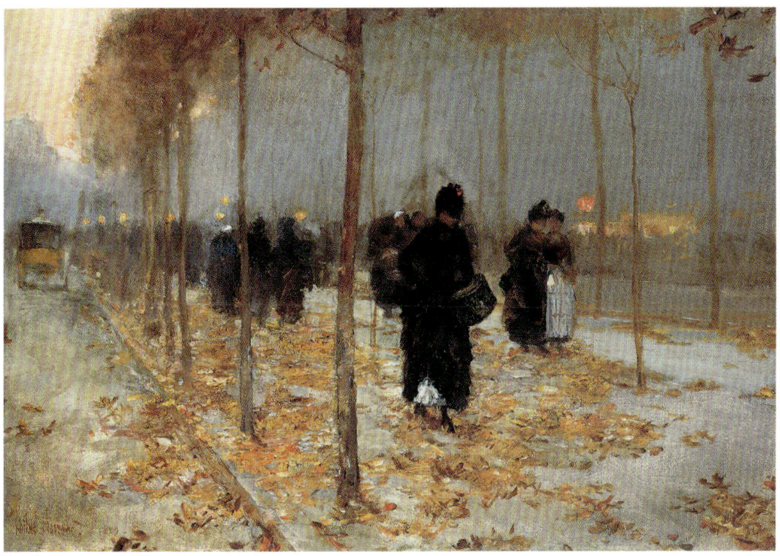

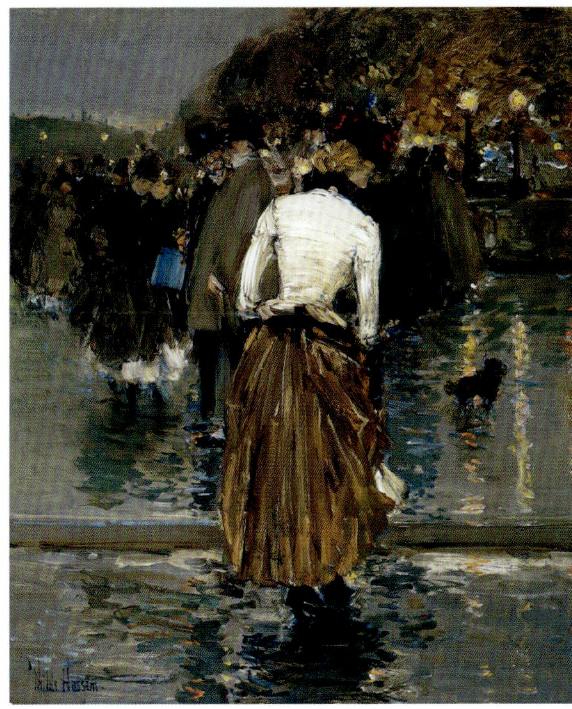

5.8 Childe Hassam, *Le Crépuscule* (*Twilight*), c. 1888. Oil on canvas. 127 × 195.6 cm (50 × 77 in.). Private collection.

5.9 Frederick Frieske, *The Balcony*, 1904. Oil on canvas. 50.2 × 61 cm (19¾ × 24 in.). Courtesy Hollis Taggart Galleries, New York.

5.10 Childe Hassam, *Paris Street Scene, Autumn*, 1889. Oil on canvas. 33.5 × 46 cm (13¼ × 18⅛ in.). Sammlung Thyssen-Bornemisza.

5.11 Childe Hassam, *Walk at Sunset in Paris*, c. 1888–89. Oil on canvas. 46 × 38.4 cm (18⅛ × 15⅛ in.). Marlene and Spencer Hays Collection, Musée d'Orsay, Paris.

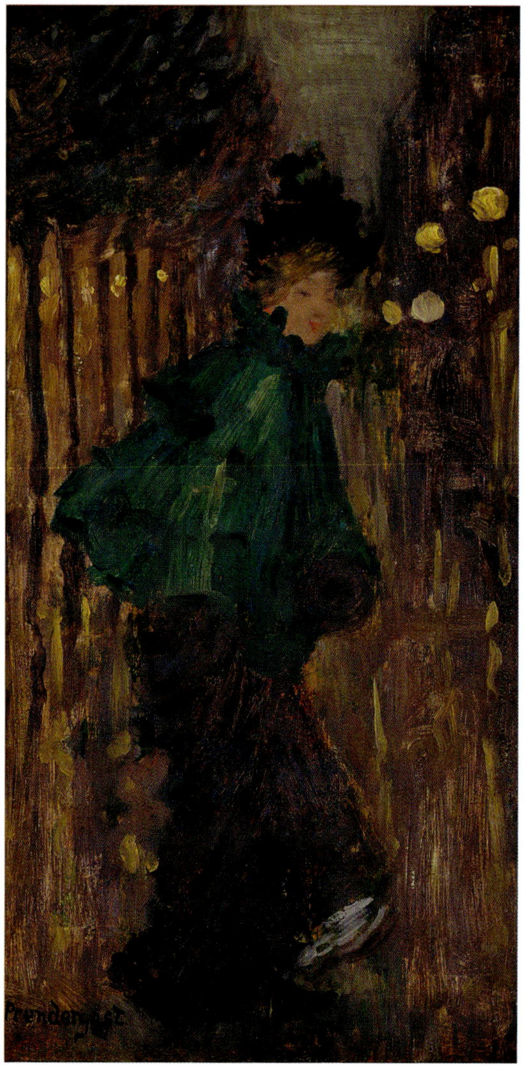

5.12 Maurice Brazil Prendergast, *Lady on the Boulevard*, 1892–94. Oil on panel. Image: 34.6 × 18.1 cm (13⅝ × 7⅛ in.); frame: 52.7 × 36.2 cm (20¾ × 14¼ in.). Terra Foundation for American Art, 1992.64. Photo: Terra Foundation for American Art, Chicago/Art Resource, NY.

Curator Stéphane Guégan took the opportunity afforded by the exhibition of the Hays Collection, the collection of a prosperous American couple, in purpose-built, extravagant galleries in the Musée d'Orsay in 2013, to encapsulate, somewhat snidely but not inappropriately, Hassam's American approach to Paris city scenes:[61]

> Hassam [and his friend the painter Frank Boggs] excelled at the night lights of Paris, still attractive in the rain. Their paintings depict city-dwellers from genteel society, and the sort of comings and goings that make up the bustle of the city. Hassam very quickly mastered the vocabulary of these modern alert observers, pretty Parisian ladies walking their dogs, and the halo of artificial lights. The painting in the Hays collection [fig. 5.11] encapsulates to perfection the by-now canonical components of Uncle Sam's Paris.

I argue that *A Paris Nocturne* (fig. 5.5), full of the "canonical components of Uncle Sam's Paris," has the qualities of a manifesto for the Paris-steeped Hassam. But his Boston twilights also defined placid and inviting evenings. A nocturne painted in Boston, *A City Fairyland* (1886; fig. 5.13), shows his early interest in crepuscular light and the faint twinkle of yellow gas streetlights, but the "temperature" of Boston street life is lower than that imagined in Paris—which is to say that the pace is relatively slow in an early evening snowfall, and the broad space chosen as the setting (probably Copley Square) is comparatively empty. Gerdts noted Hassam's lack of interest in painting major monuments in Boston. His "primary concern here and later was the city as a setting for contemporary life."[62] City monuments were missing from the Paris nocturnes as well.

While Hassam's Boston and Paris nocturnes were of a piece, then, his subsequent night pictures set in New York City and Chicago had a more daring tenor.[63] Recall that Hassam arrived in Paris in 1886, when the future of electric lighting in the streets of the City of Light was still up for grabs; when the gingerly transition to electric light from the ubiquitous gas was still unfolding both politically and visually. "Gingerly" is a polite word choice given that the transition was full of the conflicts, graft, failures of nerve, and incertitude that made the Paris of the 1890s one of the least electrified of modern cities. Indeed, the city slowed the pace of the advance of electricity by insisting on experiments when lighting was well past the testing stage and by debating the concession endlessly.[64] Hassam would have found the debate about the predations wrought by American technology in Paris hard to miss, especially if he was affected by the xenophobic outlook of those who agreed with Jules Claretie.

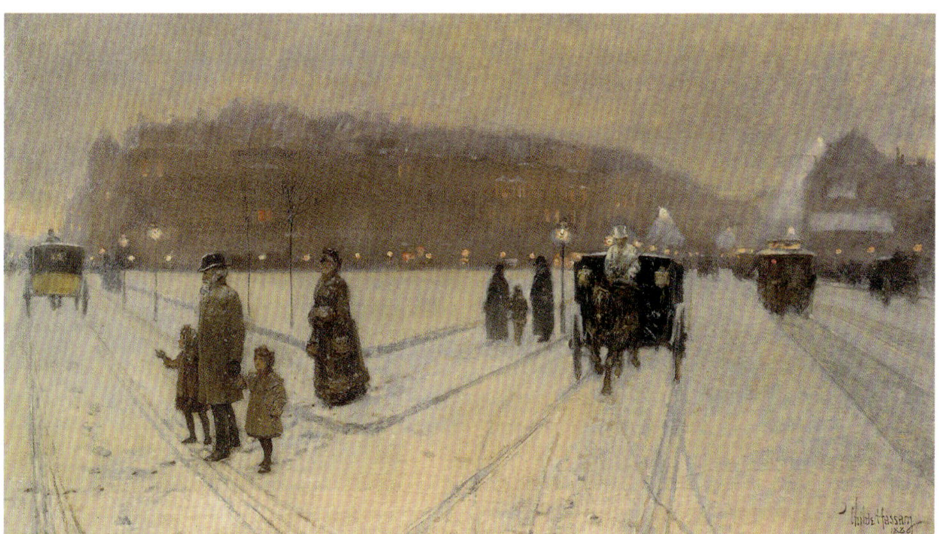

5.13 Childe Hassam, *A City Fairyland*, 1886. Oil on canvas. 50.8 × 76.2 cm (20 × 30 in.). Private collection.

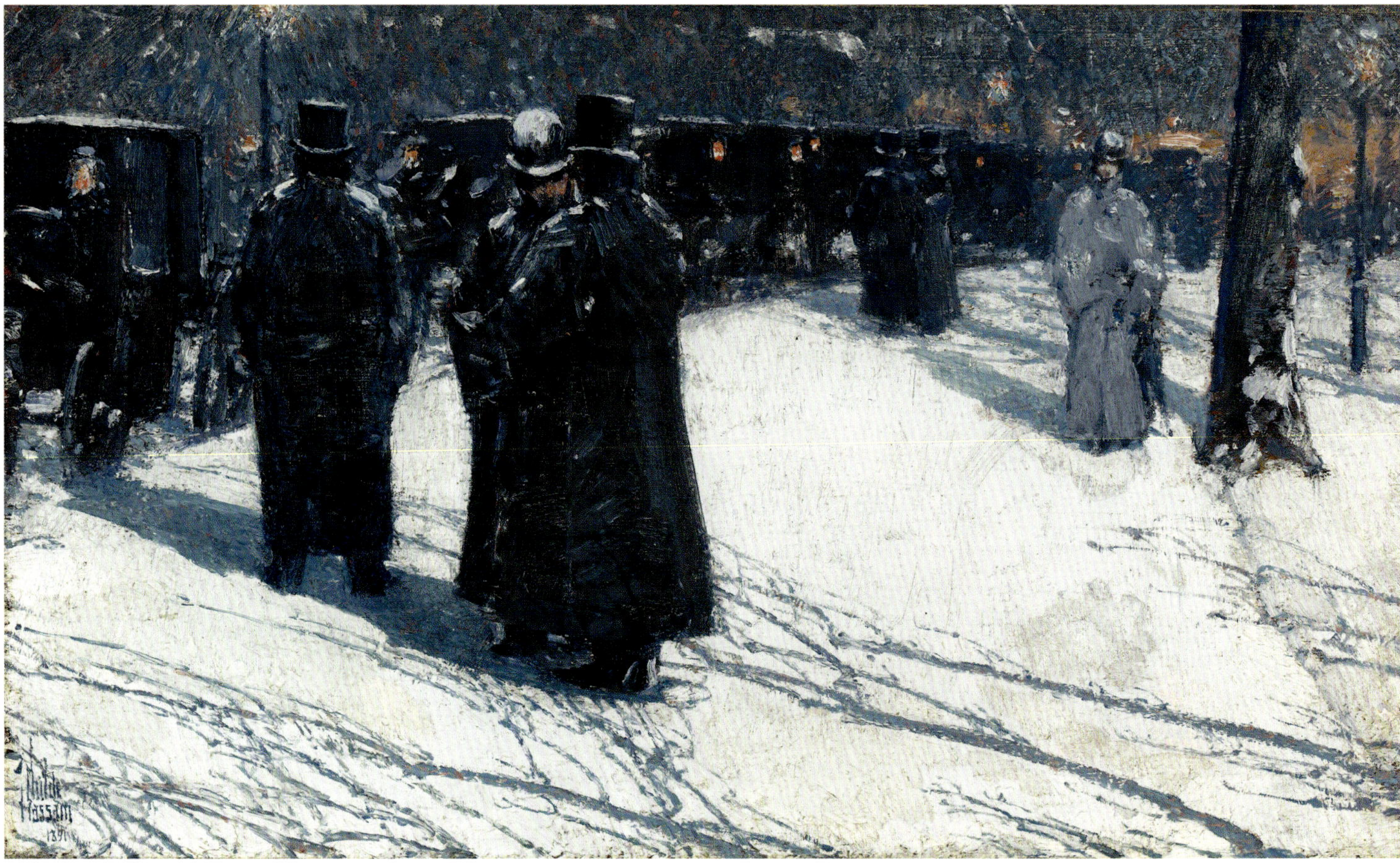

5.14 Childe Hassam, *Cab Stand at Night, Madison Square*, 1891. Oil on panel. $8\frac{15}{16} \times 14\frac{3}{8}$ in. Bequest of Annie Swan Coburn (Mrs. Lewis Larned Coburn), Smith College Museum of Art, Northampton, MA. Photo © SCMA/All rights reserved. Photo: Petegorsky/Gipe.

For reasons already suggested, Hassam and his fellow US painters kept American-style brashness and vulgar infrastructure out of their Paris nocturnes, which uniformly constructed a discreet, benign, subtly twinkling, and congenial metropolis, punctuated by gentle points of gaslight and populated by poised and respectable people. The homogeneity of the Paris nocturnes is confirmed by the discovery that the metropolitan night paintings that Hassam produced, mostly in New York City, upon his return, endowed the modern urban night with precisely the characteristics never present in the Paris pictures. In his American city nocturnes of the early 1890s, he imagined the night in jarring terms that would have been unthinkable in the Paris pictures.

Cab Stand at Night, Madison Square, NY (1891; fig. 5.14), one of Hassam's most original paintings, endows an electrically illuminated urban space with qualities, both visual and psychological, never broached in the French city pictures.[65] The atmosphere is charged with the harsh glare of punishing white electric streetlight (the canvas says "Electric Light" on the back); the solitary woman, who looks pinched and anxious, is not a figure of freedom and self-actualization; and the cabbies are hulking, conspiratorially bunched together and threatening. The composition is also daringly cropped, unlike the capacious structural approaches used earlier by Hassam. *Fifth Avenue Nocturne* (c. 1895; fig. 5.15), is the most extraordinary of the series: the avenue's electric streetlights are paired, monstrous insect eyes punctuating the turgid blue murk of moody night.[66] The solitary female stroller appears threatened by the male stroller not far behind her.

A final morsel of pictorial evidence from Hassam's nocturnes should clinch the case regarding the ways in which Paris could and could not be imagined. In 1899,

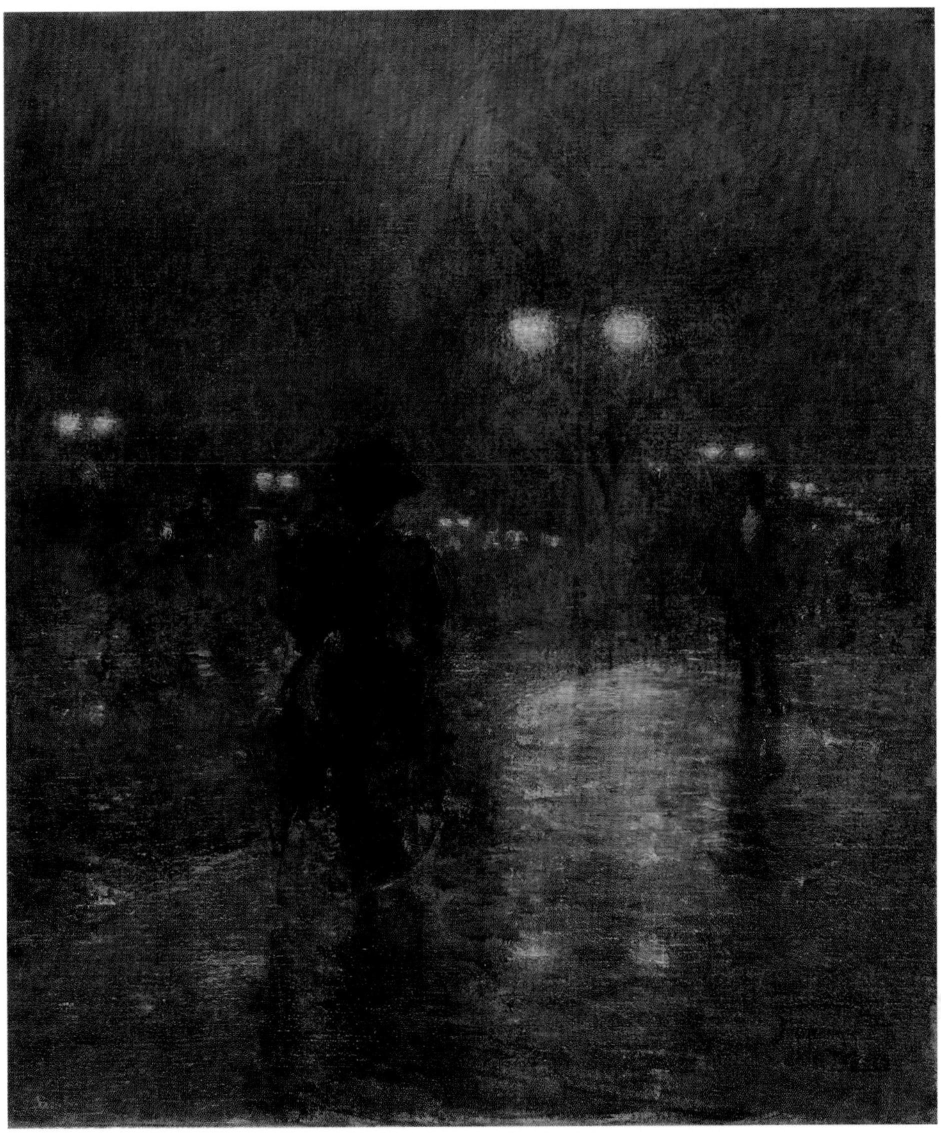

5.15 Childe Hassam, *Fifth Avenue Nocturne*, c. 1895. Oil on canvas. Cleveland Museum of Art, Anonymous gift, Bridgeman Images.

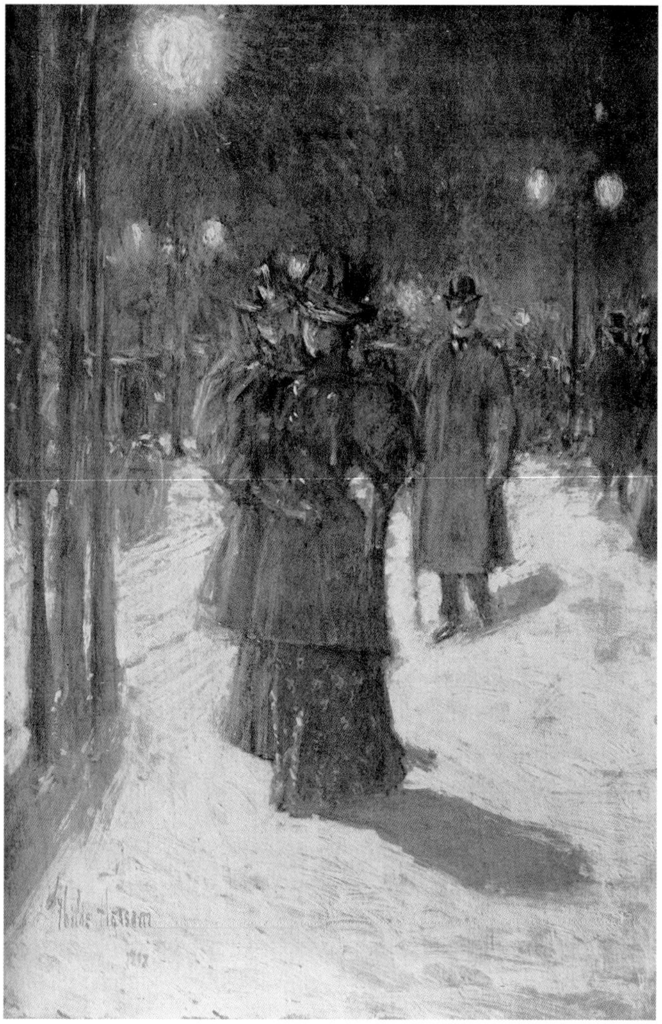

5.16 Childe Hassam, *Winter Night, Fifth Avenue, in Front of the New University Club, Looking South* (1898), from Childe Hassam, *Three Cities* (New York: R. H. Russell, 1899). Courtesy of Toledo Museum of Art.

5.17 Childe Hassam, *Nôtre Dame* (n.d.), from Hassam, *Three Cities*. Courtesy of Toledo Museum of Art.

Hassam published a collection, a folio book, entitled *Three Cities* (New York: R. H. Russell). The cities were New York, London, and Paris. New York was allotted seventeen scenes (where dated, all 1898), three of them set at night.[67] London got thirteen scenes (all dated, 1897–98), none set at night. And finally Paris was allocated nineteen scenes (where dated, all 1898), three set at night.[68] Judging by the characteristics with which he endows New York and Paris at the end of the nineteenth century, the gulf between them had only grown in Hassam's imagination and artistic practice. The New York night scenes (fig. 5.16, for example) show Manhattan as the true shock city of the era (*pace* Manchester), defined by stark contrasts between piercing white lights and the gray-black monotone of city pedestrians, umbrellas, and carriages. It is not a benign urban symbiosis, but the beginnings of an imagery of a jarring struggle for space and survival in the most modern American city. The three Paris night scenes define the French capital as a city of streetlights—one lamp is featured in each—but, by contrast to New York, Paris (fig. 5.17, for example) is now imagined as a place defined by traditional, nay old-fashioned, activities and monuments: a cathedral, chestnut roasting, and a flower-seller.[69] Paris is shorn of whatever credentials it once possessed as the erstwhile headquarters of stylish promenades and gaslit modernity, outstripped by the brasher lights and more intense forms of street life in New York City, by then dubbed "The City That Never Sleeps."

The Paris nocturnes by American artists conceptualized metropolitan modernity in ways unlike that given form by contemporaneous Parisian paintings of the nocturnal city. The perspectives of both Karl Marx and Walter Benjamin are germane here. They both argued that thinking about the future and recalling the past were interdependent in an era of hectic modernization. The cultural historian Ben Highmore's encapsulation is on target: "Karl Marx and Walter Benjamin, in their very different ways, understood that the modernizing impulse in nineteenth-century France carried with it a cargo of archaic and atavistic tendencies. Modernization, for them, was haunted by the ruins of a time it sought to surpass."[70] The works of the Parisian Painters of Modern Life were haunted by unpredictability and attempted to figure the edginess of a modernity defined by that very quality. For them, the brash city interior at night was a preferred, indeed paradigmatic, site for such explorations (think of works by Édouard Manet, as well as Edgar Degas, Georges Seurat, and Henri de Toulouse-Lautrec). Nighttime boulevard spaces did not serve their purposes. They were perhaps already considered the clichéd property of the Panglossian visitor, or found too commercial and vulgar to serve their subtle interpretative purposes.

The Americans, by contrast, sought in Paris a way out of or an alternative to their boredom and disenchantment with the cities of the United States. Paris was imagined by them as the answer to that signature ennui, and their romanticized depictions of the outdoor city at night must be understood to bracket unpredictability (along with other proverbial Paris nighttime baggage, such as crime, prostitution, vagrancy, and the like) as part of their effort to assuage their pre-Paris dissatisfactions. The painters from the United States, exemplified in this discussion by Charles Courtney Curran and Childe Hassam, who fashioned serene and congenial forms of social life suffused for the most part with friendly gaslight, endowed Paris with the capacity to please, excite, and soothe simultaneously. The American artists' invention in paint of a Janus-faced modernity for the city of Paris, modern but traditional, instances a viewpoint shaped by longing and ambition, recognizable as that of the bruised but enthusiastic foreign artist, of the storm-tossed but upbeat voluntary exile in the City of Light. *Fluctuat nec mergitur.*

· 6 ·

Man at the Window

EDVARD MUNCH IN SAINT-CLOUD

How light it was outside. One would think it was daylight. I always lie here with the curtains pulled back. It is the moon shining over the Seine; it shines into my room, through the windows, and casts a bluish rectangle of light on to the floor. As I lay there and looked out of the window, other images flashed before my eyes dim and blurred like projections from a magic lantern.

EDVARD MUNCH, 1890

Edvard Munch, a twenty-three-year-old artist from Kristiania, the capital and largest city of Norway (called Oslo after 1925), traveled to Paris for the first time in 1885 on a three-week stay that included visits to the Louvre and the annual Salon exhibition. Between 1889 and 1895, he returned to France on three long study visits. In those years, the increasingly peripatetic artist enlarged the span of his aesthetic compass by absorbing the state of play in advanced art circles in Berlin, Brussels, and Copenhagen, as well as Paris and Kristiania, securing his claim to cosmopolitanism in the process.[1] Compared to the American artists studied in chapter 5, Munch (1863–1944) became more widely traveled and less Paris-centered. The present chapter homes in, however, on his Paris work, concentrating upon his breakthrough, light-besotted 1890 painting campaign in Saint-Cloud, a quiet commune on the Seine six miles west of central Paris.

Munch was awarded a bursary by the Norwegian state in 1889 for art study in Paris, and moved to the French capital in October to study with Léon Bonnat.[2] Like other young male Norwegian art students who uprooted for instruction, his primary goal was the mastery of academic drawing.[3] His arrival was heralded by one of his own paintings, *Morning* (1884; Bergen Kunstmuseum), on display at that year's Exposition Universelle.[4] He enriched his aesthetic knowledge by also seeing the work of Gauguin and his circle at the Café Volpini exhibition that year. At the very end of 1889, he moved out of the city proper to Saint-Cloud, in order, it appears, to find serene working conditions, and to steer clear of an influenza outbreak in the French capital.[5] He remained headquartered there, with a view of the close-by river, until his return to Norway at the end of May 1890.

Distant from the residences and studios of other artists in the French capital, Saint-Cloud was a rather peaceful spot when Munch settled there. It was a destination for pleasure excursions, thanks to its elegant royal park that survived the destruction of the palace during the Franco-Prussian War, and was easily accessed from Paris via steamer on the Seine.[6] Like most foreign painters in Paris in the later nineteenth century, Munch did not know French and remained rather isolated, confined more or less to the company of those comprising the little Norwegian Paris colony.[7]

To fill out the picture of Munch's subsequent sojourns in France, in November 1890 he won another art study stipend, returned to Paris, and went on to Nice early in 1891. He spent a month in Paris at the end of his stay (from late April through the end of May 1891) before departing again for Norway. It was during that month that he painted his well-known plunging view of the *Rue Lafayette* (Oslo, National Gallery), much indebted to French Impressionist street views, especially those of Gustave Caillebotte. A third Norwegian art study award sent him back to France in autumn of 1891, when he traveled to Nice via Paris and Copenhagen. He remained in Nice until he returned home via Paris in early April 1892. The years from 1893 to 1895 inaugurated his deep engagement with the art worlds of Germany, mostly those of Berlin.

The present chapter stays close to the implications of the trailblazing picture that the young Norwegian painted during his fecund if in some ways fraught stay in a rented room above a café overlooking the Seine in Saint-Cloud in 1890. *Night* or *La Nuit* (or *Night in Saint-Cloud*) (1890; fig. 6.1), set in his own rented rooms, was the first of about fifteen painted, pastel, and etched artworks that Munch set in threshold space at night, all of them made in Paris or Nice between 1890 and about 1895. Despite his knowledge, already adumbrated, of developments in advanced art in various parts of Europe, his window "series" is singular; not beholden to earlier or contemporaneous specimens of art, whether made in Paris or elsewhere. That said, it is not the placement of pictures of modern life in interior space adjacent to a parlor's glassed French doors that was the telltale move. On the contrary, Parisian examples of structurally similar "threshold pictures" of everyday life are numerous.[8] What distinguish the pictures in Munch's series are their recurrent nighttime settings, the intense character of their "threshold people," and the use of *éclairage* whose brightness and hue stand out against the tones and colors of the evening environments.[9] The moodiness of the pictures, all of whose figures are featureless or entirely faceless, is likewise indisputable.

6.1 Edvard Munch, *Night in Saint-Cloud* (*Natt i Saint-Cloud*), 1890. Oil on canvas. 64.5 × 54 cm (25⅜ × 21¼ in.). Nasjonalmuseet (National Gallery), Oslo, NG.M.01111.

Three distinct clusters make up the Munch window corpus. One group features a top-hatted man seated close to an illuminated window at night; the set was inaugurated by the bellwether Saint-Cloud canvas (fig. 6.1) and sustained through subsequent declensions in diverse media, including an 1893 painted replica. In a second group, a bareheaded, standing woman wearing a shift gazes determinedly but furtively out a similar, if not identical, windowed door bathed in night lights. (Figure 6.2 is a prime example.) In the final group of paintings and prints (fig. 6.3, for example), Munch conjoins a man and a woman in a besotted kiss before the same kind of window. So in cluster three, a key diagnostic trait of the first two sets of window pictures—segregation by sex—is dropped, and the threshold scenario aligns with sexual activity, deploying a metaphorical alignment between the light outside and the sex inside.

An adjacent, closed French window, only partially curtained, is the shared, obliquely angled or parallel far wall of the *mise-en-scène* of all the pictures, and in each picture a rectangle, point, zone, or wedge of artificial light is, to borrow Mark Ledbury's splendid expression, "the structural focus of a drive to narrative density and episodic energy."[10] In Munch's window pictures, night light is either the object

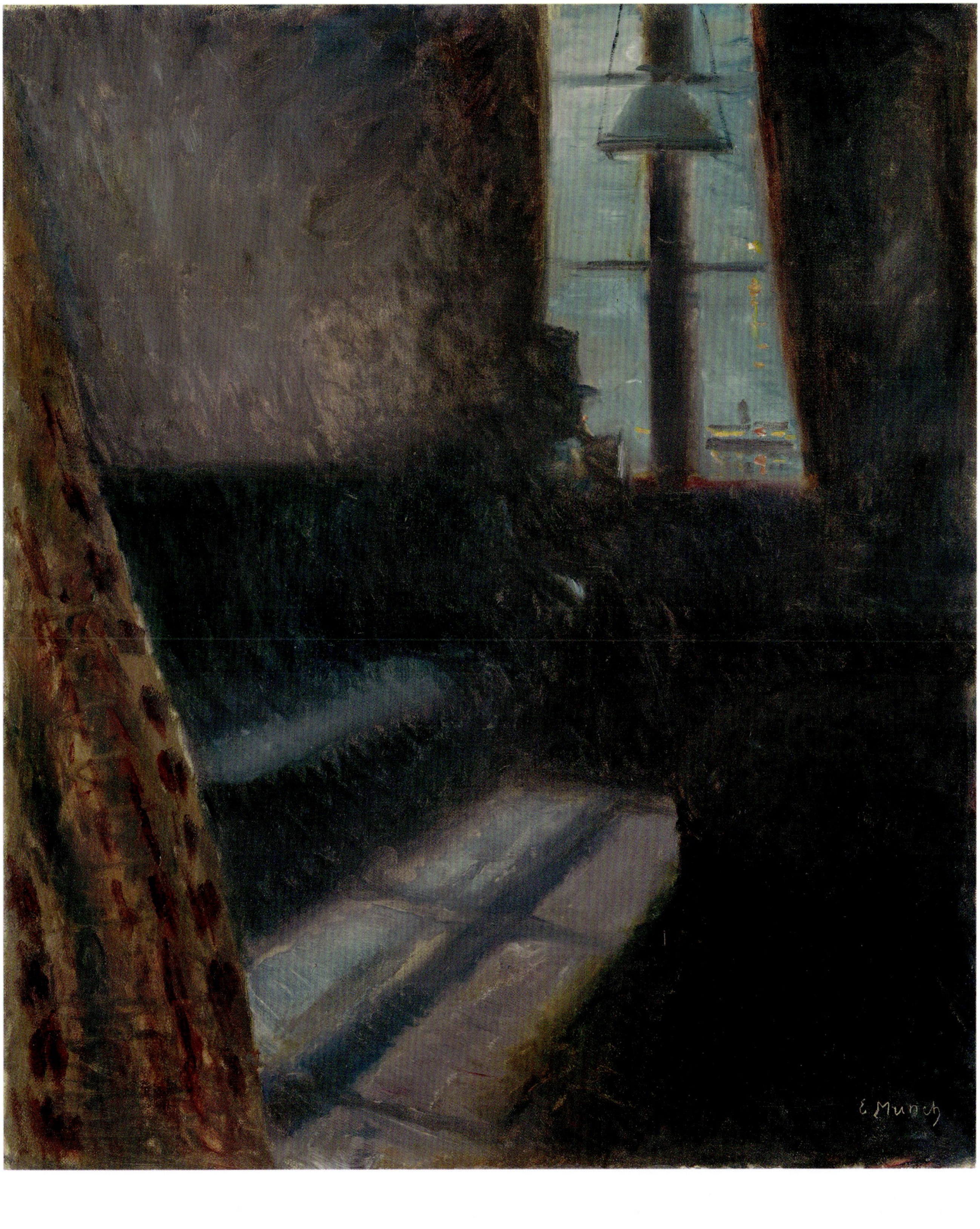
E Munch

of longing and fascination or a technologized metaphor for erotic entanglement. I argue, to put it bluntly, that artificial light, sometimes electric, seen through the window often stands for sex in these pictures.[11] The diverse artificial illuminations of the modern urban night—gas, electric, and riverboat lighting—are recruited to do both atmospheric and metaphorical service in Munch's window pictures. The corpus delivers us back onto Christoph Asendorf terrain, referenced in chapter 3: "A part of the fascination with electricity might have arisen out of the closely related analogy to erotic attraction (or repulsion), so nicely characterized as tension between the sexes."[12]

In view of the arguments made in earlier chapters, the proposal that the counterpoint created by artificial illumination in Munch's nocturnal window pictures was an aspect of their up-to-dateness will not be unexpected. Each of the pictures acknowledges the quarrel between darkness and illumination that characterized the nighttime urban landscape of the nineteenth century, and served to define its modernity.[13] When Munch began spending time in Paris in 1885, the city's preoccupation with the effects of new technologies of illumination remained intense. It appears that he made good use of the lingua franca of the new *éclairage* by drawing upon and activating the rich symbolic and allegorical resonance of the new night light in the French capital. A small oil entitled *Cabaret* (1885–86; fig. 6.4) bears obvious traces of Munch's familiarity with the Degas-Seurat Paris nexus and its use of the "moon eggs" of bright light in the spaces of commercial recreation.[14] The picture noticeably deploys a modernist semiotics of brash commercial Paris via the coexistence of caricatured male visages, and a female stage entertainer who adjoins a zone of glaring illumination at the front of the *salle,* framed and punctuated in turn by an array of bright globes. An 1889 diary entry, which is said to record "Impressions from a ballroom, New Year's Eve in Saint-Cloud," testifies to Munch's savvy regarding streetlights. At the end, he wrote: "Once again I was out on the Boulevard des Italiens—with the white electric lamps and the yellow gas jets—with the thousands of strange faces that looked so ghostly in the electric light."[15]

As the art-historical record demonstrates in abundance, innovative visualizations of modern life were not only staged in public spaces or commercial interiors. Many inventive artists working in later nineteenth-century Paris gravitated to the *mise-en-scène* of private space. Michael Fried, for example, argues that interiority and its "absorptive worlds or cloisters" defined the thematic or spatial modernity of much nineteenth-century French painting.[16] Munch himself set an important pre-Paris picture in a domestic interior: *Spring* (1889; fig. 6.5), a canvas that he called his "definitive farewell to impressionism or to realism," terms defined within a Norwegian frame of reference.[17] In a strictly Paris context, it would certainly have been called a "naturalist" picture, insofar as it does not use the curtained window border as a tense membrane, and it closely echoes the work of Jules Bastien-Lepage. The contrast between the young invalid and her attentive knitting companion gives the blond, airy, and sun-filled room a subtly despondent narrative undercurrent.

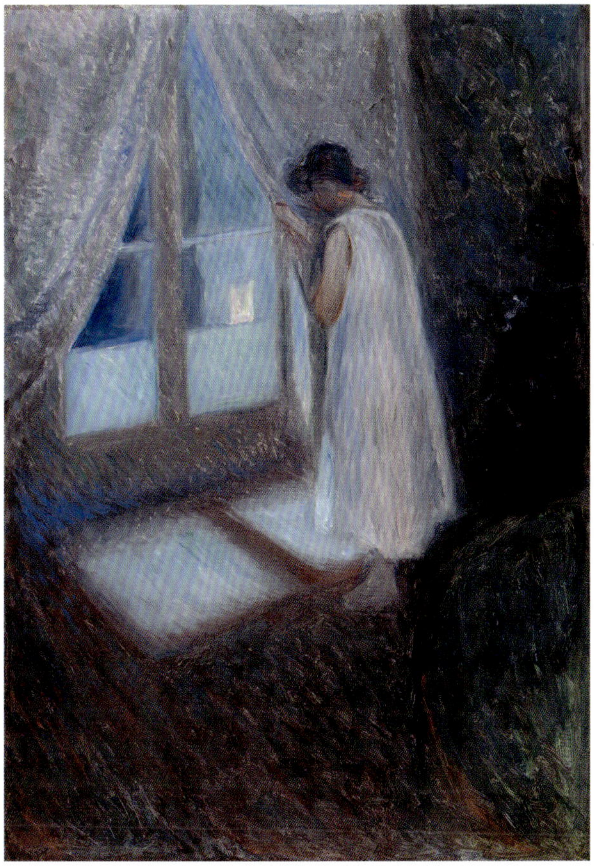

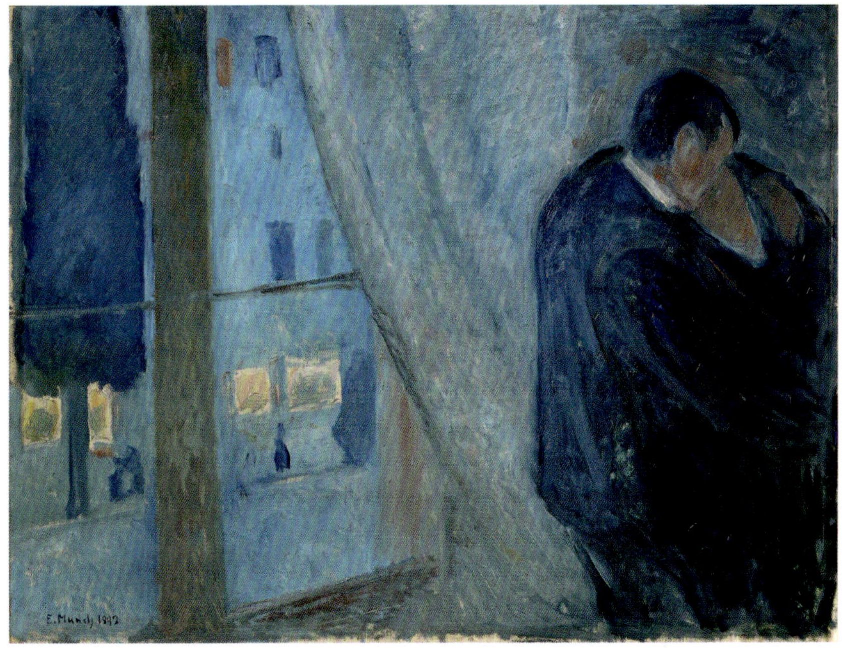

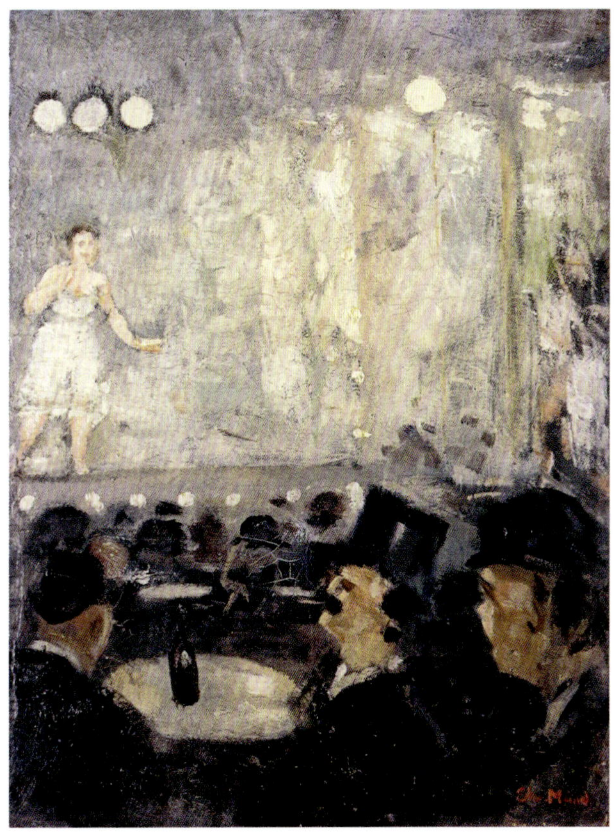

6.2 Edvard Munch, *The Girl by the Window*, 1893. Oil on canvas. 96.5 × 65.4 cm (38 × 25¾ in.). Inscribed lower right: E—Munch. Searle Family Trust and Goldabelle McComb Finn endowments; Charles H. and Mary F. S. Worcester Collection, 2000.50; Art Institute of Chicago. Photo: Art Institute of Chicago/Art Resource, NY.

6.3 Edvard Munch, *The Kiss* (*Kyss*; *The Kiss by the Window*), 1892. 73 × 92 cm (28¾ × 36¼ in.). National Gallery, Oslo, NG.M.02812. © 2018 Artists Rights Society (ARS), NY. Photo: Erich Lessing/Art Resource, NY.

6.4 Edvard Munch, *Cabaret*, 1885–86. Oil on cardboard, 60 × 44 cm (23⅝ × 17⅜ in.). Stenersenmuseet, Oslo. © 2018 Artists Rights Society (ARS), New York.

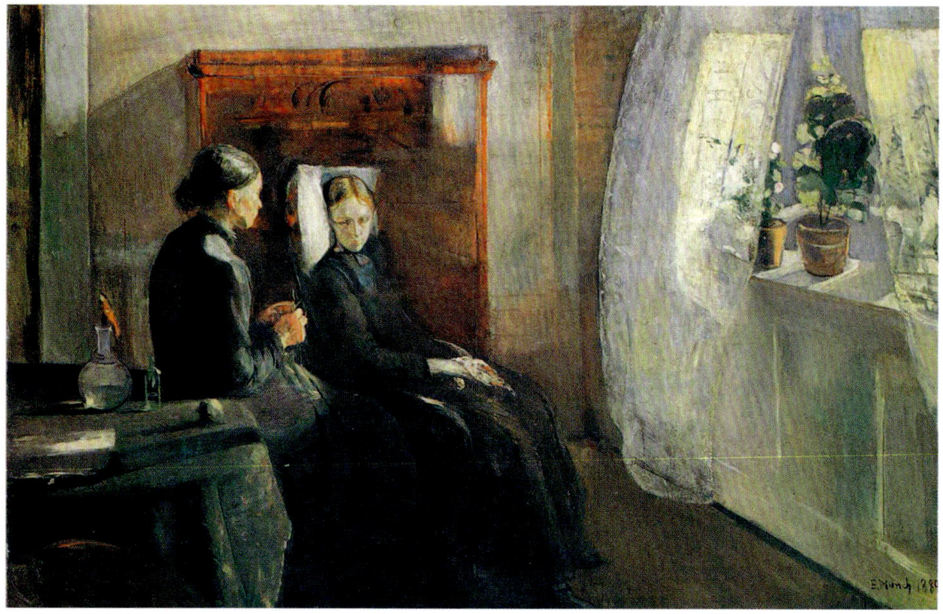

6.5 Edvard Munch, *Spring* (*Vår*), 1889. Oil on canvas. 169.5 × 263.5 cm (66¾ × 103¾ in.). Nasjonalmuseet (National Gallery), Oslo, NG.M.00498.

Modernist artists set pictures in interiors to explore the sheer difficulty of making sense of modern life, and I will read Munch's French nighttime window pictures through a lens defined by issues of space, interiority, and urban social life. Understanding Munch's extraordinary suite of pictures as investigations of contemporary life conducted in the often perplexing and stressful conditions of modernity (mostly metropolitan) during the continuing heyday of illumination discourse, the present account, while not denying personal investments, will part company with readings of Munch's French interiors set at night that discuss them exclusively or even preponderantly within the explanatory frame of autobiography.[18]

But Munch did not paint exclusively night pictures in Saint-Cloud, nor were they all interiors.[19] The non-French-speaking and often gloomy Norwegian (his father, Dr. Christian Munch, unexpectedly died on November 28, 1889) did at least eight *plein air* paintings of the Seine in his neighborhood during the day, from his window or down on the quai.[20] The sheer undertaking of such riverine views indexes his knowledge of the currency and apparent indispensability of Paris-based outdoor landscape painting.[21] He worked mostly from his side (the west bank) of the river, but actually painted from both banks of the Seine. The contrasting faces of the Pont de Saint-Cloud (built in 1810; destroyed in 1940) distinguish the views. See, for example, *The Seine at Saint-Cloud* (1890; fig. 6.8), a view from the east bank; it shows the south face of the bridge and Saint-Cloud proper punctuated by the distinctive tower of the Second Empire church, Saint-Clodoald. The smooth face of the stone bridge seen from the Saint-Cloud side of the river is, for example, delineated with geometricized precision in Munch's illustrated January letter (fig. 6.12) as well as in figure 6.8.

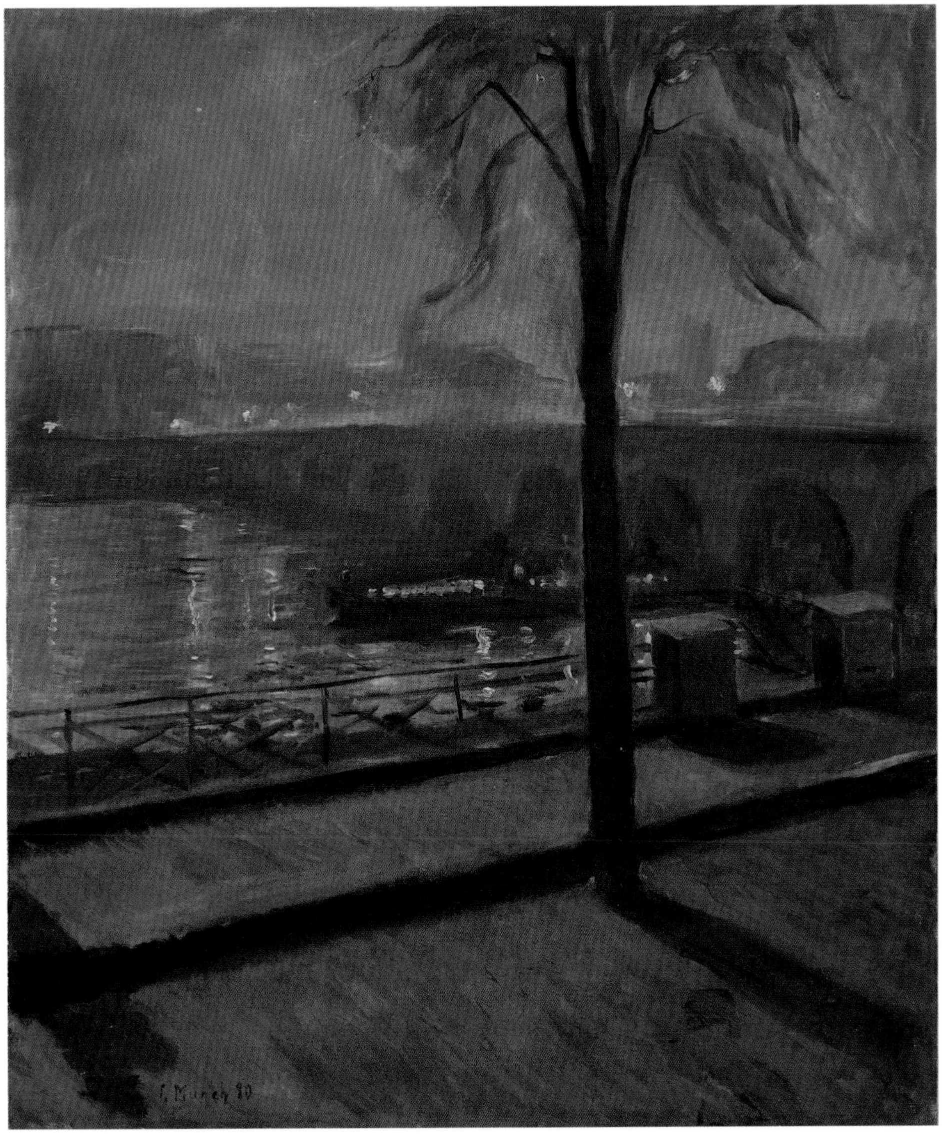

6.6 Edvard Munch, *The Seine at St. Cloud*, 1890. Oil on canvas. Gift of Mrs. Morris Hadley (Katherine Blodgett, class of 1920), 1962.1, Loeb Art Center, Vassar College. © 2018 Artists Rights Society (ARS), NY. Photo © Frances Lehman Loeb Art Center, Vassar College. Charles Porter, photographer/Alex Kalish, assistant.

The varied paint handling and blond palette of the riverscapes showcase his attraction to Impressionist facture and hues, and even Neo-Impressionist divisionism, by the spring of 1890. Another aspect of the bridge may also—this is entirely speculative—have attracted his interest. The very same bridge was the fabled location of the legendary "Filets de Saint-Cloud," nets intended to catch cadavers as they floated away from Paris. The nets were a frequent presence in nineteenth-century plays, novels, and popular histories, albeit mostly in the first half of the century. The perpetuation of the ghoulish legend of the nets is especially remarkable because they were not there after Napoléon replaced the old bridge with a new one in 1810.[22] While there is no explicit ghoulishness in Munch's river pictures, we must consider the remote possibility that the young artist may have gotten wind of Saint-Cloud's macabre reputation.

Two nighttime views of the Seine instance the artist's fascination with evening illuminations along the river, an interest that was to prove vital to the conception of *La Nuit*. *The Seine at Saint-Cloud* (1890; fig. 6.6), with its multiple specks of white, yellow, and red—denoting pale stars and bolder streetlights and steamer lamps, the latter answered and enhanced by wavy striations of reflection on the river's surface—possesses intricacies of hue and light that are very reminiscent of those in the 1888

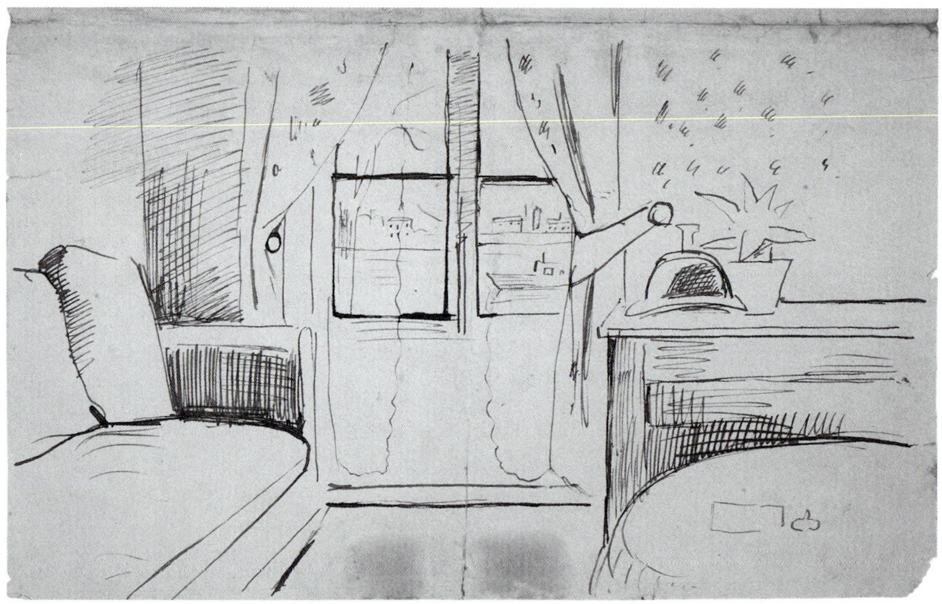

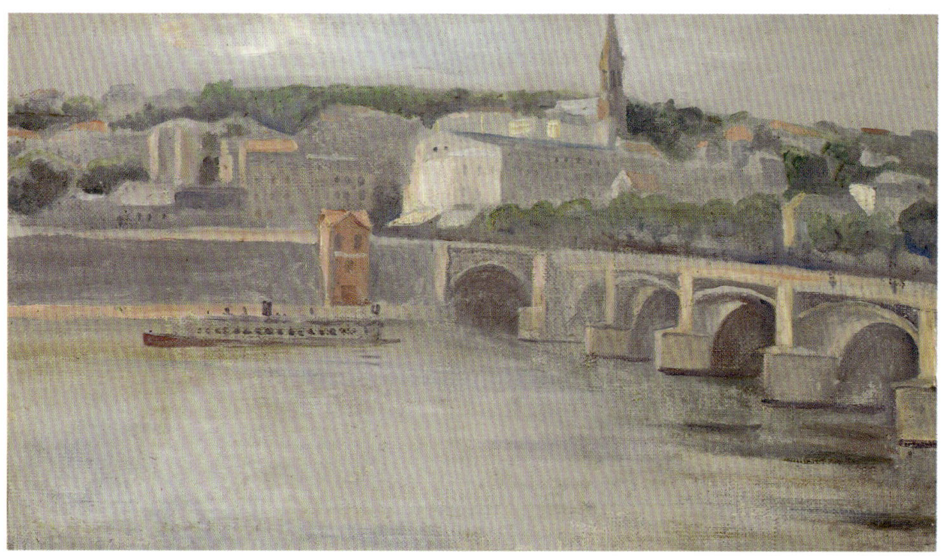

6.7 Edvard Munch, Letter to Karen Bjölstad from Saint-Cloud, January 24, 1890, recto. Munch Museum, Oslo, T2810. © 2018 Artists Rights Society (ARS), NY. Courtesy of Munch Museum.

6.8 Edvard Munch, *The Seine at Saint-Cloud*, 1890. Oil on canvas. Private collection. © 2018 Artists Rights Society (ARS), NY. Courtesy of Munch Museum, Oslo.

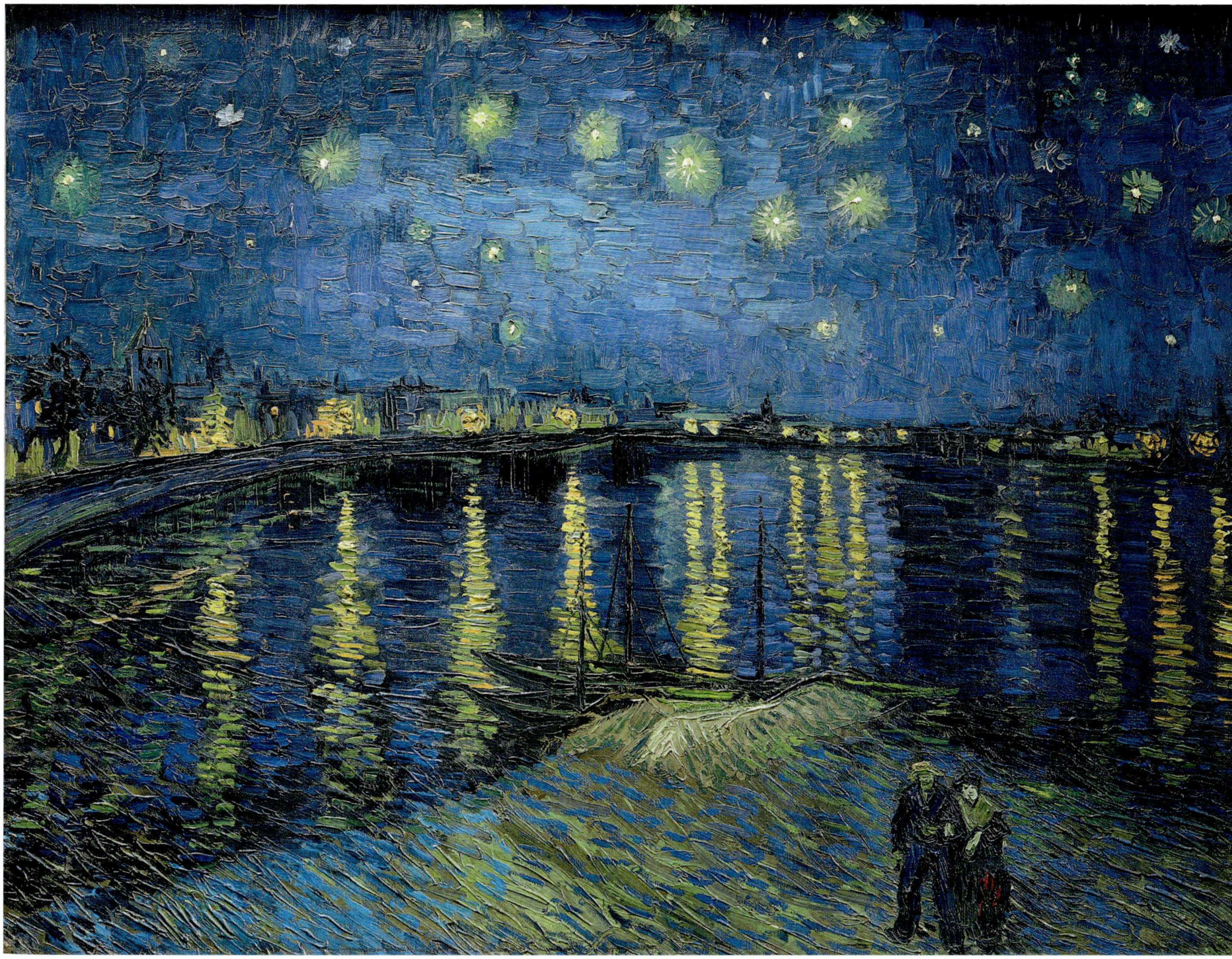

6.9 Vincent van Gogh, *Starry Night over the Rhône*, 1888. 72.5 × 92 cm (28½ × 36¼ in.). Paris, Musée d'Orsay.

Arles nocturne by Vincent van Gogh (fig. 6.9). The exception to the otherwise striking similarity between the two is Munch's abundant use of the illusion of moonlight pouring down and casting shadows upon the quai pavement rendered lavender by the luminosities of evening. Munch's moonlight is the natural light counterpart of the plentiful starlight in Van Gogh's picture. However, gaslight outshone celestial light in the Arles canvas. In Munch's Saint-Cloud evening river canvas, moonlight, the faint stars, and the specks of artificial light (the steamer's running lights and the streetlamps whose radiance is recorded in both white and muddy yellow daubs of paint) are instead harmonized. They coexist; they do not compete.

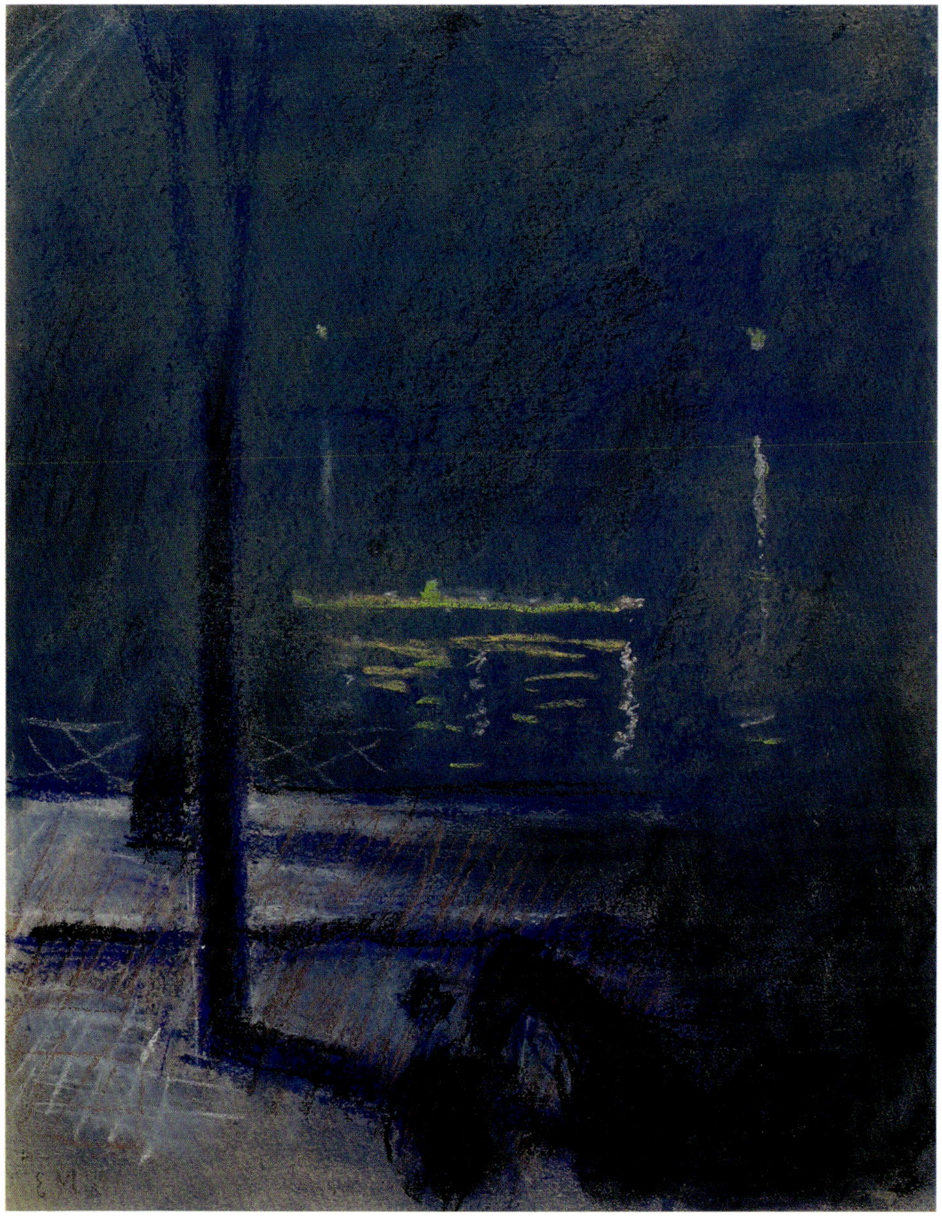

6.10 Edvard Munch, *Meudon* or *Banks of the Seine at Night*, 1890. Pastel on cardboard. 34.9 × 26.9 cm (13¾ × 10⅝ in.). Bern, Kunstmuseum. © 2018 Artists Rights Society (ARS), NY. Courtesy of Kunstmuseum Bern, Verein der Freunde, Inv. Nr. A 9546.

The pastel, *Meudon* or *Banks of the Seine at Night* (fig. 6.10), is an even bolder meshing together and flattening out of the blues of water and sky, with its suave adjustment of the visualities of *l'heure bleue* to the counterpoint of two streetlights on the bridge (emanating white light, though the lights there were gas lamps) and their very expansive reflections. The horizontal yellow waterborne strokes are so expansive so that they pull the handling of the river's surface toward the sphere of abstraction. It is the most patently Whistlerian of Munch's pictures of the river and its lights, whether or not he had actually seen a nocturne authored by the London-

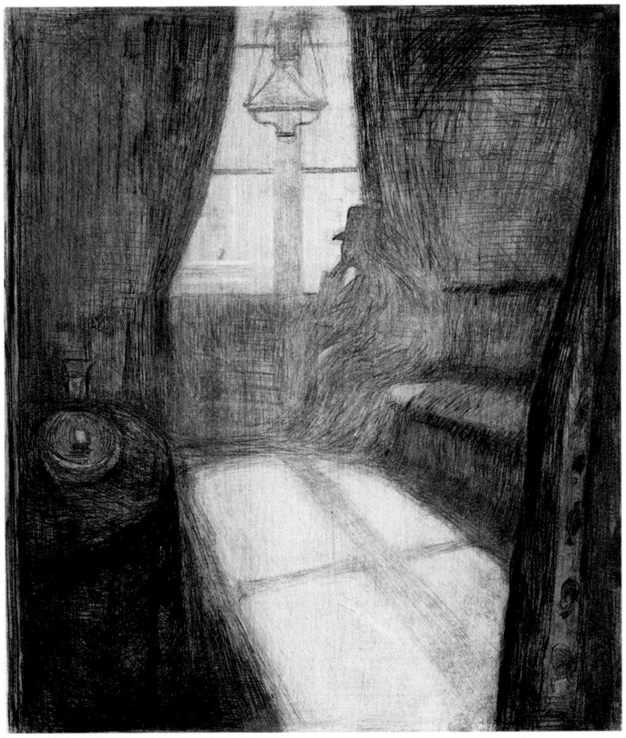

6.11 Edvard Munch, *Moonlight. Night in St Cloud*, 1895. Drypoint, with open bite and burnishing, on cream wove paper. Plate: 35.5 × 26.5 cm (14 × 10⅜ in.); sheet: 48 × 34.5 cm (18⅞ × 13⅝ in.). Clarence Buckingham Collection, 1963.315, Art Institute of Chicago. © 2018 Artists Rights Society (ARS), NY. Photo: Art Institute of Chicago/Art Resource, NY.

based artist before he drew the pastel. It also possesses a surprisingly archaizing touch in the profile of a man leading a horse in the center foreground. True to course, moonlight pours down upon the pavement, silhouetting the beast and its master as well as the tree.

The key work in the "Man at the Window" cluster of cognate pictures is, as we have stressed, the 1890 oil on canvas (fig. 6.1).[23] The 1893 painting and pastel were replicas commissioned after a successful exhibition of the 1890 canvas in Berlin in 1892. Likewise, the 1895 intaglio prints (fig. 6.11, for example) are returns to the earlier successful and decisive endeavor. Munch scholars are united in considering it the breakthrough picture of his young maturity.[24] Some have argued that dejection and melancholy are the keynotes of the picture.[25] My account does not deny the moodiness of the painting's scenario but stresses instead its structure—especially the picture's thematization of a virtual (visual) breaching of the glass threshold boundary. The indistinctness of the shadowy man serves paradoxically to lend force to his yearning toward or for what lies beyond.

Thanks to an illustrated letter that Munch wrote to his aunt, Karen Bjölstad, on January 24, 1890 (figs. 6.7 and 6.12), we plainly see that he set *Night in Saint-Cloud* in his own rented room.[26] He reported in the letter that the room he rented above a café "is quite pleasant, especially at dusk, when the fire in the fireplace illuminates it and the steam boats pass by the window with their red and green lights."[27] In a draft of another letter, written to an unidentified person during the same Saint-Cloud stay,

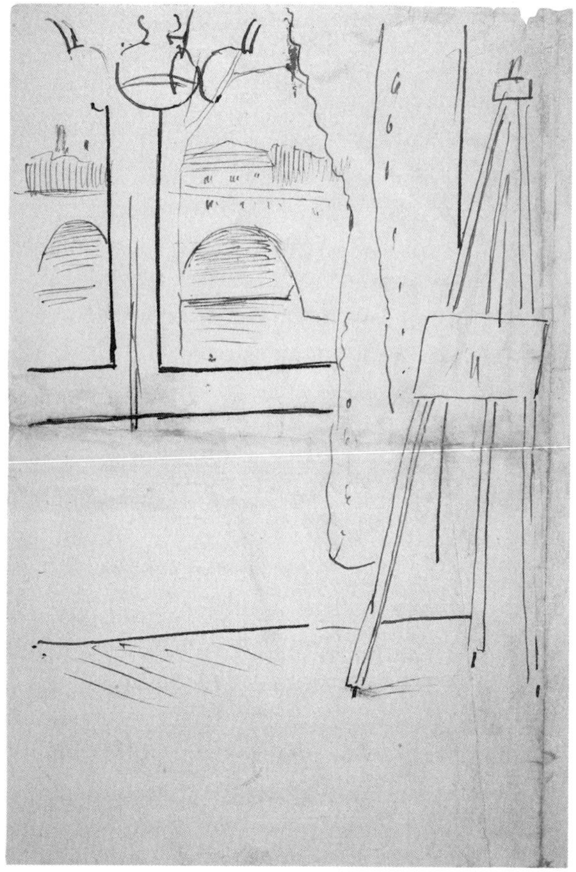

6.12 Edvard Munch, Letter to Karen Bjölstad from Saint-Cloud, January 24, 1890, verso. Munch Museum, Oslo, T2810. © 2018 Artists Rights Society (ARS), NY. Courtesy of Munch Museum.

Munch called attention to the blend of evening lights before him out the window and pouring into the room through the glass: "How many evenings have I not sat alone by the window, and regretted that you were not here, so that together we could admire the setting outside in moonlight—with all the lights on the other side and the gas lamps outside in the street—and all the steamers with their green and red lanterns, and yellow lanterns. And then the strange semi-darkness inside the room—with the bright bluish square that the moon sheds on the floor—."[28]

Munch's repeated mentions of the steamers align with the boats he drew out the window in the letter to his aunt (fig. 6.12), and in the painting the tiny dabs of intense red delineate one light at least on a passing boat and its reflection. Green lantern lights did not end up in the picture; they became yellow spots of illumination instead, better suited presumably to standing out against the field of blue. There is one more spot of red, which adjoins the sitter's hand where it melds with the face. Just a tiny dab: is it another spot of light out on the Seine, or a Sargent-style glowing cigarette end?

Rapetti has noted that Munch's 1890 painting (and its descendants) are all winter pictures according to the artist's own writings. *Night in Saint-Cloud* was thus very likely undertaken long before the spring Salon of 1890 afforded exposure to Whistler, and the agitated facture is, moreover, distinctly un-Whistleresque.[29] The pictures' shared *mise-en-scène*—a closed French door, low-hanging lamp, and a doubled set of heavy curtains—also attests to their cold-weather setting; details that also signal the accord between the furnishings in the illustrated letter and those in the painting (though the heavy foreground spotted curtain is an aspect of the décor unheralded by the illustrated letter).

That they are hivernal pictures problematizes the long-held view that they are first and foremost Scandinavian Blue pictures, based upon a regional palette connected at its root to the blue of the lengthy, often moonlit, summer twilight in Norway, a Land of the Midnight Sun.[30] To the degree that the chroma is descriptive in the case of the moody blues in the three color versions of *Night in Saint-Cloud*, taking into account the wintertime setting of the pictures, the color bespeaks meteorological as well as (or instead of) emotional climate, and deserves to be considered a species of icy blue.

The conspicuously unlit lamp, silhouetted against the window, was also prominent in Munch's illustrated letter to his aunt (fig. 6.7). Although modified somewhat in the painting, it remains identifiable as the same device, a hanging or pendant reflector lamp that would have burned "pétrole" ("paraffin" as it was termed in Britain, or kerosene in the United States). As shown in the drawing, usually three chains running from a smoke deflector to the shade ring or brackets that extended from the shade ring suspended such lamps. (In America, these lamps were called "library" or "hanging parlor" lamps.)[31] In the illustrated letter, the lamp is shown schematically but very clearly to have three suspension chains or metal brackets, but the painting delineates only two chains holding up the lighting fixture. The lamp would wobble hopelessly under those conditions, but Munch's painting needed the hominess, volumetric gravitas, and strong form of the lamp, and the two diagonal uprights flanking the shade make a more legible and symmetrical silhouette against the window than a trio of chains would have done. The simplification with regard to the lamp in the letter illustration was a move in the direction of clarity and elegance. As integrated with the seated figure, the lamp appears large and pulled down very low—so low that the absorbed man would hit his head were he to snap out of his reverie and stand up.

Then there is the triangle of very bulky, spotted brown-burgundy curtain in the left foreground. This jungle cat accessory speaks to or rather simply evokes wintertime preparations that have apparently been taken in this home: the kerosene lamp hung right up against a window rather oddly combined with the thick drape and the very large, barely discernible armchair at right (or table, as in figure 6.7) suggest the ad hoc creation of a reading alcove. The French doors, for the time being, are not in use; their closure staunches the flow and exchange of air. The closed, windowed doors are merely transparent sheets of glass, exposing and framing a sparkling riverine evening world that evidently engages the attention of the shadowy man. But is it rapt? Or is the effect accidental—a nonsignificant byproduct of the cant of his head?

That ambiguity notwithstanding, that he ponders, looks out, and does not read in this spec-built micro-environment helps to underscore the attention he pays to the twinkling riverine-cum-street scene beyond the glass doors.[32]

According to Charles Rice, the interior as a cultural form always bears its "newly illuminated impossibility as a space of retreat and immersion."[33] At the same time, the porosity of the dwelling and its susceptibility to invasion by the world outside were also understood as threats to its serenity and further characteristics of modernity. Threshold space, the terrain of Munch's pictures and the spatial *ne plus ultra* of permeability, had special purchase on thinking the interior, which was literally "on edge"—both unclear and a site of inherent tension because of its composite identity. Threshold space was where the strain intrinsic to various moments and sites of modernity, including the new domesticity sequestered in the interior, could be seen and examined. To quote Victoria Rosner: "A way of life built around separation and specialisation encounters difficulty when faced with transitional or in-between states that resist categorisation. Such states are architecturally embodied in the threshold, the space that forms a bridge between two discrete rooms."[34] Or between the inside and the outside. Thus the threshold, a superlative theater of everyday private life in the Parisian modernist imagination, was the perfect anchor of critical scenarios of modernity at its most concentrated and fraught.

Anthropology has led the development of "threshold theory." Victor Turner, the British symbolic anthropologist, for example, advanced the notion of "liminality" and argued that it is a "fundamentally interstructural mode."[35] The applicability of the tensions and potentialities of the betwixt and between condition of liminality to Munch's scenario here, oriented toward and bordered by a window, a supremely interstructural device, should be apparent and is explored in what follows.[36] The degree to which Munch's liminal people do or do not connect the interior to the *ailleurs* beyond the glass frontier is key to grasping the complexities of the pictures. As Georg Simmel observed, "In the immediate as well as the symbolic sense, in the physical as well as the intellectual sense, we [humans] are at any moment those who separate the connected or connect the separate."[37] When vision is the only means to link inside and outside—the difference between a window and a door, only the latter being breached by the body in motion—the act of connecting can always and easily be refused. The window is the site par excellence of display or its concealment. One can look as well as not look. Walter Benjamin's frequently cited observation is germane here: "The intoxicated interpenetration of street and residence . . . comes about in the Paris of the nineteenth century."[38] Charles Baudelaire also considered the importance of the window as a site of reverie; as a screen of imaginary transport and creative displacement; as an opportunity for release from the dark hole in which life otherwise lives.[39]

Claude Gandelman's insights regarding the role of doors in paintings also pertain to Munch's window as threshold. Gandelman writes, "Doors play a role of paramount importance in seeing. They either permit the gaze of a specific character to connect with its object; stand for the impossibility of seeing what is represented; express the irresistible force of vision, the impossibility 'not to see'; or signify the freedom of seeing or refusing to see."[40] Obviously, both the door and the window are intermodal structures, but the opacity and functionality of doors set them apart from the transparency and relatively passive functionality of windows. The uncurtained window—whether open or closed—permits the gaze, and its prominence in a picture tends to inaugurate a rumination on the irresistibility of looking out, while, at the same time, initiating reflection upon the parameters of that action and its power via thoughts centering upon the interconnectedness or disconnection between the zones demarcated by the window. Because windows are the exterior membranes of dwellings, they also mark the boundaries between contrasting climate or temperature zones. (An open window, not Munch's structure, would perforce blend and meld the climate zones, and, like the closed window, weather permitting, would also allow the gaze through to the outside.) Munch's handling complicates the issues of freedom and accessibility by muddling the clarity of the man in the picture. Is he embracing freedom and access wholeheartedly or only begrudgingly? Is his access mere happenstance, and thus only a coincidental product of his perch on the bench? The question naturally arises: is it a motivated gaze? Is he paying attention, or is he in a state of absentmindedness?[41] The difficulty of nailing that down enriches the complexity of the threshold status of the blue windowed doors. That complexity dovetails in turn with Isobel Armstrong's identification of the window as "*the* disputed space of the century."[42]

But in the register of the effects of painterly illusion, things are otherwise, because we must reckon with the picture's illusion of light that has passed through the window into the interior shining upon the floor. The muted natural light of evening—twilight plus moonlight—cascades into the room, and the windowpanes leave a cross-shaped shadow at the center of the trapezoid of mottled blue-mauve light. The panel of light streams diagonally across the floor in such a dramatic way (the drama and mystery are exacerbated by the broad paint handling) that it bids fair to evoke the illusion of a solid volume (a catafalque) rather than a flat pool of light crisscrossed by the equally flat shadow of the window mullions. The rectangle of light is indeed so enormous and attention grabbing that it effectively collaborates with the light-filled squares of the upright windowed doors to bracket and hem in the shadowy, dreamy man in profile. He is flattened and skewered by the flanking rectangles of pale light that border him like deformed wings.

The strip of illumination that falls upon the bench, which supports him against the wall, points to his darkened body as a diagonal directional arrow would aim for a center of interest. But it is an indicator that points to the frustrating heart of the picture: a shadowy, pensive, and, in fact, illegible cipher. Inasmuch as the silhouetted, top-hatted man in profile is, in the end, a narrative disturbance and psychic void (or nullity), the picture comes into focus as a disquisition on the environmental and personal consequences (or effects) of the modern conjunction of lights and darks at night; about the ways that the lights of modernity can be startlingly switched on to expel the dark, but not in the cases of Munch's quirky, archaic darkened interior. Not even the conjoined efforts of twilight, gas or electric streetlights, and the running lights of the passing boats on the Seine can dispel the darkness of the night in Saint-Cloud. Thus, the crossing of the threshold—the gaze directed out, the lights streaming in—may be an interpenetration or exchange, thanks to the transparency of the glass, but the commingling is unequal. The light does not illumine the man in the dark; the interior is gloomy regardless of the light-conducting properties of the transparent window ushering in the lights despite the cold temperatures beyond. As Wolfgang Schivelbusch has argued, unrelenting visibility expels the magic from the night. Munch, like other late-century partisans of the dark, may have believed that Paris (and its environs) were revealed more fully at night than in the day.[43] That the mediated lights of modernity do not affect or disenchant Munch's thinker underscores the depth and inscrutability of his old-fashioned absorption. We might argue that Munch's "man at the window" pictures instate longing that coexists with anxiety about the penetration of the home by outside forces. The subtext is a fascination with night lights.

The room's size and strong diagonals, not to mention its top-hatted man perched on a bench at a window, cause the space to resemble the interior of a railway car.[44] After all, the solitary figure is dressed for the out-of-doors. The similarity between Munch's painted room and the interior of a train carriage affirms a duality at the heart of the painting: stasis (the immobile man) versus mobility (the world beyond of boats, water, and lights). Schivelbusch has written of the ways the nineteenth-century railway passenger, as a result of a train's swift movement, felt that he "ha[d] lost contact with the landscape."[45] Munch's man, like a train passenger, albeit from a stable position, also looks out, at least in part, to see and to seize visually aspects of what is glimmering (changing), sparkling (on the water's surface), and passing by outside (steamers with their lights and movement on the quai). His serene aspect, poised between attention and inattention, forecloses any hint of stress in his demeanour; there is no trace whatsoever of panic that he cannot keep up with, or keep track of, the varied ingredients of the illuminated scene outside.

Among roughly contemporary cognate paintings set in Paris, Munch's may call to mind Gustave Caillebotte's *Young Man at the Window* (1875), among his exhibits in his first "Impressionist Exhibition," the second show held in 1876. The painting broached for the first time in Caillebotte's work the theme and *mise-en-scène* of looking out the window at the city of Paris. Though a daylight picture, it warrants comparison with Munch's Saint-Cloud night painting.[46] Gustave's brother René posed on the second floor of the family apartment at the corner of rue de Miromesnil and rue de Lisbonne in a fashionable bourgeois quarter of the Right Bank.[47] This standing threshold protagonist, who unhesitatingly commands his surroundings before an open window, contrasts strongly with Munch's comparatively restrained and seated onlooker, poised reticently on the brink; on the interior face of the glass membrane.

Caillebotte's observer, seen from behind, forthrightly addresses the view beyond the open glazed doors and stone balustrade assuming a paradigmatically forceful masculine pose: his firmly planted, widely spaced legs exude somatic and emotional command and control. His body appears to convey authority and ease, which successfully breaks through the barrier separating inside from outside. With the effectiveness of an electrified fence, this same invisible wall implies the sequestration of Munch's top hatted man, or at least a strong sense of his separation from the beyond despite being dressed for the out of doors.

That said, a contrary reading of the man looking out the window in Caillebotte's picture is actually in order because more persuasive. And it serves to bring the two window protagonists into closer proximity. Apropos of Caillebotte's man at the window, the authors of the 1994 Caillebotte exhibition catalogue observed: "These openings [of Caillebotte's windows onto streets, rooftops, or courtyards] provide views of what is, in effect, another interior: that of the city."[48] This statement helps to characterize the hemmed-in quality of the man at the window; the ways in which his commanding, manly posture addresses and controls nothing but a cave-like expanse of almost deserted stone and metal. The tiny woman in profile walking through the center of his field of vision is a meager catch indeed in the wide net cast by his unimpeded, even impertinently posed body. The ostensible restlessness that propelled him from his nearby chair to the open window resembles the condition of a sequestered invalid more than that of an empowered and unconstrained *flâneur*.[49] Indeed, this Parisian man looking for something out the window—his body registers determination, not dreaminess—is the exemplary thwarted *flâneur*. Or perhaps "thwarted *flâneur*" was an oxymoron by the 1870s. Following the indispensable formulation of Priscilla Parkhurst Ferguson, Caillebotte's man at the window would be read instead as a highly legible image of the consummate *flâneur* of his day—a figure of alienation and detachment, of being in the city without being of it.[50] Placing a man like this in threshold position generates a probing assessment of urban masculine ambition face to face with its limits and discontents.

Scene through this lens, Caillebotte's and Munch's men at the window share ideological territory as well as indistinctly defined ocular desire. The link between the two Paris paintings, albeit set at contrasting times of day, helps us further to gloss the richness of threshold space.[51] Munch's nighttime setting and its welter of sparkling lights beyond the glass membrane considerably enhance the male at the threshold scenario. *Night in Saint-Cloud* is an outsider nocturne entangled in the matrix of illumination discourse of exceptional complexity.

CONCLUSION

Art Fueled by Lights

Switch on your electric light.
Then we can get down to what
is really wrong.

VAN MORRISON, 1970[1]

Disappointment and Discovery

This book studies the entanglement of the visual arts with artificial illumination in Paris in the later 1800s. But my readings do not add up to a synthetic account of the subject, nor do my assessments of sundry practices lead easily to a conclusion. I launched the study to investigate artists' "reactions" to the transition from gaslight to electric light in the streets, shops, and homes of the French capital, and I intended to title it *Electric Paris*, but the project never arrived at the targeted destination. The situation that my research brought to light differed from what I had anticipated, since the shift from the dominion of one lighting regime to another was far from smooth, let alone rapid. A battle for supremacy between gaslight and electric light defined the city's illumination environment as well as its energy economy throughout the Belle Époque. So the artworks and situations that I scrutinized could only be unique events rather than cases.[2] As a result, generalizing about these singular instances would have been forced.

Also, the contest between older and newer lighting technologies did not erase ancient flame equipment—candles and oil lamps—from the modern interior until the end of the nineteenth century and the early twentieth century.[3] In fact, gas was not much used in the urban home in the decades we have studied. The historian Lenard R. Berlanstein reports that "no more than five percent of all Parisian residents were customers of the Parisian Gas Company. Far from being a necessity of life, domestic gas was a curiosity."[4] The main *éclairage* event of the era—the designation of Paris as the Electric Capital in 1881 notwithstanding—was the resilience of gaslight in the public sphere. To wit, approximately 80 percent of the 93,000 workshops and stores in Paris used gas for lighting in 1880.[5]

That gaslight maintained the upper hand much longer than I had expected—the contest with electricity continued well into the twentieth century—fundamentally altered and threatened to undermine my undertaking. But a luminous realization dawned above the dark cloud of disappointment over electric light's delay. The illumination-conscious artworks that attracted my interest shared something fundamental. They had in common a critical determining technological circumstance, the French capital's unprecedented motley lighting environment. My postdespair insight helps to explain why the art at the center of this study of maximum illumination stress and hybridity in the French capital was chiefly of two kinds: (1) art that vouchsafed its maker's awareness of the erstwhile headline-grabbing electric arc lights, the Jablochkoff candles, that challenged the ubiquity of the gas flame; and (2) art that disavowed or ignored the brash electric contenders or acknowledged them indirectly or furtively. Another commonality is that in virtually all instances the primary unit of exploration was night and pleasure.[6]

The chapters in this book argue that nocturnal culture is definitive of metropolitan modernity, and while lamp-centric artworks emphasize *éclairage*, they define the night's modernities in varied ways. Chapter 1 studies works of art enmeshed in but subtly distant from the glory days of gas street lighting: Marville's uncanny Second Empire daylight streetlamp photos, and Caillebotte's painted reference (in 1877) to the dominion of the *réverbère* of an earlier era. The art examined in chapters 2, 3, and 4—by Sargent, Cassatt, Degas, and the caricaturists—indirectly and directly heralded a multifaceted lighting environment, including the novel electric lamps that dazzled between 1878 and 1882. The centerpieces of chapters 5 and 6—outsider nocturnes by Curran, Hassam, and other Americans, and the 1890 campaign by Munch—are singular responses of attraction mingled with avoidance with respect to a still variegated but somewhat more electrified scene in the Paris streets.

As electric light became more of a presence in late-century art practices, my interest in it flagged. The gingerly rise of electric light did shape the compass of the book after all, if in a negative way. As its visual presence in art became a matter of descriptive fastidiousness and a telegraphic sign of brashness and hypertechnologized modernity—in painted and graphic works by Louis Anquétin (1861–1932), Jean Béraud (1848–1935), Pierre Bonnard (1867–1947), Jean-Louis Forain (1852–1931), Théophile-Alexandre Steinlen (1859–1923), Félix Vallotton (1865–1925), and Édouard Vuillard (1868–1940), for example—and less a phenomenon to resist and displace, or to denounce and laugh at, my curiosity waned.

What exactly dampened my enthusiasm? In appraising the outcomes of transactions between aesthetic choices and illuminations in the visualization of nighttime Paris in the late 1800s, I observed that the embrace of electric light threatened to dumb down depictive modernism, otherwise focused complexly, often ambiguously and anxiously, on the nocturnal city. Like Gustave Kahn writing in 1928, I find electric light "contre-impressionisme," or rather, to put it more broadly, its embrace discouraged subtlety and complexity.[7] This perspective provides a cogent explanation for the Impressionists' avoidance of outdoor city spaces at night, a pattern noted in chapter 1. The artists seem to have intuited that urban modernity at its most pungent was a matter of mixtures and not homogeneous classifications, whether of persons or illuminations. And gaslight was too old-fashioned and benign, while Jablochkoff candles and other experimental electric lights were too vulgar and brazen. For this reason, social and environmental compounds and composites, structures of indistinctness and multiplicity in the realms of technology and environment, dominated the vocabulary of the paintings and prints of modern life studied here (with the exception of the caricatures).[8]

Paths Not Traveled

This may help to explain why I did not single out certain practices that leap to mind when *fin-de-siècle* art and brilliant lights are evoked, such as the brightly colored posters of Henri de Toulouse-Lautrec (1864–1901), the inventor of Wicked Paris. His garish graphic work conveys the indelible impression that it is the art and electric light corpus par excellence, and thus should have been indispensable to my quest as originally defined. His mature *œuvre* was a prime candidate for examination as the book first took shape. You may think that Lautrec should have been grouped with those gathered in the problematic "explicit electric light group," but once his art's surprising ingredients were unpacked, it turned out that his work presented challenges that I was not inclined or equipped to tackle.

My expectations were shaped by my longstanding angle on Lautrec's light fixture-filled art, which has been to understand the artist at the height of his powers in the 1890s working at the leading edge of explicitly commercial visual culture. Lautrec's work advertised brash corridors of the entertainment fast lanes of Montmartre in the eighteenth arrondissement of Paris in those late-century decades when the city itself was called the café of Europe. As I endeavored to unearth the aesthetic representation of lighting fixtures used to maximize the commercial crassness of entertainment *milieux*, I took my cue from lighting scholars Blühm and Lippincott: "In 1880s Paris, it would seem that almost anything viewed under electric light became a desirable commodity."[9] Following the logic of *fin-de-siècle* Parisians rather than Van Morrison (in the epigraph to this conclusion, above), I expected to find a riotous assembly of dazzling electric lights in Lautrec's work of around 1890 because I presumed that he would have used the widely accepted semiological alliance between electric light and vulgarity to secure the commercial vividness and allure of his art. I got it wrong.

A career milestone for Lautrec occurred in 1886, when his artwork as well as his life came to center on the dives of Montmartre—most important, a seedy cabaret known as Le Mirliton. But his subsequent works were affiliated with the Moulin Rouge, an altogether different kind of dance hall, built specifically to appeal to middle-class and high-society Paris.[10] In 1891, the nightclub itself was the subject of Lautrec's first poster, *Moulin Rouge, La Goulue* (fig. c.1), a work that made him famous all over Paris. The Moulin Rouge was created to siphon off posh clientele from older dance halls. As Reinhold Heller has observed, the Moulin Rouge offered "a retreat from the everyday into a Disneyesque hyper-reality whose supreme revelation was a fictionalized demimonde . . . [with] 'natural' lower-class performers in the excited choreographed display of their sexual allure."[11]

Lautrec's agreement to make a lithographed poster for the club fused his art ever more tightly with the dance hall as a commercial enterprise.[12] He was not only continuing to dance the dance of advanced art linked to mass culture, which had been a vanguard dyad and dialectic of embrace vs. refusal since the mid-century era of

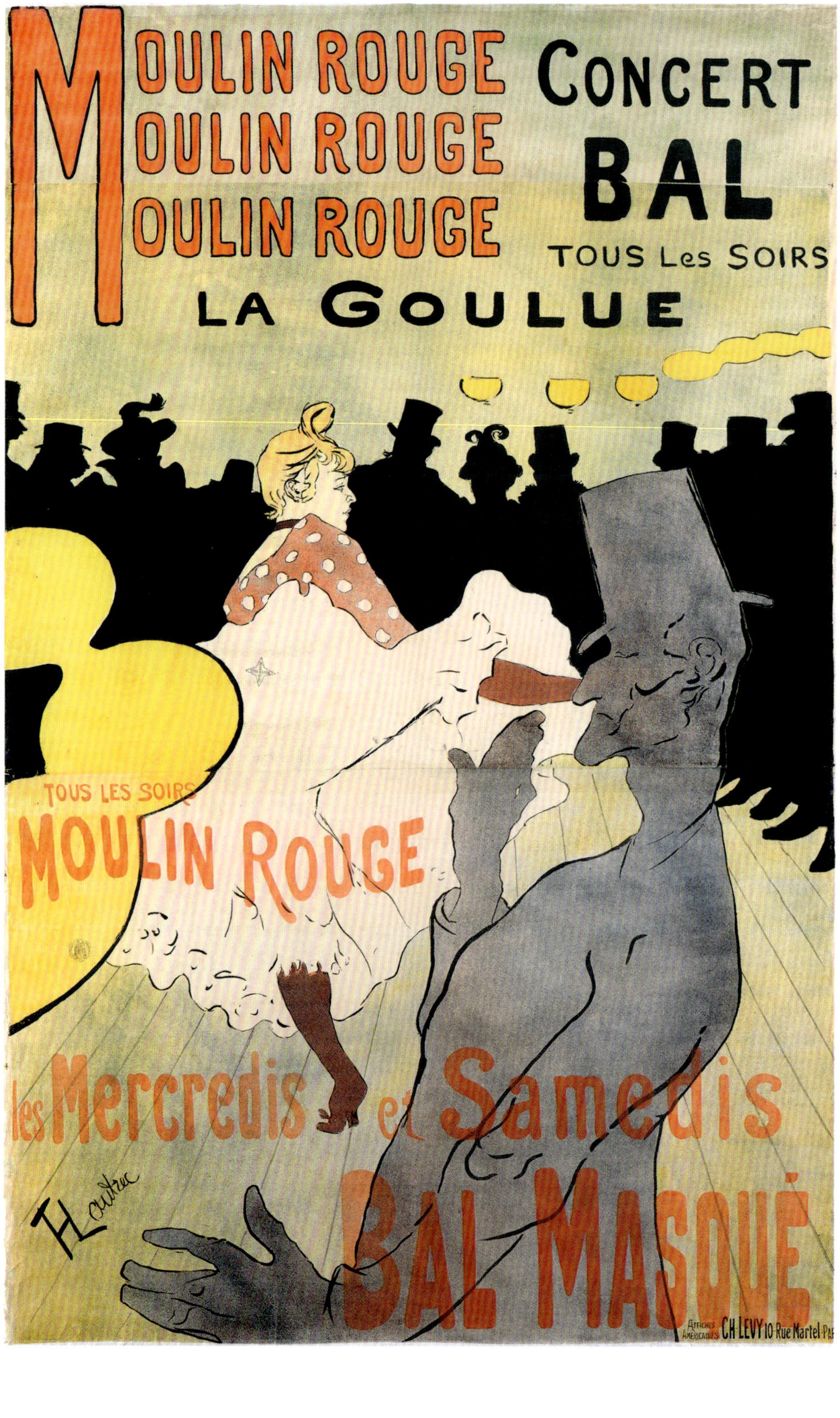

C.1 Henri de Toulouse-Lautrec, *Moulin Rouge: La Goulue*, 1892. Lithograph printed in four colors; 3 sheets of wove paper. 190 × 116.5 cm (74 13/16 × 45 7/8 in.). Printed by Affiches Américaines, Charles Lévy (Paris). Metropolitan Museum of Art.

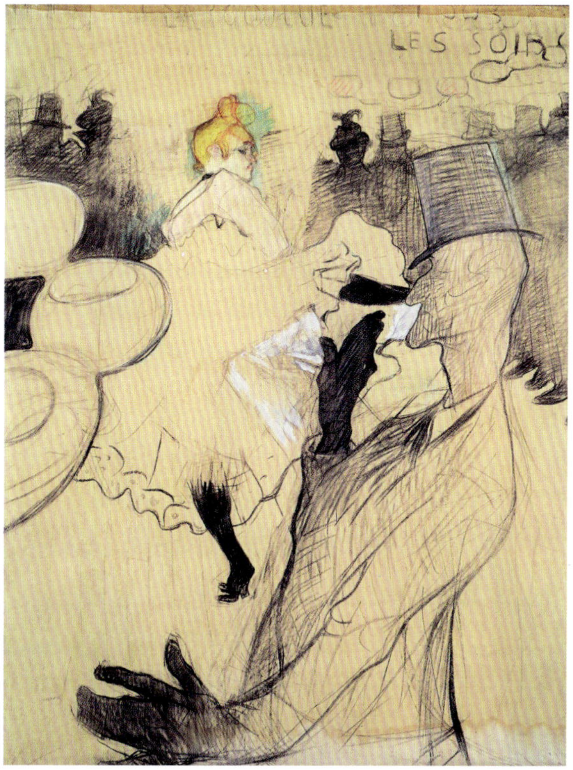

C.2 Henri de Toulouse-Lautrec, *La Goulue et Valentin le Désossé*, 1891. Pen and ink and colored pencil on paper. Musée Toulouse-Lautrec, Albi, France/Bridgeman Images.

Gustave Courbet. Lautrec was shortening the circuit between the two spheres by literally crafting an advertisement for a business. With the attraction of customers in mind, Lautrec emphasizes the spectacular and magnetic nature of La Goulue's daring dance by surrounding her with silhouetted bourgeois admirers. Valentin le Désossé (the boneless one) is likewise a key presence in the poster; his monochromatic and caricatured profile and eccentric hand gestures remove him from the action per se but define him as the presiding dance hall deity. The bourgeois character of the clientele (no ambiguity about those hats) makes plain that Lautrec had adjusted his vision and aesthetic language, with an accent on high chroma, to the upmarket élan of his new clientele.

A large preliminary drawing, *La Goulue et Valentin le Désossé* (1891; fig. C.2), served as a kind of laboratory in which Lautrec initiated the startling process of simplification and synthesis that characterizes his bellwether poster. Looking at them side by side enabled me to identify the objective referent of the flat yellow shape made of three linked blobs at the left edge of the poster, a seemingly organic form that accentuates the audacity of the dancer's body posture. It is actually a stylization of the three spherical globes of a low-hanging gas chandelier or *torchère*, both using *becs papillon* (batwing gas burners).[13] What deserves emphasis here is that the most daring abstraction or at least cheekily flattened form in the entire poster is a reference to gaslight, not electric light as I had anticipated. And those flame-heavy, large gas fixtures were Belle Époque nightclub signatures, despite the pressure upon theaters and dance halls to switch to safer electric lighting.

An anarchist like Paul Signac associated Lautrec's poster with degradation; he saw it as a graphic rendition of the pleasure enjoyed by a society in decay. Its wide distribution (contributing to the fame of the artist) also fed the growing "affichomania" of the early 1890s, and placed a commercial intervention squarely in the sphere of the fine arts. Lautrec's invention of the visual iconography of Wicked Paris drove demimonde values into the visual mainstream and reinforced the viability of gaslight, not the even brasher competitor electric light, as a visual sign of decadence.[14] The manifold paradoxes and exaggerations in Lautrec's use of the industrial light of modernity distanced me from his artwork.[15]

The example of Lautrec's work might serve to model directions that a future scholar of Art in the Era of High Nocturnalization, the banishment of darkness from cities, may wish to pursue.[16] Coined by Craig Koslofsky and reused by Tim Edensor, nocturnalization denotes "the expansion of social and economic activity into the night and the subsequent spread of illumination, a process that continues to be informed by moral and modernist discourses," the development denied or glossed by the pre-1890s artworks examined in this book.[17]

Also absent from my consideration and deserving of recognition was the work of the allegorists of night light. A future study of Art and Night (or Art and Electricity) should give them their due. First, Jean-Léon Gérôme, *Night* (c. 1850–55; oil on canvas, Paris, Musée d'Orsay). Next, Pierre Puvis de Chavannes and his panels done for the Boston Public Library (1895; oil on canvas). The panel on telegraphy, on the left wall, reads: "Physics: By the wondrous agency of Electricity, Speech flashes through Space and swift as lightning bears tidings of good and evil." Finally, Henri Fantin-Latour, *Night* (1897; oil on canvas, Paris, Musée d'Orsay).

Art and Lighting

One of my central arguments is that qualities of urban lighting, and a related sensitivity to contrasts of light and darkness, shaped the definition of the telltale visual and social qualities of the French capital on the part of a number of innovative artists oriented toward modern life between about 1860 and 1890. Surveying a broader swath of modern art, it becomes clear that a visibility/invisibility dialectic, rooted in an attentiveness to the valences of darkness and conflicting luminosities, has been the spine of many of the most intelligent and enduring artworks made in illuminated cities, especially Paris, the first City of Light, but also in Berlin, New York City, London, and Chicago, for over a century and a half. But concerning the French capital in particular, I will not now try to explain two antithetical developments in art that may appear relevant to the stakes of this book. Neither the rise of explicit references to electric light in some corners of visual culture, nor when and why certain

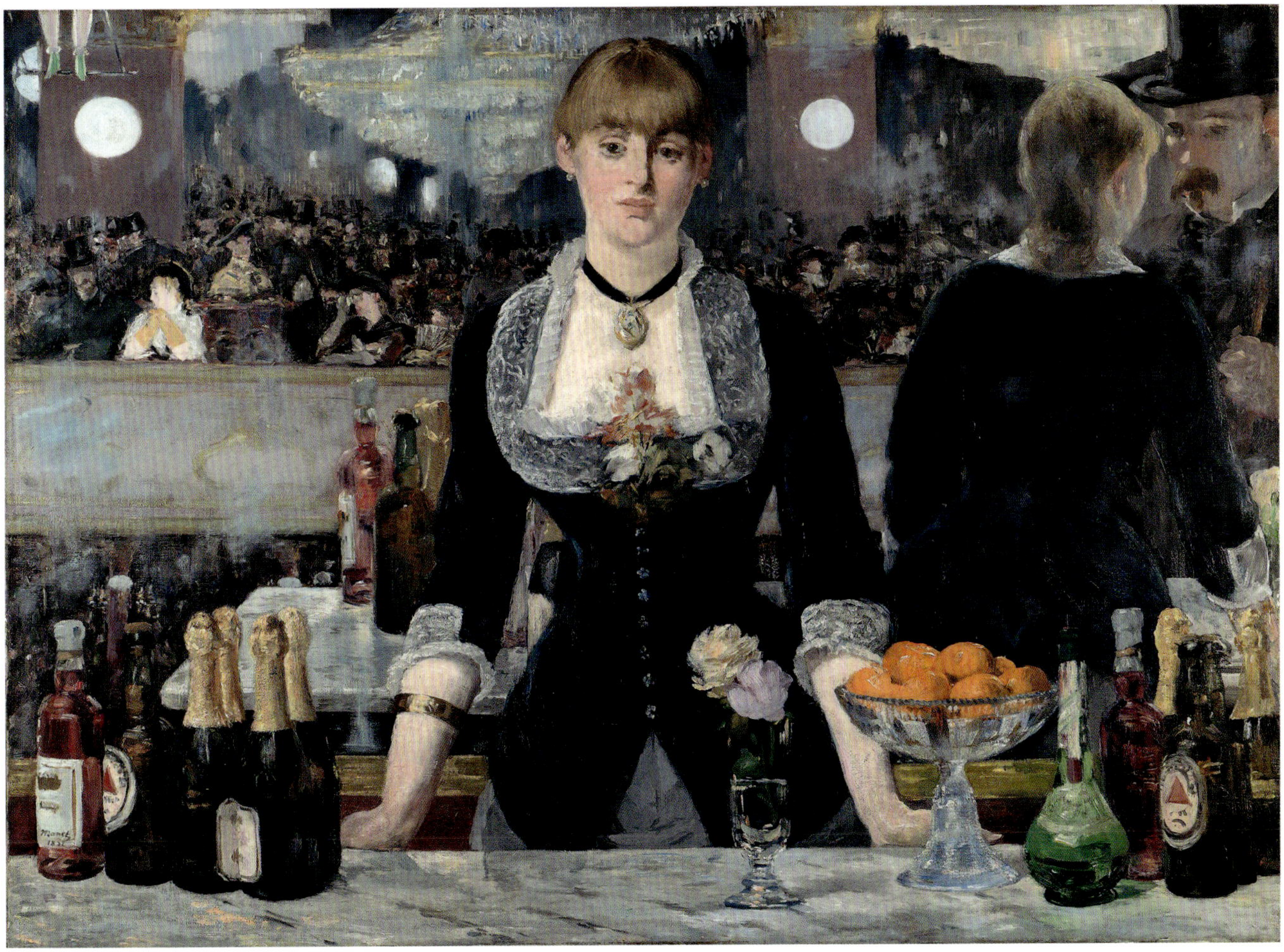

C.3 Édouard Manet, *A Bar at the Folies-Bergère*, 1882. Oil on canvas. 96.5 × 129.5 cm (38 × 51 in.). Courtauld Gallery, London.

innovative, Paris-based painting and printmaking practices shed their attentiveness to the infrastructure of illumination is addressed. Grappling with those issues would surpass the scope of the book.

I close instead with an especially vivid example of illumination discourse, one that graphically reaffirms the Parisian art world's obsession with lighting in the later 1800s; the preoccupation that links all my chapters. It is a little-known outcry by a seasoned art critic and lion of French letters, J.-K. Huysmans (1848–1907)—one that helped to inspire my project in the first place. Writing about the Salon of 1882, he erupted in a fit of fussiness about Édouard Manet's inadequate depiction of indoor lighting in *A Bar at the Folies Bergère* (1881; fig. C.3), his well-known late painting set in a Parisian nightclub:[18]

> The subject is thoroughly modern and M. Manet's idea to put his female figure in her own milieu is ingenious, but what does this lighting [*éclairage*] mean? That is gaslight and electric light? Come on, it is a hazy *plein air*, a bath of pale daylight!—

From that, everything collapses—the Folies-Bergère only exists—and could only exist—in the evening; thus understood and sophisticated, the club is absurd. It is really deplorable to see a man of M. Manet's worth sacrifice [his art] by such subterfuges and make paintings [*tableaux*] that are as conventional as those by others.

Huysmans's surprising denunciation of Manet's illusionistic abilities targets the sphere of lighting effects. The thrust and ferocity of his diatribe demonstrate a broad awareness of and sensitivity to, even touchiness about, the visual properties of lighting apparatuses and lamps; about the differences between arc-light globes and gas chandeliers. Huysmans inveighs against Manet by averring noisily that the meaning of the club is erased (the club becomes absurd) minus an accurate record of its unsophisticated lights. Huysmans seems to be saying that Manet's descriptive insufficiencies blunt the club's tawdry edge by inadvertently softening its lights. In the critic's terms, Manet's triviality as an artist is indexed by his conversion of an artificially illuminated nightspot into a daylight tableau.

As evidence that the visual and semiological properties of conflicting forms of lighting were taken deadly seriously in the art-critical community of later nineteenth-century Paris, Huysmans's flabbergasting text cannot be improved upon. His denunciation of an unintentional *plein air* demonstrates an art critic's unwillingness to judge evasion as a positive representational strategy. Huysmans does not credit the avoidance of depictive certainty in the sphere of lighting fixtures and lights as a potentially shrewd assessment of the heterogeneous visual character of technologized urban modernity. The standpoint he refuses is an argument about modernism and the night that this book embraces. The most compelling paintings and prints (graphic satire excluded) of the modernities of nocturnal Paris grapple with darkness and lights, but forswear descriptive closure.

Notes

Introduction

1. "Bien que l'art de l'éclairage ne soit le monopole d'aucun pays, d'aucune capitale, il est certain que c'est à Paris qu'il doit principalement son essor": Henri Maréchal, *L'Eclairage à Paris: Etude technique des divers modes d'éclairage employés à Paris sur la voie publique, dans les promenades et jardins, dans les monuments, les gares, les théâtres, les grands magasins, etc. et dans les maisons particulières . . . gaz, électricité, pétrole, huile, etc. . . .* (Paris, 1894), v.
2. Wolfgang Schivelbusch, *Disenchanted Night: The Industrialization of Light in the Nineteenth Century*, trans. Angela Davies (Berkeley: University of California Press, 1988).
3. Walter Benjamin, "Paris, Capital of the Nineteenth Century," 1935 and 1939 versions, in *The Arcades Project*, trans. Howard Eiland, ed. Rolf Tiedemann (Cambridge, MA: Harvard University Press, 1999), 3–26, 955–58.
4. "L'éclairage est en effet un des éléments clées de la féerie parisienne qu'ausculte Benjamin, le moyen le plus primaire mais le plus efficace d'attirer le regard. . . . L'alternance de la lumière et de l'ombre est à la base du spectacle visuel au XIXe siècle" Martin Bressani and Marc Grignon, "La Fantasmagorie du gaz d'éclairage à Paris au XIXe siècle," in Libero Andreotti, ed., *Spielraum: W. Benjamin et l'architecture* (Paris: Éditions de la Villette, 2011), 57.
5. "Les nuits parisiennes du XIXe siècle se font objet de discours et lieu d'expériences au moment où les moyens techniques destinés à vaincre l'obscurité gagnent en sophistication" Simone Delattre, *Les douze heures noires: La nuit à Paris au XIXe siècle* (Paris: Albin Michel, 2004), 115.
6. Christopher Prendergast, *Paris and the Nineteenth Century* (Oxford, 1995); Hollis Clayson and André Dombrowski, eds., *Is Paris Still the Capital of the Nineteenth Century? Essays on Art and Modernity* (New York: Routledge, 2016).
7. Followed quickly by "la place Vendôme, la rue Castiglione, la rue de la Paix, la rue et la place de l'Odéon, and les galéries du Palais Royal" (Marc Gaillard, *Paris Ville lumière* [Amiens: Martelle, 1994], 32). Gas did not eclipse oil lighting immediately. In 1834 the city was still using much more oil lighting than gaslight (Gaillard *Paris Ville lumière,* 32). In 1855, all the competing and proliferating gas companies were fused into one sole company by Haussmann (Gaillard, *Paris Ville lumière,* 33).
8. Marie de Thézy, *Charles Marville Réverbères* (Paris: Tête d'Affiche, 1993), 61.
9. Prendergast, *Paris and the Nineteenth Century*; Joachim Schlör, *Nights in the Big City: Paris, Berlin, London, 1840–1930*, trans. Pierre Gottfried Imhof and Dafydd Rees Robers (London: Reaktion, 2016), passim.
10. de Thézy, *Charles Marville Réverbères,* 62.
11. François Loyer, *Paris Nineteenth Century: Architecture and Urbanism*, trans. Charles Lynn Clark (New York: Abbeville Press, 1988), 312.
12. Loyer, *Paris Nineteenth Century,* 312.
13. "Est-ce à dire . . . que la lumière électrique fera disparaître l'éclairage par le gaz? Il faut écarter bien loin cette pensée": Louis Figuier, *L'Art de l'éclairage* (Paris: Jouvet et Cie., 1882), 256.
14. Eugène Defrance, *Histoire de l'éclairage des rues de Paris* (Paris: Imprimerie Nationale, 1904), 117.
15. Schivelbusch, *Disenchanted Night*, 118. He adds: "In arc-light, the eye saw as it did during the day, that is with the retinal cones, while in gaslight, it saw as it did at night, with retinal rods."
16. Jonathan Crary, "Attention and Modernity in the Nineteenth Century," in Caroline A. Jones and Peter Galison, eds., *Picturing Science, Producing Art* (New York: Routledge, 1998), 475–99.

17. The foundational study of the antagonism is Lenard R. Berlanstein, *Big Business and Industrial Conflict in Nineteenth-Century France: A Social History of the Parisian Gas Company* (Berkeley: University of California Press, 1991).
18. "En réalité, malgré les progrès de l'électricité, les dernières décennies du XIXe siècle et la Belle Époque constituent l'apogée du gaz d'éclairage." Gaillard, *Paris Ville lumière*, 44.
19. Maréchal, *L'Eclairage à Paris*, 18.
20. "Au moment où j'écris ces lignes, une révolution radicale semble être sur le point de se produire dans l'éclairage de la Voie Publique de Paris: la substitution de la lumière électrique à celle du gaz" (quoted in Gaillard, *Paris Ville lumière*, 43).
21. I plead guilty to using "Electric Paris" to title two exhibitions I helped to curate: first, at the Clark Art Institute in Williamstown, Massachusetts, in 2013 (http://clarkart.edu/exhibitions/electricparis/content/exhibition.cfm); and second, at the Bruce Museum in Greenwich, Connecticut, in 2016 (https://brucemuseum.org/site/exhibitions_detail/electric-paris).
22. According to electricity scholar Charles Malégarie, the eighteen-year period was not sufficient to encourage serious investment, and the permissions granted did not anyway guarantee a monopoly to the entity assigned a sector. Charles Malégarie, *L'Électricité à Paris* (Paris: C. Béranger, 1947), 9. According to A. N. Holcombe, the resultant prices were high, "but that was the price the city must have expected to pay when in 1888 it shifted the risk of initial experimentation in the electric-lighting industry to private enterprise." More to the point, he observes: "The leading cause of the failure of the municipal electric-lighting undertaking at Paris to become a model of public enterprise seems to have been the indifference of the majority in the municipal council." See A. N. Holcombe, "The Electric Lighting System of Paris," *Political Science Quarterly* 26, no. 1 (March 1911): 124, 126.
23. "Les critères d'appréciation de la luminosité sont instables. Ainsi, le gaz, une révolution indissociable du climat du Paris nocturne baudelairien ('les lueurs enfumées'), devient progressivement insupportable à partir des années 1880: irrégulière, nauséabonde, dangereuse, cette technique d'éclairage est perçue comme aussi archaïque que l'huile." Marc Armengaud, *Paris La Nuit: Chroniques Nocturnes* (Paris: Pavillon de l'Arsenal, 2013), 93–94.
24. Ernest Freeberg, *The Age of Edison: Electric Light and the Invention of Modern America* (New York: Penguin, 2013), 110. Armengaud, *Paris La Nuit,* 1008, provides a map of the brightly lit "emblematic places of festivity of the Belle Époque."
25. "La marche fut cependant assez lente pendant les cinq années qui suivirent l'exposition; on ne peut guère citer, comme réellement intéressantes, que les installations faites par la Société Edison dans divers théâtres, dont le principal est le Grand Opéra. . . . La terrible catastrophe de l'Opéra-Comique, ce désastreux incendie [dû au gaz d'éclairage, 27 mai 1887], fut la poussée déterminante. A la suite de cet événement s'éleva un cri général: 'il faut mettre l'éclairage électrique dans tous les théâtres' et naturellement on alla jusqu'au bout de l'idée en réclamant la lumière électrique pour tous. La réalisation de ce vœu pour les théâtres ne tarda pas." Charles Malégarie, *L'Électricité à Paris* (Paris: Librairie Polytechnique Ch. Beranger, 1947), 7–8.
26. The term "electro-enthusiast" was coined by Sasha Archibald, "Blinded by the Light," *Cabinet* 21 (Spring 2006): 1.
27. "Il en est ainsi de la lumière électrique: elle éblouit et n'éclaire pas." Jules Luquiens, ed., *French Prose: Popular Science* (Boston: Ginn and Co., 1893), 171.
28. Christoph Asendorf, *Batteries of Life: On the History of Things and Their Perception in Modernity*, trans. Don Reneau (Berkeley: University of California Press, 1993), 99.
29. Henry James, *Parisian Sketches: Letters to the New York Tribune, 1875–1876*, ed. Leon Edel and Ilse Dusoir Lind (Westport, CT: Greenwood Press, 1957), no. 19, "Summer in France" (written July 22, 1876, and published August 12, 1876), 189–90.
30. Schivelbusch, *Disenchanted Night*, 148–49.
31. Richard Harding Davis, *About Paris* (New York: Harper and Bros., 1895), 76–77.

32. "Toute lumière, pour être convenablement employée à des services généraux et publics, doit pouvoir s'éparpiller, se fractionner à l'infini; sans cela, elle reste un foyer restraint, éclatant, mais impropre à satisfaire aux exigences d'une grande ville. Il en est ainsi de la lumière électrique: elle éblouit et n'éclaire pas; dans bien des circonstances elle peut être utilisée, mais on n'est pas encore parvenu à en faire un agent d'éclairage régulier." "L'éclairage," in *Paris: Ses organes, ses fonctions et sa vie dans la seconde moitié du XIXe siècle*, 3rd ed. (Paris: Hachette, 1875), vol. 5, 308.
33. Edmondo de Amicis, "The First Day in Paris, 28 June, 1878," in *Studies of Paris* (New York: G. P. Putnam's Sons, 1886), 29. Credit goes to the late Charles Harrison for the reference.
34. Robert Louis Stevenson, "A Plea for Gas Lamps" (1878), in *Virginibus Puerisque and Other Papers* (Harmondsworth, Middlesex, UK: Penguin, 1946), 153–57.
35. See Archibald, "Blinded by the Light."
36. "Nous sommes habitués à nous représenter la lumière électrique sous forme de foyers éblouissants, scintillants, durs à l'oeil, bruyants, changeant sans cesse d'intensité, aux tons variables et blafards. Ici, au contraire, on a devant soi une lumière qui a été en quelque sorte civilisée, accommodée à nos habitudes, mise à notre portée; chaque bec éclaire comme du gaz, main comme un gaz qu'il eût fallu inventer, un gaz donnant une lumière d'une fixité parfait, gaie et brillante sans gêner la rétine." *L'Électricité et ses applications: Exposition de Paris* (Paris: G. Masson, 1882), 354.
37. Photography is the light-dependent art par excellence, but I do not discuss it here because it is not a neglected area of study. Alongside abundant study of the daylight practice of a pioneer like Gustave Le Gray and his contemporaries, urban night photographs (especially those made during the genre's heyday, from the 1890s to the 1930s) have also drawn scholars' interest. See Katharina Menzel, "Night Photographs," in *Nightscapes: Photographs of the Night* (Ulm, Germany: Stadhaus Ulm, 2001), 9–17, 19. My remit—Belle Époque figurations in the painted and graphic arts of the artificially illuminated night, whether indoors or out—has not attracted similar attention.
38. A recent exhibition catalogue (Iris Schaefer, Caroline von Saint-George, and Katja Lewerentz, *Painting Light: The Hidden Techniques of the Impressionists* (Florence: Palazzo Strozzi, 2008) did not even mention night light as paintable, as a form of light. But Impressionist night pictures set outdoors are in fact rare, as discussed in what follows.
39. "Mais l'archéologie des représentations de l'électricité . . . reste encore à faire. Pensons que nous sommes presque incapables de dire si dans certains des plus grands tableaux impressionnistes la lumière est celle du gaz ou déjà de l'électricité!" Bruno Foucart, "Histoire de l'art et histoire de l'électricité: L'histoire de l'art face à l'électricité," in *L'électricité dans l'histoire: Problèmes et méthodes: Actes du Colloque de l'Association pour l'histoire de l'électricité en France, Paris, 11–13 octobre 1983* (Paris: Presses universitaires de France, 1985), 148.
40. "Il faudrait que l'historien d'art fasse essentiellement appel à ses éventuelles facultés d'historien et recueille les informations qui lui permettraient de savoir si tel lieu peint par l'artiste était alors éclairé au gaz ou à l'électricité. Par exemple dans *le Bal masqué à l'Opéra* de Manet, on est à peu près sûr que l'éclairage des appliques que l'on voit sur les colonnes portant la galerie était au gaz. Mais très vite l'électricité devait succéder au gaz dans cet opéra de la rue Lepelletier [*sic*] comme dans celui de Charles Garnier. Les tableaux de Degas posent des interrogations comparables." Foucart, "Histoire de l'art et histoire de l'électricité," 151.
41. It is not worth unpacking Foucart's confusion about Parisian theaters in detail, but while urging art historians to get their facts straight, he mixed things up. Here are the bare facts:

La Salle Le Peletier was destroyed by fire on 28 October 1873.

Charles Garnier's Opéra opened in 1875. Electric arc lights were installed on the façade in 1877 and on the avenue in 1878. Its interior was supplied with electric lights in October 1881.

L'Opéra Comique (the second Salle Favart) was destroyed by a gas fire on May 27, 1887. Thereafter, the Paris city council required a change to electric light in all public establishments.

Chapter One

Epigraph: Martin Bressani, "Paris," in Sandy Isenstadt, Margaret Maile Petty, and Dietrich Neumann, eds., *Cities of Light: Two Centuries of Urban Illumination* (London: Routledge, 2015), 28.

1. Sarah Kennel, "Charles Marville: Hidden in Plain Sight," in *Charles Marville: Photographer of Paris* (Washington, DC: National Gallery of Art, 2013), 37.
2. Kennel, *Charles Marville,* 29. The richest troves of Marville's streetlamp photos reside in the collections of the Musée Carnavalet (at least 38 *tirages*) and the Bibliothèque Historique de la Ville de Paris.
3. We know, for example, that the urinal campaign, his series on the *vespasiennes*, was engaged in 1875.
4. Marie de Thézy, *Charles Marville Réverbères* (Paris: Tête d'Affiche, 1993), 57.
5. Marc Gaillard, *Paris Ville lumière* (Amiens, 1994), 178.
6. Marie de Thézy, *Marville* (Paris: Éditions Hazan, 2015), 206.
7. Marie de Thézy, *Charles Marville, Collection Photo Poche* (Paris: Centre National de la Photographie, 1996), n.p.; Kennel, *Marville*, 28, 31 and following; Françoise Reynaud, "Marville and Old Paris," in Kennel, *Marville*, 188–205; Ruth Fiori, *L'Invention du vieux Paris: Naissance d'une conscience patrimoniale dans la capitale* (Paris: Éditions Mardaga, 2013).
8. By emphasizing the photographs' fuzzy backgrounds, de Thézy's description is not incompatible with mine. "Le plan rapproché laisse voir l'environnement, mais souvent l'arrière-plan apparait noyé dans une sorte de flou qui rehausse l'objet et lui confère une étrange présence" (de Thézy, *Marville* [2015], 206–7).
9. I saw this pattern emerge as I pored over the photographs in Sarah Kennel's superb exhibition, *Charles Marville, Photographer of Paris*, National Gallery of Art, September 29, 2013–January 5, 2014. I started thinking along the lines fleshed out here when Sarah Kennel invited me to speak in the December 6, 2013, public symposium, "Old Topographics: Photography and Urbanization in Nineteenth-Century Paris"; http://www.nga.gov/content/ngaweb/audio-video/audio/marville-old-topographics.html.
10. François Loyer, *Paris Nineteenth Century: Architecture and Urbanism*, trans. Charles Lynn Clark (New York: Abbeville Press, 1988), 306.
11. Quoted by Loyer, *Paris Nineteenth Century*, 312.
12. Loyer, *Paris Nineteenth Century*, 306.
13. The factual question naturally arises: how tall were the streetlamps that Marville photographed? Gaillard reports that the earliest generation of lamp stood between 2.30 and 2.75 meters high (between 7.5 and 9 feet) to allow lighting by hand, with or without a ladder. He notes, but without specifying the exact height, that "their height increased during the Second Empire" ("Leur élévation s'accrut sous le Second Empire"). Importantly for this discussion of Marville's photos, it was not until after 1878, when lamps could illuminate automatically, that it became possible for them to be much taller—higher than 7 meters (23 feet) along the avenue de l'Opéra, the *grands boulevards*, the Tuileries, the quais, and in the *places*. Gaillard, *Paris Ville lumière*, 78.
14. On the modernity of the alternation between light and dark, see Martin Bressani and Marc Grignon, "Les Fantasmagories du gaz d'éclairage à Paris au XIXe siècle," in Libero Andreotti, ed., *Spielraum: Walter Benjamin et l'architecture* (Paris: Éditions de la Villette, 2011), 51–71.
15. On the science behind the first night exposures, see C. Klary, *La Photographie nocturne: Applications de la lumière-éclair obtenue par la combustion du magnésium* (Paris: Société d'Éditions Scientifiques, 1893). Klary (n.p.) observes that the practical use of the combustion of magnesium in powder form "pour la production des photographies instantanées" was discovered at the end of 1887. Stereographic faux photographic nocturnes prospered in the 1860s. There are many examples in the collections of the Musée Carnavalet (reproductions purchasable through the Photothèque des Musées de la Ville de Paris) and the J. Paul Getty Museum, Los Angeles. These "vues stéréoscopiques avec éclairage" were made possible by piercing holes in the print made of tissue paper, sometimes hand colored, and viewing the works in an illuminated stereoscopic viewer. Frequently "shot" in this format around 1860 (called "tissues" or "vues stéréoscopiques transparentes effet jour/nuit" in French) were the Place de la Concorde and the Bal Bullier. I owe my familiarity with this medium to Françoise Reynaud, the

very generous former curator of the photographic collections of the Musée Carnavalet, Paris. On the genre of night photography (including recognition of the barely known night photos of Gabriel Loppé, taken in Paris between 1880 and 1900), see Quentin Bajac, "The Latent Images of the Night," in Sylvie Aubenas and Quentin Bajac, *Brassaï: Paris Nocturne* (London: Thames and Hudson, 2013), 190.

16. Wolfgang Schivelbusch, *Disenchanted Night: The Industrialization of Light in the Nineteenth Century*, trans. Angela Davies (Berkeley: University of California Press, 1988), 152.
17. My discussion here relies on Herbert Muschamp's review of the 1999 Belknap Press of Harvard University Press publication of *The Arcades Project*: "The Passages of Paris and of Benjamin's Mind," *New York Times*, January 16, 2000; and Charles Rice, "Architecture's Capital? Revisiting Benjamin's Paris," in Hollis Clayson and André Dombrowski, eds., *Is Paris Still the Capital of the Nineteenth Century? Essays on Art and Modernity, 1850–1900* (London: Routledge, 2016), 19–34.
18. Rice, "Architecture's Capital?" 20.
19. Eric Hazan, *The Invention of Paris: A History in Footsteps*, trans. David Fernbach (London: Verso, 2010), 348.
20. The *locus classicus* of this scenario in *Paris Spleen* is Baudelaire's "The Eyes of the Poor." No one discussed it more penetratingly than the late Marshall Berman in *All That Is Solid Melts into Air: The Experience of Modernity* (New York: Simon & Schuster, 1982). Not wanting to give the impression that Eric Hazan is my entrenched and constant antagonist, I will not quote anything more from his 2010 book in the text, but I would like this discussion to serve as a critique of Hazan's belief in Marville's machinic objectivity. Hazan, *The Invention of Paris*, 343–47: "Marville showed the silent charm of what others liked to see as disturbing and unhealthy. Without any quest for the picturesque, without the least resort to an aesthetic of poverty, he simply used the resources of photography in a way that much later would be described as 'objective.'"
21. Peter Sramek undertook a campaign to rephotograph the sites recorded by both Marville and Atget. Peter Sramek, *Piercing Time: Paris after Marville and Atget, 1865–2012* (Bristol, UK: Intellect, 2013). For his 2011 photo of the rue Soufflot, see Sramek, *Piercing Time*, 313. A summer 2010 intern at the Musée Carnavalet, Ellen Hlozan, guided by Françoise Reynaud, rephotographed every lamp (or the milieu of a now vanished lamp) photographed by Marville between 1858 and 1871; prints of these are in the collection of the Carnavalet. The surviving lamps outnumber those that have been destroyed. Ellen Hlozan, "Réverbères de Marville: Marville's Gas Lamps," unpublished manuscript, 2010. She was an undergraduate art history student at York University, Toronto, at the time.
22. Paul Sébillot, "Exposition des impressionistes," *Le Bien Public* (April 7, 1877), reprinted in Ruth Berson, ed., *The New Painting: Impressionism 1874–1886: Documentation* (San Francisco: Fine Arts Museums of San Francisco, 1996), vol. 1, 190.
23. Literature on Monet's electric London nocturnes is extremely scarce. They are neither mentioned nor reproduced in Richard Thomson's majestic 2011 Monet exhibition catalogue, *Claude Monet, 1840–1926* (Paris: Réunion des Musées Nationaux and New York: Abrams, 2010). Ditto for John House and colleagues, *Monet's London: Artists' Reflections on the Thames 1859–1914* (St. Petersburg, FL: Museum of Fine Arts and Ghent: Snoeck, 2005), and Katharine Lochman, *Turner Whistler Monet* (Toronto: Art Gallery of Ontario in association with Tate Britain, 2005). One was included in *Claude Monet: A Tribute to Daniel Wildenstein and Katia Granoff* (New York: Wildenstein & Co., 2007).
24. Nancy Forgione's otherwise brilliant essay, "Everyday Life in Motion: The Art of Walking in Late-Nineteenth-Century Paris" (*Art Bulletin* 87, no. 4 [December 2005]: 664–87), did not, for example, deal with the differences between daytime and nocturnal circulation.
25. Varnedoe's game-changing work on this previously neglected artist started with his *Gustave Caillebotte: A Retrospective Exhibition* (Houston, 1976). See also his updated *Gustave Caillebotte* (New Haven, CT: Yale University Press, 1987). Michael Marrinan's essay in the 2015 exhibition catalogue sheds new light on the painting by arguing that an understanding of human movement through space permeates the structure of the canvas. Marrinan, "Caillebotte's Deep Focus," in Mary Morton and George T. M. Shackelford, eds., *Gustave Caillebotte: The Painter's Eye* (Washington, DC: National Gallery of Art, 2015), 23–38.

26. Something very much worth noting is that the prominent lamp was an old-fashioned one in 1877. The 1830s-era lamp marks the foreground, the rue de Turin, as pre–Second Empire terrain. Kirk Varnedoe explained that this kind of gas lamp was the standard fixture before the adoption of the floreate-patterned Oudry lamps of 1865. Varnedoe, *Gustave Caillebotte: A Retrospective Exhibition* (1976), 113n2; and Varnedoe *Gustave Caillebotte* (1987), 213–14n2. Julia Sagraves argues that the old lamp serves to identify the painting as "a view of Paris only recently and partially cleansed of its history [the events of 1870–71]." Julia Sagraves, "The Street," in Anne Distel et al., *Gustave Caillebotte: Urban Impressionist* (Chicago: Art Institute of Chicago with Abbeville Press, 1995), 96. Juliet Wilson-Bareau informs us that two circular pedestrian islands, each sporting a five-branched candeladrum of "the latest design," were built in 1878 in what was the middle ground of Caillebotte's 1877 painting. Juliet Wilson-Bareau, *Manet, Monet, and the Gare Saint-Lazare* (Washington, DC: National Gallery of Art, 1998), 92. The two additional green streetlamps in the middle ground of the painting (to the right of the umbrella handle [fig. 1.9]) have, however, eluded mention by modern scholars, perhaps because they were less visible before the 2014 cleaning of the painting in Chicago. The one at left is short and overlapped by a woman on the sidewalk, and the second is very thin and tall, ascending above the umbrella's edge. Neither makes much sense.
27. Another, similarly composed modern-life picture is *La Seine à Charenton* (1875), by Jean-Baptiste-Armand Guillaumin, an artist with Impressionist connections. His painting (sold by Sotheby's in New York, March 16, 2011, lot 29) juxtaposes a peasant woman to riverine factories, and uses a huge isolated streetlamp on the broad *quai* as an imposing and completely decontextualized intermediary. Another painting that makes streetlamps more consequential than people in a Paris setting is Vincent van Gogh's *Terrace and Observation Deck at the Moulin de Blute-Fin, Montmartre* (early 1887; Art Institute of Chicago).
28. Adolphe Alphand, *Les Promenades de Paris: Histoire, description des embellissements, dépenses de création et d'entretien des Bois de Boulogne et de Vincennes, Champs-Elysées, parcs, squares, boulevards, places plantées. Etude sur l'art des jardins et arboretum* (Paris: J. Rothschild, 1867–73). The relevant page showing the rue Soufflot fixtures, "Voie Publique. Candélabres," is reproduced in Andreas Blühm and Louise Lippincott, *Light! The Industrial Age, 1750–1900: Art and Science, Technology and Society* (Pittsburgh: Carnegie Museum of Art, 2000), 182.
29. Marrinan, "Caillebotte's Deep Focus," 34.
30. T. J. Clark, *The Painting of Modern Life: Paris in the Art of Manet and His Followers* (1984; rev. ed., Princeton, NJ: Princeton University Press, 1999), 28.
31. Clark, *Painting of Modern Life*, 29.
32. Clark, *Painting of Modern Life*, 29.
33. Marville's photos of urinals surmounted by gas streetlights prompt a further thought about the links between the photographer's project and the paradoxes of *le mobilier urbain* of Paris when seen through the lens provided by literary scholar Michael Rubenstein. Rubenstein calls our attention to the productive ambiguity in a suitably complex definition of public works. On one hand, they are prerequisites for the existence of the great society. On the other, they are things that result from the existence of the great society. In other words, necessity fosters excess. Marville photographed the double-duty, composite pieces of street furniture par excellence, light-surmounted *vespasiennes.* Two examples appear in Kennel, *Marville*, cat. no. 93, *Urinoir à 3 stalles* (chaussée du Maine) (1875–76, albumen print from collodion negative, 26.5 × 34.5 cm [10½ × 13⅝ in.], Musée Carnavalet, Paris); and cat. no. 94, *Urinoir (système Jennings)* (plateau de l'Ambigu) (1876, albumen print from collodion negative, 26.7 × 36.4 cm [10½ × 14⅜ in.], Musée Carnavalet, Paris). They are daytime views of structures that house brute male bodily need, illuminated both inside and out after dark in pools of gaslight. Public necessity is appropriately served by purely instrumental illumination, valenced nonetheless in the ideological conditions of Paris in the 1870s as the light of enchantment. Michael Rubenstein, *Public Works: Infrastructure, Irish Modernism, and the Postcolonial* (Notre Dame, IN: University of Notre Dame Press, 2010).

Chapter Two

1. "Le crepuscule commence à tomber. L'ombre coule des arbres qui s'étagent en masses sombres. . . . L'ombre gagne encore noyant toute la terrasse, effaçant les blancs contours des marbres." J.-K. Huysmans, "A travers le jardin du Luxembourg," in *Les Chefs-d'œuvre d'art au Luxembourg* (Paris: Librairie Ludovic Baschet, 1881), n.p.
2. This moment of the day, twenty to thirty minutes after sunset, is also called civil twilight (by the US Naval Observatory)—the limit at which twilight illumination is sufficient for objects to be seen. Gratitude goes to E. David Luria for teaching me this term; see also Christopher Dewdney, *Acquainted with the Night: Excursions through the World after Dark* (Toronto: HarperCollins, 2004), 27 and following.
3. In 1924, the artist wrote: "I am sorry I cannot remember which is the original sketch [*sic*] and which is the replica of two versions of "In the Luxembourg Gardens." I think I did them both at about the same time, and one was exhibited and bought by Mr. John G. Johnson, and the other I gave to McKim." John Singer Sargent to Mr. Alan Burroughs, February 4, 1924. A photocopy of the letter (on Copley-Plaza Hotel, Boston, letterhead) is in the object file in the Minneapolis Institute of Art. Sargent retained the Minneapolis version in his own collection, until he gave it to the architect Charles Follen McKim, probably at some date in the 1890s. The Philadelphia version, more finished at the level of facture, was shown in New York. See note 12 for a complete history of the PMA version's early history.
4. E. S., "The Hart-Sherwood Collection," *Art Amateur* 2, no. 2 (January 1880): 30.
5. *Handbook of the Minneapolis Institute of Art* (Minneapolis: Society of Fine Arts, 1917), 81.
6. Kelsey Brosnan is not alone in mistaking the moon for the sun in the Minneapolis picture: "Sargent firmly situates the viewer in time and place: the descent of the icy yellow sun indicates that it is twilight, and the conclusion of that day's fashion 'parade.'" Kelsey Brosnan, "The Female Body in the Jardin du Luxembourg," *Collegiate Journal of Art* 7 (Spring–Summer 2011): 13.
7. In their discussion of Sargent's technique, Lance Mayer and Gay Myers note: "In the 1880s [he] devised a methodical plan to improve his skills by painting quick outdoor sketches." That does not prove that the 1879 pictures were done at least in part outside, but it is an indirect corroboration to learn that he moved systematically in that direction. Mayer and Myers, *American Painters on Technique, 1860–1945* (Los Angeles: Getty Publications, 2013), 176. John Esten's perspective is also suggestive: "Sargent . . . associated summer afternoons with the pleasure of painting outdoor places (13)." John Esten, *Sargent Painting Out-Of-Doors* (New York: Universe, 2000), 18–27. Esten's book includes many examples of Sargent's pre-1879 outdoor sketches and paintings, though not done in Paris itself.
8. Lunar calendar for 1879 at www.rodurago.net (consulted August 30, 2012). Credit and warm thanks to my sister Laurel Clayson who started helping me figure this out years ago using that classic of period software, Starry Night.
9. Charles Baudelaire, at the end of the section on landscape in "The Salon of 1859," quoted in Eric Hazan, *The Invention of Paris: A History in Footsteps* (London: Verso, 2010), 341.
10. Patricia Hills, "The Formation of a Style and Sensibility," in Patricia Hills et al., *John Singer Sargent* (New York: Whitney Museum of American Art in association with H. N. Abrams, 1986), 31.
11. The summer of 1879 was, in fact, the one summer he spent at home, though in August he went to Mme. Pailleron's estate at Ronjoux in Savoy to paint commissioned pictures of the lady and her children. Richard Ormond and Elaine Kilmurray with Richard H. Finnegan, *John Singer Sargent: Figures and Landscapes, 1874–1882, Complete Paintings,* vol. 4 [hereafter *CP* v4] (New Haven, CT: Yale University Press, 2006), 181; Albert Boime, "Sargent in Paris and London: Portrait of the Artist as Dorian Gray," in Patricia Hills et al., *John Singer Sargent*, 83; and Stanley Olson, "Chronology," in Patricia Hills et al., *John Singer Sargent*, 277.
12. Geographic, social, and aesthetic reasons for choosing Paris sites as painting subjects are discussed in what follows. Not mentioned are market considerations. According to the *Catalogue raisonné*, Sargent made the Philadelphia canvas as "a picture to sell," and it was the only Paris painting that he sent for exhibition. *CP* v4, 181. The authors argue that its apparent failure to sell when displayed at the

National Academy of Design late in 1879 and when put up for auction later may "have contributed to Sargent's reluctance to undertake any further Parisian subjects . . . but he was also under pressure from portrait commissions." *CP* v4, 182. Further research (by Joseph Hammond), however, shows that the Luxembourg canvas was *not* included in the May 1879 exhibition at the National Academy of Design. Reviews for that exhibition described Sargent's contribution as the *Neapolitan Children Bathing* (1879, oil on canvas, 16.8 × 41.1 cm, Sterling and Francine Clark Art Institute, Williamstown, MA). *The Luxembourg Gardens* was displayed at the National Academy of Design only prior to being successfully auctioned at the nearby Chickering Hall on December 17, 1879. Probably because it was a short sales viewing rather than a formal gallery exhibition, the display in December appears to have gained few notices. The painting is described in the sales catalogue and in two articles on the auction. Literature: "The Academy Exhibition," *New York Times*, May 2, 1879, 5; "Fine Arts. Exhibition of the Academy of Design. II," *Nation* 28, no. 725 (May 22, 1879), 359; Edward Strahan, "The Art Gallery: The National Academy of Design," *Art Amateur* 1, no. 1 (June 1879), 4–5; "The Society of American Artists," *Scribner's Monthly* 18, no. 2 (June 1879), 311; *Valuable Paintings, the Collections of Mr. John H. Sherwood and Mr. Benj. Hart* (New York: George A. Leavitt & Co., 1879), lot 61; "Paintings on the Block: Selling the Sherwood and Hart Collections," *New York Times*, December 18, 1879, 5; ES [Edward Strahan], "The Hart-Sherwood Collection," 30; "John S. Sargent," *Art Amateur* 2, no. 6 (May 1880): 188. Additionally, Marc Simpson observes that in 1879 the gallerist George Petit "had been involved in getting [the PMA *Luxembourg Gardens*] into the hands of its first owner." Marc Simpson in Richard Ormond et al., *Sargent and His People* (London: National Portrait Gallery, 2015), 31. He cites as evidence a letter by Sargent written from Madrid (now lost) dated October 20, 1879 (247n69).

13. Later nineteenth-century Luxembourg pictures that defined the park as a space of children and child-minding include Edgar Degas, *A Nurse in the Luxembourg Garden* (1875); Fernand H. Lundgren, *Gardens of Luxembourg* (c. 1882); Pierre-Auguste Renoir, *At the Luxembourg Garden* (1883); Charles Courtney Curran, *Luxembourg Gardens* (1889); Maurice Prendergast, drawings (1890–94); James McNeill Whistler, etchings and lithographs (1892–95), including *Nursemaids: Les bonnes de Luxembourg* (lithograph, 1894); and Félix Vallotton, *Jardin du Luxembourg* (1895).
14. Sargent made eight related drawings. *CP* v4, 190–92, figs. 103–8.
15. Despite the garden's aggravated and troubling history, H. Barbara Weinberg, for example, views the immediately pre-Sargent transformations of the Luxembourg favorably. "Its parks, including the Luxembourg Gardens, had been built or reconfigured in response to urban growth and overcrowding, thus epitomizing the French capital's new energy." *American Stories: Paintings of Everyday Life, 1765–1915* (New York: Metropolitan Museum of Art, 2009), 118. Greg M. Thomas is of the same opinion: "Although Haussmann and [Adolphe] Alphand [chief engineer for the parks] met some fierce criticism, the parks for the most part succeeded spectacularly; tourists and Parisians alike flocked to them and they were widely heralded as shining emblems of modernization." Greg M. Thomas, "Women in the Parks of Paris," in Aruna D'Souza and Tom McDonough, eds., *The Invisible Flâneuse? Gender, Public Space, and Visual Culture in Nineteenth-Century Paris* (Manchester, UK: Manchester University Press, 2006), 36.
16. The street's prolongation had a long history. It was announced in 1805, but only begun in 1846. Jacques Hillairet, *Dictionnaire historique des rues de Paris* (Paris: Éditions de Minuit, 1963), vol. 2, 528. The present-day small Place Edmond-Rostand, situated at the intersection of the rue de Médicis and the boulevard Saint-Michel, which now prevents the rue Soufflot from connecting directly with the boulevard Saint-Michel as it did in Sargent's day, was not built until 1924. Hillairet, *Dictionnaire historique*, vol. 1, 468. On Marville's photos of this locale, see Peter Sramek, *Piercing Time: Paris after Marville and Atget, 1865–2012* (Chicago: Intellect and University of Chicago Press, 2013), fig. 5.30, 312; and Sarah Kennel, *Charles Marville: Photographer of Paris* (Washington, DC: National Gallery of Art, 2013), 123.

17. Eugène Defrance, *Histoire de l'éclairage des rues de Paris* (Paris: Imprimerie Nationale, 1904), 118. "On vit donc briller la lumière électrique sur la place du Château-d'Eau (aujourd'hui place de la République); puis, en mars 1879, aux Halles; ensuite place de la Bastille et rue Soufflot, depuis le Luxembourg jusqu'au Panthéon." While the first arc lights in the world appeared in France (1875), they were in use in the United States as well. Charles F. Brush developed the first commercially successful arc light systems in North America, first used in Cleveland in 1879. http://www.edisontechcenter.org/ArcLamps.html#growth.
18. Charles Barnard, "Electric Candle: The Light of the Future—Practical Working of the New Light in Paris," *Chicago Daily Tribune*, June 29, 1878, 12. Warm thanks to Ernie Freeberg for the reference.
19. "A Concise Chronology," in Elaine Kilmurray and Richard Ormond, eds., *John Singer Sargent* (Washington, DC: National Gallery of Art, 1999), 273.
20. Kilmurray and Ormond, eds., *John Singer Sargent*, 84, no. 15.
21. *CP* v4, 183.
22. The American painter William A. Coffin recalled going to the concerts with Sargent. See http://www.mfa.org/collections/object/rehearsal-of-the-pasdeloup-orchestra-at-th-cirque-dhiver-31960. See also Kilmurray and Ormond, eds., *John Singer Sargent*, 84, no. 15; and Ellen Roberts, "John Singer Sargent: An Impressionist and an Intransigeant," in *The Age of American Impressionism: Masterpieces from the Art Institute of Chicago* (Chicago: Art Institute of Chicago, 2011), 70. Coffin recalled that the sheets of music were lit by "little lamps on the racks," but the ghost chandelier in the Chicago version assures us that large lamps hung from the ceiling.
23. Andreas Blühm and Louise Lippincott, *Light! The Industrial Age, 1750–1900: Art and Science, Technology and Society* (Pittsburgh: Carnegie Museum of Art, 2000), 188. James Tissot's renderings of the arc lights in the Hippodrome exist in both painted and etched form: *Les Femmes de Paris: Ces dames des chars*, oil on canvas (c. 1883–85, Rhode Island School of Design Museum), and etching and drypoint (1885, private collection).
24. The Art Institute of Chicago dates the picture to 1876–78, but others, including me, are sure that it must be later; certainly not before 1879. The case for a later date is discussed in http://www.mfa.org/collections/object/rehearsal-of-the-pasdeloup-orchestra-at-th-cirque-dhiver-31960; and in Kilmurray and Ormond, eds., *John Singer Sargent*, 84, no. 15.
25. One is considered the final canvas, two as studies, and one a sketch after. See *CP* v4, nos. 763–66.
26. Lacey Taylor Jordan, *After Whistler: The Artist and His Influence on American Painting* (Atlanta: High Museum of Art, 2003), cat. 56, p. 220.
27. The ingredients of this part of the garden constitute its bona fides as "un jardin à la française." The boat basin is the critical feature, since the Frenchness of the garden is secured by the presence of "a central island occupied by a vast octagonal basin upon which paths converge and surrounded by balustraded terraces" ("Un terre-plein central vers lequel convergent les allées est occupé par un vaste bassin octagonale environné de terrasses à balustrades"). Gilles Marchand, *Dictionnaire des monuments de Paris* (Paris: Éditions Gisserot, 2003), 440. The staircase, which gives access to the elevated terrace defined by balustrades, greenery, and the exit gates beyond, is in plain sight in both pictures. Notably, Sargent's painted balustrades are entirely parallel to the top and bottom edges of the paintings and parallel to the picture plane as well. He straightened out and simplified the actual garden balustrade's slow curve in the process.
28. The pair's placement was painstakingly settled upon. According to John Davis, far from having been an impulsive record of an observed event, "that strolling couple is actually the American artist J. Carroll Beckwith and the famous Capri model, Rosina Ferrara. Beckwith relates that Sargent had the two of them stroll across that scene again and again until he got it right." This observation shores up the possibility that Sargent worked *sur place*. Davis continues: "Hardly the composition 'à l'improviste' that we've come to know!" Personal communication, Fall 2005; confirmed February 21, 2014. James

Carroll Beckwith was the American student that Sargent and his father had met the first day in the atelier of Carolus-Duran. The two men also studied drawing at the École des Beaux-Arts together and became fast friends. Carter Ratcliff, *John Singer Sargent* (New York: Abbeville Press, 1982), 40.

29. Note the difference between the renderings of the woman's pink dress from picture to picture. In the Minneapolis canvas, more of the outer dress is shown hitched up to aid her movement, revealing the pale underdress or crinoline beneath in the process, but the woman lifts her skirt too high. The couple's ensembles and behaviors are actually eccentric. It was not correct in 1879 to combine a black suit with a straw boater, and it was unseemly to be bareheaded and smoking in the presence of a woman. The cloud of pale gray cigarette smoke in the MIA picture is indeed sizable. The woman's gauze-wrapped straw hat, like the man's boater, is for daytime not evening wear, and the black belt that interrupts her pink dress is a move against the dominant long lines of that year's fashionable dress, though pink was certainly a very popular dress color that year. In other words, the two strollers are not, strictly speaking, fashionable. Warm gratitude goes to Justine DeYoung for appraising the character of the clothing in the paintings. Personal communication, March 2, 2014.
30. Nancy Forgione, "'The Shadow Only': Shadow and Silhouette in Late Nineteenth-Century Paris," *Art Bulletin* 81, no. 3 (September 1999), 497.
31. Forgione, "The Shadow Only," 490.
32. Susan Stewart, "Out of the Darkness: Nocturnes," in *Poetry and the Fate of the Senses* (Chicago: University of Chicago Press, 2002), 291. Thanks to Caro Fowler for recommending this text to me.
33. Hills, "The Formation of a Style and Sensibility," 32.
34. Jonathan Crary, *24/7: Late Capitalism and the Ends of Sleep* (London: Verso, 2013), 65–66.
35. I am referring here to the circular blobs of bright orange just above the balustrade railing at the far left, unmistakably intended to suggest the lights on the rue de Médicis poking through the greenery. And the cognate if fainter balls of orange pigment (streaky red and yellow, actually) employed to denote lights on the boulevard Saint Michel at the top of the staircase in both pictures.
36. Patricia Hills, *The Painters' America: Rural and Urban Life 1810–1910* (New York: Praeger for the Whitney Museum of American Art, 1974), 97.
37. John Siewert, "Rhetoric and Reputation in Whistler's Nocturnes," in *After Whistler: The Artist and His Influence on American Painting* (Atlanta: High Museum of Art, 2003), 64; and Margaret F. MacDonald and Patricia de Montfort, "Nocturnes," in *An American in London: Whistler and the Thames* (London: Dulwich Picture Gallery, 2013), 124–31.
38. Siewert, "Rhetoric and Reputation," 66.
39. Note that the moon is slightly to the left (north) of the basin's edge in the PMA picture, but being off axis diminishes neither the intensity nor the lateral extent of the water's moon glow. The moon in the MIA canvas (cradled symmetrically by a curved notch in the dark foliage) is on axis with the boat basin below; the reflected light is correspondingly intense but whiter and less expansive. Defrance narrates the sequence of lighting experiments that dotted Paris on both Right and Left Banks in 1878 and 1879. In February 1878 came the infamous implantation by the Syndicat d'Étude de la Lumière Électrique of Jablochkoff arc lights in the Place de l'Opéra, in the Avenue de l'Opéra, and in the Place du Théâtre Français. In December, the Conseil Municipal proposed that those experiments be extended for one year. Thereafter, the Place du Château-d'Eau (today the Place de la République) was added to the list, and in March 1879, Les Halles was followed by the Place de la Bastille and the rue Soufflot, leading from the Luxembourg to the Panthéon. Defrance, *Histoire de l'éclairage*, 117–18.
40. Another artist fascinated by the coexistence of natural and industrial light, Camille Silvy (1834–1910), was an ancestor of the dissonant nocturne in the world of photography. He made an extraordinary, heavily manipulated photo in 1859, *Studies on Light: Twilight* (private collection). A sidewalk-lined street is represented in twilight with two male figures in the foreground alongside a glowing gas streetlamp. There are three more lamps blurred in the gloaming in the background. This illusion re-

quired four separate negatives. Further, "the light in the lamp may have been drawn on the negative as, unlike the lamps farther away, it has no aureole." According to Haworth-Booth, "the distant figure is probably the first intentional use of blur in the history of photography." Mark Haworth-Booth, *Photographer of Modern Life: Camille Silvy* (Los Angeles: J. Paul Getty Museum, 2010), 59. Thanks to Sarah Kennel for telling me about Silvy. See also an apposite, if less vivid urban nocturne by the American artist, Louis Rémy Mignot, *Bal de Nuit* (1867; Paris). Mignot became acquainted with Whistler in 1867. Katherine E. Manthorne with John W. Coffey, *Louis Rémy Mignot: A Southern Painter Abroad* (Washington, DC: Smithsonian Institution Press for the North Carolina Museum of Art, 1996), cat. nos. 96, 159, 205.

41. The paintings of Joseph Wright of Derby (1734–97) are also germane, but knowledgeable readers asked me to cut my discussion of them from this chapter. Warm gratitude goes nonetheless to Jason La Fountain, Dominique Poulot, and Susan Wager for encouraging me to compare Wright of Derby's nocturnes to Sargent's. Though set far away, Wright of Derby's scenarios have much in common with Sargent's. He sometimes juxtaposed celestial illumination to worldly luminosities in the form of moonlight coexisting with factory lights in, for example, *Arkwright's Cotton Mills by Night* (c. 1782; Derby Museum and Art Gallery, UK). The modern glowing, red-brick factory nestles cozily in the hills, well below the crest of the hilly horizon, posing no threat to the natural order and thoroughly outdone by majestic white clouds and a glowing moon that grandly dominate the scene. Jonathan Crary uses *Arkwright's Cotton Mills by Night* to make a significant point about modernity and its illuminations that bears directly upon my analysis of the works by Sargent. He notes that while Wright of Derby's painting conveys the physical nearness of one natural and one invented sphere, the key thing is that "this relatively early image of an irreconcilable adjacency nonetheless counters the notion of an 'industrial revolution' that devastated the countryside and quickly herded rural laborers into cities and factories. Instead there was a protracted and piecemeal deterioration of older forms and spaces" (24/7, 63). To his delineation of the coexistence and clash of two orders, Crary adds a crucial observation: "Wright of Derby's image is an early revelation of modernity's concurrence and contiguity of ultimately incompatible systems" (24/7, 66). Jason La Fountain also called my attention to a little-studied United States–based nocturnist, Johann Mengels Culverhouse (American, born Holland, 1820–95), who specialized in night scenes lit by competing forms of light, usually firelight or torchlight versus moonlight. See, for example, *Moonlight Boating Party* (1876; Brooklyn Museum of Art), and *At the Market* (1872).

42. For another Paris nocturne by Jongkind closer in date to Sargent's paintings, see *La rue Saint-Séverin* (1877, oil on canvas, 47.5 × 34.5 cm; Paris, Musée Carnavalet, P1995). It goes even further than *Notre Dame de Paris* (1854) in moodily archaizing the simultaneously gaslit and moonlit capital city.

43. Hannah Feldman suggested that the gleaming basin functioned as a zone of displacement. Good idea.

44. Defrance, *Histoire de l'éclairage*, fig. 33. It is dated 1880 by Defrance, but Cham died in 1879.

45. A third connotation is a remote possibility: a reference to the silver nitrate solution used in photography, though night photos were still in the future.

46. "La Chaste Phoebé. Pourquoi chaste ? Plus nombreux que les étoiles du firmament sont les criminels auxquels elle a prêté sa lumière complaisante, depuis que le monde est monde, et que Phoebé l'éclaire."

47. "L'éclairage électrique, d'une couleur blafarde, comme celle de la lune, projetait des rayons que ne fatiguaient nullement la vue. . . . On a pu lire très facilement à l'aide de ce rayon lumineux, à 150 mètres de distance." Quoted in Defrance, *Histoire de l'éclairage*, 114.

48. Julian Berthold Jongkind's sketchbook drawing dated August 10, 1887, *La Terrasse du Luxembourg* (Louvre, Département des Arts Graphiques, RF 11164, verso), also features a newspaper-reading man seated on a bench, but in daylight.

49. Joachim Schlör, *Nights in the Big City, Paris, Berlin, London, 1840–1930*, trans. Pierre Gottfried Imhof and Dafydd Rees Roberts (London: Reaktion, 1998), 95.
50. Kudos to Sarah Kennel for "il s'argent."

Chapter Three

1. Baudelaire, "De l'essence du rire et généralement du comique dans les arts plastiques" (published 1855), in Henri Lemaitre, ed., *Curiosités esthétiques. L'Art romantique et autres œuvres critiques* (Paris: Éditions Garnier Frères, 1962), 241–63. Translated as "On the Essence of Laughter," in Charles Baudelaire, *The Mirror of Art: Critical Studies*, ed. and trans. Jonathan Mayne (Garden City, NY: Doubleday, 1956), 133–53.

2. William Henry Preece complained that very little was done to improve the illumination of the streets of Paris before the 1881 International Exposition of Electricity, but mentioned the eccentric lighting of the Boulevard des Italiens as an exception. "[B]ut prior to the opening of the Exhibition, a portion of the Boulevard des Italiens was lit up by four De Mersanne [*sic*] lamps, suspended high up, at wide intervals, over the center of the road. The effect was very fine, but the lamps were very bad." Comte Th. du Moncel and Wm. Henry Preece, *Incandescent Electric Lights, with particular reference to the Edison Lamps at the Paris Exhibition* (New York: D. Van Nostrand, 1882), 147. Obviously, the plate reproduced in figure 3.2 called them "Million Lamps." Nomenclature was slippery in these days of constant innovation.
3. The original *La Caricature* was published from 1830 to 1843. On Robida's magazine, see Laurent Bihl, "La Caricature (1880–1904)," March 1, 2009, http://www.caricaturesetcaricature.com/article-33576190.html. Some of the ideas in this chapter were published in Hollis Clayson, "Bright Lights, Brilliant Wits: Caricature and Electric Light in Later Nineteenth-Century Paris," in Alain Beltran, Léonard Laborie, Pierre Lanthier, and Stephanie Le Gallic, eds., *Electric World: Creations, Circulations, Tensions and Transitions, 19th–21st Centuries*, History of Energy, Comité d'Histoire de l'Électricité et de l'Énergie (Brussels: Peter Lang AG, 2016), 17–38.
4. Philipon also edited *Le Charivari*, the other major journal of political caricature of the era.
5. Edison's appearance as a grave and respectable young worthy on the cover of the October 19, 1881 issue of *L'Illustration* testifies to the rise of his standing in the French capital city, especially once his contributions to the 1881 International Electricity Exhibition grabbed the headlines. Historian Alain Beltran's assessment of Edison exemplifies the post-1881 opinion amongst the French of the American as a shrewd businessman: "Le grand inventeur Thomas Alva Edison était doublé d'un très efficace tempérament d'entrepreneur." (The great inventor Edison was endowed with the very efficient temperament of an entrepreneur.) Alain Beltran, *La Ville-Lumière et la Fée Électricité: Service public et entreprises privées: l'énergie électrique dans la région parisienne* (Paris: Éditions Rive Droite, 2002), 116. See also Charles Bazerman, *The Languages of Edison's Light* (Cambridge, MA: MIT Press, 1999), who explains that Edison gave up his public role as a man of science, 132ff.
6. The "discussion" of the Fidèlimètre continues on pages 2–6 of the same issue.
7. Richard Terdiman, *Discourse/Counter-Discourse: The Theory and Practice of Symbolic Resistance in Nineteenth-Century France* (Ithaca: Cornell University Press, 1985). Bertrand Tillier is ambivalently in this camp: *À la charge! La Caricature en France de 1789 à 2000* (Paris: Éditions de l'Amateur, 2005), 16. "Le plus souvent, la caricature est . . . une *charge* qui outre pour ridiculiser." Eirwen Nicholson critiques the problematic tendency to regard caricature as a mass medium, which in turn hinges on defining it as "an inherently democratic and potentially subversive medium." Eirwen Nicholson, "Consumers and Spectators: The Public of the Political Print in Eighteenth-Century England," *History: The Journal of the Historical Association* 81, no. 261 (January 1996): 5–21.
8. Todd Porterfield aligns with this point of view. Porterfield, "The Efflorescence of Caricature," in Porterfield, ed., *The Efflorescence of Caricature, 1759–1838* (Farnham, UK: Ashgate, 2011), 1–9. Mike Goode

argues that caricature is "essentially conservative." Goode, "The Public and the Limits of Persuasion in the Age of Caricature," in Porterfield, ed., *The Efflorescence of Caricature,* 117–36. "In many cases, the primary purpose of an image [in caricature] seems to have been to specify or recycle a particular character type or set of types." Mike Goode, in Porterfield, ed., *The Efflorescence of Caricature,* 120.

9. "La jeune américaine voyage beaucoup; pendant que son mari siège à son comptoir ou dans les assemblées politiques, elle court le monde en toute liberté. Mais le Fidèlimètre est un témoin et une garantie. Toutes les semaines, une photographie du Fidèlimètre est envoyée au mari, et, à la moindre déviation, celui-ci envoie, par télégraphe ou téléphone, l'ordre de rentrer *at home*."
10. "L'oncle Sam est un malin, chacun sait ça. Dès la première apparition du Fidèlimètre, toutes les épouses de l'Union en ont été pourvues. Quel objet d'ailleurs est d'une utilité plus incontestable? À l'ouverture de la dangéreuse saison des bains de mer, qui font courir tant de risques aux maris retenus à la ville par les affaires, le Fidèlimètre a été adopté avec enthousiasme par toutes les épouses de la fashion américaine."
11. Roger Jouan et al., *Voyages Très Extraordinaries dans le Paris d'Albert Robida* (Paris: Paris bibliothèques éditions, 2005). Robida also imagined an electric future in book form, Albert Robida, *Le Vingtième Siècle: La Vie Électrique* (Paris: La Librairie Illustrée, 1890, facsimile, Lexington, Ky.: Adamant Media Corporation, Elibron Classics, 2006). See also Sandrine Doré, "Entre caricature et anticipation, la Parisienne définie par Albert Robida (1848–1926)," in Ségolène Le Men, ed., *L'art de la caricature* (Paris: Presses Universitaires de Paris Ouest, 2011), 211–32, and Caroline Dubbs, "Cultures of Time in fin-de-siècle France: The Popular Literature and Graphic Art of Albert Robida (1848–1926)," PhD diss., Department of French, University of Pennsylvania, 2015.
12. Laura Kipnis, "The Lothario," in *Men: Notes from an Ongoing Investigation* (New York: Henry Holt, 2014), 86.
13. The new arc lamp streetlights were Jablochkoff candles and were mostly put in place between 1877 and 1879. They continued at some locations through 1880, 1881, and the early spring of 1882. On Cham, see David Kunzle, "Cham, The 'Popular' Caricaturist," *Gazette des Beaux-Arts* 96 (December 1980): 213–24. Kunzle, a Cham partisan, argues that while Honoré Daumier, the better-known figure, spoke mostly to "the wealthier upper-middle-class art-conscious Republican public," Cham spoke to the other end of the scale, "those less concerned with 'enduring art' than the irritations and follies, large and small, of the socio-political day-to-day" and that Cham reached a much bigger audience. Kunzle, "Cham, The 'Popular' Caricaturist," 214.
14. Thank goodness Jean-Louis Cohen identified these figures correctly.
15. Christopher Curtis Mead, *Charles Garnier's Paris Opéra: Architectural Empathy and the Renaissance of French Classicism* (Cambridge, MA: MIT Press, 1991), 196.
16. The Jablochkoff experiment on the avenue de l'Opéra was terminated on April 1, 1882, putting an end to the first attempt to place electric light in Paris on a permanent footing. Alain Beltran, *La Ville-Lumière et la Fée Électricité: Service public et entreprises privées: L'Énergie électrique dans la région parisienne* (Paris: Editions Rive Droite, 2002), 71. A period characterization of the difference between the electric light on the avenue de l'Opéra and the gaslight on the nearby rue du Quatre-Septembre is too good not to quote: "Sur un refuge, nous avons entendu un mot d'un titi qui résume assez bien l'impression général: C'est curieux tout de même, il fait lune avenue de l'Opéra et soleil rue du Quatre-septembre." "On a traffic island, we heard a word from an apprentice that sums up the general impression: it's strange all the same that the avenue de l'Opéra is moonlit while the rue du Quatre-septembre is lit by the sun." This use of celestial metaphors very cleverly records the difference in hue between a white light and a yellow one. So while the arc light's blaze is often found to rival sunlight in intensity, there is no doubt that the light was white (and how!). *Liberté*, May 3, 1879, quoted in Beltran, *La Ville-Lumière et la Fée Électricité,* 62–63.
17. Charles Garnier, "L'Éclairage électrique à l'Opéra," *La Lumière électrique: Journal universel d'électricité* 1, no. 8 (October 15, 1879): 150 and following.

18. Mead, *Charles Garnier's Paris Opéra,* 196. Mead also reports that on November 25 the minister of fine arts agreed that Garnier could install electric lighting in the Opéra on a permanent basis.
19. The two of them attended a monthly dinner of twenty men, all friends ("vingt convives, vingt hommes d'esprit"), chez Brébant, that they called the "diner du Canard aux navets." (the dinner of duck with turnips). Félix Ribeyre, *Cham: Sa vie et son œuvre* (Paris: Librairie Plon, 1884), 239–41.
20. Baudelaire and Freud agree on this. Baudelaire, "De l'essence du rire," 241–63; Sigmund Freud, "On Humour" [1927], in *The Complete Psychological Works of Sigmund Freud*, trans. James Strachey, vol. 21 (1961), 160–66.
21. This runs counter to Kunzle's definition of Cham's down-market audience.
22. Dr. N. Th. Klein, *De l'influence de l'éclairage sur l'acuité visuelle* (Paris: G. Masson, Libraire de l'Académie de Médecine, 1873), 59. "Faut-il admettre, avec tous ceux qui ont écrit sur ce sujet, que l'acuité visuelle diminue lorsque la lumière est trop forte? " "[U]n éclairage très-fort, comme celui de la lumière électrique, produit une excitation douloureuse que peu de personnes sont à même de supporter." "Mais nous avons toujours vu, à la suite du premier moment d'éblouissement, une augmentation de l'acuité."

23. Andreas Blühm and Louise Lippincott, *Light! The Industrial Age, 1750–1900: Art and Science, Technology and Society* (Pittsburgh: Carnegie Museum of Art, 2000), 150.
24. https://sites.google.com/a/plu.edu/paris-salon-exhibitions-1667-1880/salon-de-1879. Consulted January 14, 2014. Patricia Mainardi writes that in 1879, the year of the Republican accession to power, Edmond Turquet, an official in the Beaux-Arts administration, decided to light the Salon electrically so that "workers could attend in the evening." Patricia Mainardi, *The End of the Salon: Art and the State in the Early Third Republic* (Cambridge, UK: Cambridge University Press, 1993), 72. Also reporting electric lighting used at night in 1879 is Viktorya Vilk, "Laying It on the Line: Experimenting with an Aesthetic Hang at the Parisian Salon c. 1872–1883," MA thesis, Courtauld Institute of Art, London, 2010, 18. On how it looked and what observers thought, she concludes: "Despite record numbers [drawn to the Salon thanks to evening hours], public perception of the installation proves impossible to establish. Critics, in support of their own assessment, alternatively claimed public delight or outrage" (20–21). See also F. G., "La Lumière électrique au Salon," *La Lumière électrique: Journal universel d'électricité* 1, no. 3 (June 15, 1879): 58.
25. *Nouveau Larousse illustré, Dictionnaire universel encyclopédique*, ed. Claude Augé (Paris: Librairie Larousse, 1898–1904), vol. 1, 492. While the *Larousse* defines artificial light as "lumière produite par la combustion d'un corps gras ou du gaz hydrogène," we will track only postcandle artificial lighting: oil light, gaslight, and arc and incandescent electric light.
26. Paul Tucker, "1874 The First Exhibition: The First Impressionist Exhibition in Context," in Charles S. Moffett et al., *The New Painting: Impressionism 1874–1886* (Washington, DC: National Gallery of Art, 1986), 106.
27. "Ainsi, la première manifestation impressionniste a bénéficié d'un éclairage artificiel en 1874, qui permettait de regarder les œuvres après la nuit tombée." Pierre Pinchon, *La Lumière dans les arts européens, 1800–1900* (Paris: Éditions Hazan, 2011), 149.
28. Richard Kendall, *Degas by Himself: Drawings Prints Paintings Writings* (London: Macdonald Orbis, 1987), 116. Kendall writes "Jablochkof." I have changed it to "Jablochkoff" to align with Degas's original spelling. Did arc light ever blaze in an Impressionist exhibition? Martha Ward, the leading expert on Impressionist exhibition installations, is on the case. "La Cie Jablochkoff nous propose de nous éclairer à la lumière électrique," quoted in the original in Ronald Pickvance, "1879, The Fourth Exhibition: Contemporary Popularity and Posthumous Neglect," in Charles S. Moffett et al., *The New Painting: Impressionism 1874–1886* (Washington, DC: National Gallery of Art, 1986), 263n28.
29. Pickvance, "1879, The Fourth Exhibition: Contemporary Popularity and Posthumous Neglect," 250. Pickvance dates Degas's letter to early April insofar as it mentions artwork being deposited in the exhibition rooms (263n33).

30. Charles S. Moffett, "1880, The Fifth Exhibition, Disarray and Disappointment," in Charles S. Moffett et al., *The New Painting: Impressionism 1874–1886* (Washington, DC: National Gallery of Art, 1986), 294.

31. Blühm and Lippincott, *Light!* 150; and Michael Compton, "The Architecture of Daylight," in *Palaces of Art, Art Galleries in Britain 1790–1990* (London: Dulwich Picture Gallery, 1991), 42. I learned from Jo Briggs, assistant curator of eighteenth- and nineteenth-century art, Walters Art Museum, Baltimore, that William T. Walters illuminated the 1880s gallery that he built behind his townhouse with rows of open gas jet flames near the ceiling to supplement the room's skylight. Personal communication, March 25, 2014. A photograph dated 1884 exists of this arrangement.

32. Martha Ward, "Impressionist Installations," *Art Bulletin* 73, no. 4 (December 1991): 605 and fig. 5, 606.

33. Vilk, "Laying It on the Line," 22.

34. Ward, "Impressionist Installations," 615, and see figs. 9, 10.

35. ". . . fit de sa luxueuse et mondaine galerie située près de la Madeleine un lieu obligé de la vie nocturne parisienne." Pinchon, *La Lumière dans les arts européens, 1800–1900*, 152.

36. Pickvance, "1879, the Fourth Exhibition: Contemporary Popularity and Posthumous Neglect," 263n39.

37. This paragraph relies upon Compton, "The Architecture of Daylight," 43.

38. "Plan de l'Exposition Internationale d'Électricité," *La Nature* 428 (August 15, 1881): 161, shows that room no. 1 on the first floor was a "Galerie de tableaux éclairée le soir à la lampe Soleil." The American Gustave Glaser evaluated the gallery before the lamps were installed: "Hall 1 contains a gallery of paintings but it is to be hoped that the light of the "Lampe-Soleil" which is here exhibited will be better than the pictures, which are wretchedly bad." One wonders what they were. Gustave Glaser, Ph.D. (dateline Paris, August 17, 1881), "The Paris Electrical Exhibition," *Science* 2, no. 63 (September 10, 1881): 433.

39. Blühm and Lippincott observe that Americans were the leaders in domestic illumination and in lighting *tout court*.

40. See the discussion of Ernest Chesneau's opinions at the end of the chapter.

41. "L'éclairage du palais de l'Industrie est un succès, le public s'y porte déjà en grand nombre et tout annonce que l'affluence ne fera que s'accroître. Cet empressement nous paraît entièrement justifié. L'aspect du palais ainsi illuminé est agréable.

"La sensation générale est celle d'un jour discret et un peu adouci.

"Dans les galeries de tableaux, l'effet est complètement satisfaisant; les globes sont placés assez haut, leur lumière est concentrée par des réflecteurs sur les parois; on oublie volontiers l'heure et l'on se livre à l'examen des peintures avec la même quiétude que dans le jour. Dans le jardin, où sont les statues, la lumière n'est pas tout à fait assez abondante; il est vrai que le problème était autrement compliqué; une nef immense, point de plafonds ni de surfaces réfléchissantes, de la lumière à distribuer partout jusque dans les derniers recoins; le résultat n'est pas parfait, mais il est certainement remarquable.

"Quelques remarques particulières peuvent être faites sur cette expérience à grandes proportions.

"Elle nous semble démontrer que la lumière n'a pas l'éclat éblouissant et exagéré qu'on lui repproche [*sic*] quelquefois. Les globes du jardin sont à médiocre hauteur et ne gênent point. Nous signalons particulièrement les foyers établis sur le palier d'arrivée au 1*er* étage, ils sont presque à hauteur d'homme ; on les regarde en face de très-près et personne n'en paraît incommodé. En compensation, les scintillations et les changements de couleur de la lumière sont très-apparent. Dans les galeries de tableaux, cet effet est peu sensible, soit que les couleurs mêmes de la peinture empêchent de l'apercevoir, soit que les globes plus rapprochés fondent leurs lumières en une moyenne plus égale; mais dans le jardin, l'implacable blancheur des statues ne laisse échapper sans la signaler aucune des plus petites inégalités et il faut bien avouer qu'elles sont assez fréquentes.

"Malgré cette petite réserve, empressons-nous de reconnaître que cette illustration, la plus vaste sans doute qui ait été tentée encore en lieu clos, fait le plus grand honneur à ceux qui l'ont exécutée; elle montre une fois de plus ce que l'on peut attendre de la lumière électrique, et ne peut manquer d'en accélérer le progrès."

F. G., "La Lumière Électrique au Salon," *La Lumière Électrique* 1, no. 3 (June 15, 1879): 58.

42. "A la sculpture, l'éclairage donnait des ombres perlées, quelquefois un peu dures." "En somme, cette innovation a obtenu un grand succès cette année." Anon., "Le Salon à la lumière électrique," *Le Monde illustré*, July 5, 1879, 7. Warmest thanks to Justine De Young for putting this precious newspaper in my hands.

43. Another subtle account of the lighting in the 1880 Salon stresses its vexing tendency to shut off suddenly: Nélius, "Sur quelques imperfections de la bougie Jablochkoff," *La Lumière électrique* 2, no. 12 (June 15, 1880): 233–34.

44. "L'éclairage est, comme d'habitude, dément. Ce n'est ni le jour, ni le crépuscule, ni le Jablochkoff." J.-K. Huysmans, *L'Art moderne* (Paris, 1883), 67, quoted in Pickvance, "1879, the Fourth Exhibition: Contemporary Popularity and Posthumous Neglect," 263. The balance of the quotation, cut short by Pickvance, has come to light, clarifying his strictly metaphorical use of "Jablochkoff." "Ce n'est ni le jour, ni le crépuscule, ni le Jablochkoff, c'est quelque chose de vineux et de sale, une lumière passant sous des vitres brouillées et remplies de poussière. M. Bonnat a fait son petit trompe l'oeil en enlevant dans cet éclairage des chairs violacées sur du noir. " "It's neither day nor twilight nor Jablochkoff, it's something winey and dirty, a light passing through clouded windows and full of dust. M. Bonnat made his little *trompe l'oeil* by taking off (removing) the purplish fleshtones in this lighting against the black." J.-K. Huysmans, "Le Salon de 1879," in *Écrits sur l'art, 1867–1905*, ed. Patrice Locmant (Paris: Éditions Bartillat, 2006), 146. For the record, Huysmans took no notice whatsoever of the electric light at the 1879 and 1880 Salons.

45. "Les tableaux accrochés à la muraille sont parfaitement insignifiants. Ils font *tapisserie*. On les regardera plus tard. L'important pour le moment est de se montrer." Jules Claretie, *La Vie à Paris 1880* (Paris: Victor Havard, Éditeur, c. 1880), 97.

46. "Ces exhibitions n'auront, avant quelques années, plus rien à voir avec l'art." Claretie, *La Vie à Paris 1880*, 97.

47. "On annonce déjà, en grosses lettres rouges, sur les affiches officielles, que le Salon sera, tous les soirs, éclairée à la lumière électrique: c'est un premier pas vers une transformation douloureusement inévitable. Le soir où, pour charmer les loisirs des visiteurs, on ajoutera la séduction de la musique aux charmes de la peinture—de la peinture passée au gaz Jablochkoff—l'exposition des œuvres des artistes vivants remplacera agréablement les Folies-Bergère. . . . Ce serait, pour les peintres de la modernité, comme on dit, d'une originalité charmante." Claretie, *La Vie à Paris 1880,* 97–98.

48. Alain Beltran and Patrice A. Carré, "Une fin de siècle électrique," *Les Cahiers de médiologie* 10 (2000), special issue: *Lux, des lumières aux lumières*: 92.

49. "C'est adorable! Avec cette trainée de lumière sur les allées et les marronniers, on dirait la sortie des Italiens donnant sur le printemps!" Claretie, *La Vie à Paris 1880*, 98. Or is he saying even more pointedly: exiting les Italiens gives onto Au Printemps, the grand department store close at hand on boulevard Haussmann at the rue du Havre?

50. Frank Géraldy, "Les Éclairages électriques à Paris, Système Jablochkoff: Magasins du Louvre," *La Lumière électrique, Journal universel d'électricité* 2, no. 12 (June 15, 1880): 230.

51. "Ce n'est pas sans peine. Le jury s'y était opposé, dans un intérêt d'art; il pensait que les toiles étaient mal vues et mal appréciées avec cet éclairage. Ces scrupules se justifieraient, ce me semble, s'il s'agissait de ne voir le Salon qu'à la lumière électrique en supprimant la lumière du soleil; mais il s'agit de fournir un supplément: les peintres ne perdent rien et gagnent quelques heures. En tout cas, le grand juge est encore le public; il est venu en grand nombre, l'année dernière, il viendra encore cette année; il trouve donc à son gout et à sa convenance cette façon de visiter le Salon; les peintres ont-ils intérêt à l'en priver? Il est bien clair que non." Frank Géraldy, "Les Éclairages électriques à Paris, Système Ja-

blochkoff: Exposition de Peinture," *La Lumière Électrique, Journal universel d'électricité* 2, no. 12 (June 15, 1880): 227.

52. "Je sais les défauts que l'on reproche à l'éclairage électrique tel qu'il existe: scintillations, changements de couleur, reflets difficiles à éviter, rayons particuliers qui modifient les colorations. Je conviens tout le premier qu'il y a quelque vérité, mais en médiocre proportion." Géraldy, "Les Éclairages électriques à Paris, Système Jablochkoff: Exposition de Peinture," 227. He went on to discuss the effects in more detail.
53. "La scintillation et les variations de nuances sont, à mon avis, surtout sensibles à la sculpture où la blancheur des plâtres ou des marbres la dénonce impitoyablement, mais dans la masse de lumière, cela n'offre pas de désagrément trop sensible." Géraldy, "Les Éclairages électriques à Paris, Système Jablochkoff: Exposition de Peinture," 227.
54. The painstaking research was done by my RA at the National Gallery of Art in 2013–14, the incomparable Joseph Hammond.
55. No. 6219 was *Portrait de M. H. Carnot, sénateur, ancien ministre de l'instruction publique; buste, plâtre teinté*. *Livret*, 577. *Explication des ouvrages de peinture, sculpture, architecture, gravure et lithographie des artistes vivants*, Paris, 1880, known familiarly as the *Salon Livret*. *Documenting the Salon: Paris Salon Catalogs 1673–1945*, ed. John Hagood (Washington, DC: National Gallery of Art Library, 2016). Is this an inside joke about the transformation of a former French minister into an oriental man? Caricature lives and dies by its contingencies, often lost to the later researcher. To his left, the pipe-smoking bust of a man in *casquette*, no. 6691, corresponds well with the no. 6691 on display, *Travailleur, buste, plâtre*, by Georges-Gabriel Tattegrain. *Livret*, 625. There is no way of identifying the nude winged female with upraised left arm just behind him; ditto for the nude young woman with upraised arms, the swerving herm behind her, and the fleeing shadowy nude behind no. 6374. Nor can a precise identification be made for the Joan of Arc *qua* exit sign (SORTIE) just below the "D" in Draner.
56. *Portrait décoratif* (marble bust) by Ernest-Eugène Hiolle, no. 6414. *Livret*, 596.
57. *Le Professeur Brunnet, buste, plâtre*, no. 6110, by Joseph-Charles de Blezer. *Livret*, 566. No. 6630 was sculpted by Désiré Ringel and bore an extravagant title: *La marche de Rakoczy; le démon de Rakoczy accourt à Paris inspirer les tziganes; statue, plâtre*. An excerpt accompanied the *Livret* entry, taken from a poem by N. Lenau, which narrated a very intense musical performance by a fiery man. *Livret*, 619. The medieval or Elizabethan figure on the plinth in tights with a sword corresponds approximately to (and perhaps takes off from) the sculpture displayed under the number 6374 (not 6574): *Henri V, roi d'Angleterre; statue, plâtre*, by Lord Ronald Gower ("né à Londres"). *Livret*, 592.
58. "Ce pauvre Dante, il ne s'est pas méfié des effets du cidre." The drawn figure corresponds to a bronze on exhibit, *Dante Alighieri* (no. 6061) by Jean-Paul Aubé. *Livret*, 561.
59. "Onze heures sonnent! Enfin je vais donc pouvoir aller me coucher." *Mercure; statue, marbre*, by Jean-André Delorme (no. 6255). *Livret*, 581. Perhaps the evening hours extended to 11:00 p.m.!
60. The principal fold-out caricature in issues of *La Caricature* of the 1830s included a similarly dressed, jester-like guide labeled "La Caricature." Robida was perpetuating a venerable practice, though his jester is suave and handsome compared with the slightly grotesque guides of Philipon's era.
61. "Il fallait y aller carrément, comme cela, tout le monde eut été ébloui et l'on n'aurait pas pu reprocher à l'administration son absence de lumières." My gratitude for help translating this tricky text goes to Michel Hochmann, Sarah Maza, and Susan Wager. It takes a village.
62. Mainardi, *End of the Salon*, 129.
63. A photograph of the lampless interior of the nave packed with sculpture during the Salon of 1881 is available at http://www.culture.gouv.fr.
64. For a vivid account of the drawbacks of arc lighting in these early days, see Ernest Freeberg, *The Age of Edison: Electric Light and the Invention of Modern America* (New York: Penguin, 2013), 18–19. For a detailed discussion of the system that generated the light in the 1880 Salon (including images of the generators), see Géraldy, "Les Éclairages électriques à Paris, Système Jablochkoff: Exposition de Peinture," 227–29.

65. Mainardi, *End of the Salon*, 72. It is not clear from Mainardi's statement if the increase came in 1879 or in 1880.
66. Mainardi, *End of the Salon*, 72.
67. "Mille excuses, madame, je me croyais aux rondes bosses et je vous prenais pour un marbre."
68. "Une surface arrondie . . . Ce doit être la colonne que je cherche." Draner, "Effets de Brouillard," *La Caricature* 56 (January 22, 1881): 3.
69. "Le moyen de s'éclairer sur les finesses du modèle."
70. "Et moi, je vous répète que mon mari a toujours eu la vue faible et que l'éclairage électrique l'aveugle." Draner, *La Caricature*, May 29, 1880, 3, left column.
71. "—Mais, monsieur, madame abîme les sculptures! —C'est de votre faute . . . pourquoi votre lumière électrique indispose-t-elle mon Émérance." Draner, *La Caricature*, May 29, 1880, 3, right column.
72. "Faible aperçu de ce que l'on verrait si, par suite d'un accident à la machine, on n'y voyait plus." Draner, *La Caricature*, May 29, 1880, 3, right column. Other references to the quality of the light are restricted to painters joking that if they move their artwork it might be better illuminated and receive more attention. And a cheeky exchange between an older man and a very young woman speculating about certain objects of view, people included, best left unilluminated.
73. Unsigned essay, 347. For another comprehensive description of the diversity of lamps and lights on display, see Glaser, "The Paris Electrical Exhibition," 430–35.
74. "Des lunettes bleues tous les deux?" The context of the joke is the couple's preparation for a bright piece of comic theater, "L'Éclair" (The Flash).
75. Held in the Palais de l'Industrie from August 11 to November 30, 1881.
76. Alain Beltran and Patrice A. Carré, "La 'Révolution' de 1881," in *La Fée et la servante: La société française face à l'électricité, XIXe–XXe siècle* (Paris: Belin, 1991).
77. "[U]n des événements scientifiques les plus importants du dix-neuvième siècle." Quoted in François Caron and Fabienne Cardot, eds., *Histoire générale de l'électricité en France*, vol. 1, *Espoirs et Conquêtes, 1881–1918* (Paris: Fayard, 1991), 17.
78. "Près de 900 000 visiteurs se sont pressés au Palais de l'Industrie pour découvrir des techniques qui n'allaient pas tarder à engendrer et à caractériser *une nouvelle civilization matérielle*." Beltran and Carré, "La 'Révolution' de 1881," 64. "Avec l'éclairage électrique et sa lente diffusion c'est tout un système d'attitudes psychologiques qui se modifie." Alain Beltran and Patrice A. Carré, "Histoire de voir: Éclairage électrique et vie privée au tournant du siècle," in *Lumières, je pense à vous* (Paris: Centre Georges Pompidou, Centre de Création Industrielle, 1985), 39.
79. "Le téléphone est parmi les innovations électriques présentées au public l'une de celles qui semble le fasciner. Mais l'innovation majeure, celle qui retient l'attention de toutes et de tous, est sans nul doute l'éclairage électrique." Beltran and Carré, "La 'Révolution' de 1881," 67.
80. Ernest Freeberg, *The Age of Edison: Electric Light and the Invention of Modern America* (New York: Penguin, 2013), 40–41.
81. I documented red THOMAS A. EDISON, MAN OF THE MILLENNIUM banners bearing the inventor's portrait that encircled the main square in his birthplace, Milan, Ohio, in 2004.
82. Bernard S. Finn et al., *Edison: Lighting a Revolution: The Beginning of Electric Power* (Washington, DC: National Museum of History and Technology, Smithsonian Institution, 1979), 3.
83. Beltran and Carré, "La 'Révolution' de 1881," 64.
84. K. G. Beauchamp, *Exhibiting Electricity* (London: Institution of Electrical Engineers, 1997), 137.
85. William Henry Preece, "Electric Lighting at the Paris Exhibition," *Journal of the Society of Arts*, quoted in du Moncel and Preece, *Incandescent Electric Lights*, 127.
86. "Que pourrons-nous dire encore? Que le soir l'Exposition resplendit de feux que jamais aucun homme n'a pu voir briller jusqu'à ce jour. . . . Jamais on n'aura été appelé à admirer un si bel éclairage que produisent de puissantes machines." Gaston Tissandier, "L'Exposition d'électricité," *L'Illustration*, August 20, 1881, in Patrice Carré, "Documents: L'Exposition internationale d'électricité de 1881 à travers quelques textes," *Bulletin d'histoire de l'électricité* 2 (1983): 69–70.

87. "Au Palais, l'électricité avait permis de produire, avec 1,800 chevaux, une lumière équivalente à plus de 55,000 becs de gaz, soit environ 6,000 becs de gaz de plus qu'il n'en existe dans toutes les rues et les promenades de Paris. . . . Nous sommes à l'aurore d'une époque nouvelle." Henri de Parville, *L'Électricité et ses applications: Exposition de Paris* (Paris: G. Masson, 1882), 530.
88. Preece, "Electric Lighting at the Paris Exhbition," quoted in du Moncel and Preece, *Incandescent Electric Lights*,125.
89. "C'était le soir qu'il était préférable d'entrer pour la première fois à l'Exposition. Si l'on n'avait su d'avance ou se trouvait le palais, on l'aurait bien vite deviné à la lueur qu'il projetait au loin sur la ville. On aurait dit que le feu était aux Champs-Élysées ou qu'une magnifique aurore boréale resplendissait à l'occident. La lumière s'échappait par les plafonds vitres et allait éclairer les nuages. Deux puissants foyers électriques munis de réflecteurs et installés au sommet du portail de la porte d'honneur envoyaient leurs sillons étincelants sur l'Arc-de-Triomphe et la place de la Concorde." Parville, *L'Électricité et ses applications: Exposition de Paris*, 4.
90. "Des trainées de cette lumière blanche, qui est caractéristique à l'électricité, s'échappent des fenêtres du Palais de l'Industrie, plongent au loin, et semblent vouloir fouiller dans les ténèbres des sombres avenues des Champs-Elysées." Extract from H. Valette, "Exposition Internationale d'Électricité," opening of evening hours, *Cosmos—Les Mondes*, May–August, 1881, in Carré, "Documents: L'Exposition internationale d'électricité de 1881 à travers quelques textes," 75.
91. Beauchamp, *Exhibiting Electricity*, 160. Electricity had played a growing role in the immediately prior sequence of Universal Expositions, especially those in 1867 (Paris), 1873 (Vienna), and 1878 (Paris). Beltran and Carré, "La 'Révolution' de 1881," 58 and following; and Beauchamp, *Exhibiting Electricity*.
92. Beltran and Carré, "La 'Révolution' de 1881," 65.
93. Beltran and Carré, "La 'Révolution' de 1881," 58.
94. Beltran and Carré, "La 'Révolution' de 1881," 63.
95. "manifestation aussi éclatante que symbolique de la formation d'un milieu d'électriciens composite, mais prêt à mobiliser son énergie pour porter la nouvelle technologie aux premier rangs des usages industriels et domestiques, à conquérir le monde en un mot." Caron and Cardot, eds., *Histoire générale de l'électricité en France*, vol. 1, *Espoirs et conquêtes, 1881–1918*, 9.
96. "Cette Exposition d'électricité, presque née au milieu de l'indifférence générale, aura été féconde a plus d'un titre ; elle aura donné bien plus qu'elle n'avait promis; elle a surpris par son succès éclatant, même ceux qui avaient été ses plus ardents partisans des la première heure. Elle laissera peut-être plus de traces durables que son ainée, la grande Exposition universelle de 1878. Elle a fait sauter aux yeux des moins clairvoyants le rôle désormais assuré des Expositions spéciales." Parville, *L'Électricité et ses applications: Exposition de Paris*, 529.
97. It is hard even for historians who specialize in the rise of electric light in the nineteenth century to remember that it was far from evident, let alone inevitable, that Edison's bulb would outshine the competition and become the international standard. A special commission at the exhibition measured the efficiency of the lamps from the various manufacturers. Four were ranked and in this order (top to bottom): Edison, Swan, Lane-Fox, Maxim. Beauchamp, *Exhibiting Electricity*, 162. The original report itself, "Rapport de la Sous-Commission Spéciale des Lampes a Incandescence," is reproduced in Emile Allard et al., *Expériences faites à l'Exposition d'Électricité* (Paris: Gauthier-Vallars, 1883), 90–108. The other factor to remark, well studied by Robert Fox, is that while Edison himself did not set foot in Paris until 1889 (when he was treated as a celebrity and awarded the Légion d'Honneur), he had an elaborate network of agents stationed in Paris before the exhibition, currying favor with the *électricien* crowd and winning (buying) many converts to the Edison cause in the process. Robert Fox, "Edison et la presse française lors de l'Exposition International d'Électricité de 1881," in *Un siècle d'électricité dans le monde* (Paris: PUF, 1987), 223–35.
98. "From the beginning, Edison thought of his electric light as not simply a lamp, but a system. . . . While many of the inventions of the nineteenth century—from railroads to telephones—required system-building, none of them came to exemplify the systemic nature of modern technology more than

electric light and power." Bernard S. Finn, *Edison: Lighting a Revolution* (Washington, DC: Smithsonian Institution, National Museum of History and Technology, 1992), 12.

99. Beltran and Carré, "Histoire de voir," 30. Paris was extremely slow in adopting electric light both outdoors and in. It was not until 1914 that La Compagnie Parisienne de Distribution d'Électricité was created. As late as 1936, Parisians consumed half as much electricity as New Yorkers did. See Beltran and Carré, "Histoire de voir," 32 and following.

100. "L'apparition des éclairages électriques signifie—pour de nombreux témoins—un changement profond dans les modes de perception. L'électricité modifie, en effet les façons de voir, les façons de regarder. Très vite on se rend compte que dorénavant le regard porté sur les choses ne pourra plus être le même. Nouveau régime donc, qui touche tout d'abord l'organe de la vue." Beltran and Carré, "Histoire de voir," 34.

101. Christoph Asendorf, *Batteries of Life: On the History of Things and Their Perception in Modernity* (Berkeley: University of California Press, 1984, 1993), 155.

102. Glaser, "The Paris Electrical Exhibition," 102.

103. "Lorsqu'on avait franchi le grand portail, le spectacle devenait magique. Il convenait de monter immédiatement au premier étage; le regard embrassait dans son ensemble la grande nef et ses innombrables lumières; l'illumination était incomparable de splendeur. Ça et la, partout, sans ordre ni symétrie, au milieu des machines, des appareils en mouvement, brillaient comme des lanternes vénitiennes éclatantes les lampes électriques enfermées dans leurs globes opalins; la nef était comme piquée de gros diamants blancs, qui marquaient les emplacements des nations et limitaient la place des exposants: ici, des foyers a la lumière intense; là, des lampes à rayonnement doux et chaud; partout des lustres, des candélabres, des lampadaires, des bougies envoyant leurs clartés disparates sur les oriflammes, les bannières, les drapeaux multicolores. Du premier étage on eut dit une immense mosaïque d'or, d'argent et de pierres fines. . . . Le palais des Champs-Élysées était bien devenu le véritable palais de la Lumière." Parville, *L'Électricité et ses applications: Exposition de Paris,* 5. On the varying colors of electric light, see also E. Hospitalier, trans. Julius Maier, *The Modern Applications of Electricity* (London: Kegan Paul, Trench & Co., 1882), 261.

104. "Je vous l'avais bien dit que, malgré vos opinions réactionnaires, vous deviendrez rouge un jour!"

105. It may be appropriate to align the slight uptick in bravado here and more so in the November 19 issue (discussed below) with the new law on the liberty of the press that went into effect on July 29, 1881, which lifted censorship. See Richard Thomson, *Art of the Actual: Naturalism and Style in Early Third Republic France, 1880–1900* (New Haven, CT: Yale University Press, 2012), 144 and following.

106. "Promenades conférences intimes sur l'attraction des corps et les secousses des fluides."

107. "[D]e l'intensité lumineuse, l'électricité en donne tant qu'on en veut. Ce que nos yeux doivent voir, ce n'est pas le point lumineux, mais l'objet éclairé; le soleil n'a pas été créé pour qu'on regarde son disque brillant, mais bien les objets qui nous entourent et qu'il est chargé d'illuminer." Extract from Valette, "Exposition Internationale d'Électricité," in Carré, "Documents: L'exposition internationale d'électricité de 1881 à travers quelques textes," 77.

108. "Mlle. Titi, des Folies-Boulevardières, profite d'un rayon pour juger de ses effets dramatiques dans la prochaine Revue."

109. "Le maladroit! Il n'a pas vu mon mari prendre ma place, et il continue à parler de son amour!!!"

110. "Grand Dieu! Ma femme avec son cousin? . . . Ah! j'y vois clair maintenant . . ."

111. "Avec la lampe soleil! Ce serait malheureux." According to William Henry Preece, the *lampe soleil* held an intermediate position between arc and incandescent lamps. Preece, "Electric Lighting at the Paris Exhbition," quoted in du Moncel and Preece, *Incandescent Electric Lights,*139.

112. Haussmann, *Mémoires*, vol. 3, 163, quoted in Anthony Vidler, *The Scenes of the Street and Other Essays* (New York: Monacelli Press, 2011), 126n198.

113. "AVEUGLE. Ayez pitié d'un pauvre savant qui a étudié les diverses lampes électriques." This character debuted in one of Draner's responses to the electric Salon of 1880. A gallery guard in that case wore

the sign "AVEUGLE." The caption reads "Ayez pitié des pauvres gardiens du Salon, victimes des expositions à la lumière électrique." Draner, *La Caricature*, May 29, 1880, 2. (See fig. 3.19.)

114. "Honneur au courage malheureux!" Thanks to Louise Bourdua for her translation guidance. Draner's follow-up work contained numerous visits to the same humor nexus. The double-page spread, "À l'Exposition d'électricité," in *La Caricature*, November 19, 1881, contained nine small monochrome vignettes. Two of the cartoons have plots determined by the exhibition within the genre of romantic comedy that entangles unsuspecting fairgoers in the net of sex and electricity. As in his prior cartoons, men involuntarily touch the bodies of respectable female strangers because the men are dazed by the light or unhinged by electric current. On the left page (below), electricity propels a man's whole frame into the upper body of a well-dressed woman with such ferocity that he is practically horizontal. His top hat flies off, and his hyperbolic facial expression registers a mixture of surprise and delight. His outstretched hands are concealed, but it is clear that his arms encircle the very surprised woman's waist. Her mouth is a perfect O. His unstable position should be somatic prelude to falling onto the woman and knocking her down. Never mind the volatility of his posture, there is time, and he has the presence of mind, to utter an apology. "Pardon me, Mademoiselle! It's the effect of an electrical current . . . are you perhaps magnetized?" Electricity is the fundamental culprit, but its force may make the woman culpable, too; never the impertinent and ridiculous man. On the facing page, lower right, another electrically instigated, indecorous encounter unfolds. The dialogue provides an alibi for the man's groping: "Sir, what an impropriety! [what unseemliness!] —A thousand excuses, Madame, in exiting from there, one can only grope one's way along." The bourgeois man, rendered blind by the piercing lamps in the exhibition, is unable to find his way once outside the gallery until his hands land on the soft relief of a stunned woman's chest.

115. In the series entitled "À l'Exposition d'Électricité, (suite) et (fin)," two particularly spicy electric light-driven sexual encounters figure among the cluster of vignettes. The work containing the cat metaphor, which I discuss in detail in the text, is the most interesting erotically themed cartoon. In the second, which appeared in the lower left on page 382, the need for shelter from the bright lights drives a man's marauding advance upon a female friend with an obligingly wide-brimmed hat. "I beg of you, my friend [*cousine*], this light is blinding me, shelter me for a little while." "—Je vous en prie, cousine, cette lumière m'aveugle, abritez-moi quelques instants."

116. "Maintenant qu'on n'y voit plus clair ce serait le moment d'éclairer mon chat." Warm thanks to Susan Wager for helping me work this out.

117. One of the drawings from the May 29, 1880, two-page spread amplifies the cat metaphor somewhat but not entirely. A naked, laughing sculpted man on a plinth, adjoined by an attentive (real) young woman exhibition visitor, switches on a manic cat light. "Pourquoi ne pas utiliser les propriétés électriques du chat pour éclairer ce groupe?" The speaker of the caption is unclear. Parsing the electric properties of cats in the register of metaphor in this era requires further study.

118. Pierre Larousse, *Grand dictionnaire universel du XIXe siècle; français, historique, géographique, biographique, mythologique, bibliographique, littéraire, artistique, scientifique, etc.* (Paris: Administration du Grand dictionnaire universel, 1865–90), vol. 3, 1064.

119. Larousse, *Grand dictionnaire universel du XIXe siècle,* vol. 3, 1064.

120. "A côté de la merveille enfantée par Edison, le téléphone, le phonographe et la lampe sont de simples bagatelles."

121. "Avec cet ingénieux appareil, l'homme le moins susceptible d'enflammer les cœurs n'a qu'à presser un timbre pour être adoré tout de suite."

122. "Les expériences du palais de l'Industrie ont parfaitement réussi; on a vu des sénateurs, des membres de l'Institut, des hommes politiques décatis faire naître a volonté des passions soudaines parmi les innocentes visiteuses de l'exposition."

123. Goode, "The Public and the Limits of Persuasion in the Age of Caricature," 133.

124. Terdiman, *Discourse/Counter-Discourse,* 160 and following.

125. "Tout le monde va vouloir s'abonner à la compagnie parisienne des sentimentalifères électriques. Enfin, on va donc être aime pour soi-même! quel changement dans les habitudes, sociales quel bouleversement! quelle révolution!!!"

126. "La véritable et grande nouveauté, la révolution, on peut le dire, qui s'est révélée a l'exposition internationale d'électricité, c'est l'éclairage des appartements, c'est-à-dire la production d'un luminaire de petit volume, n'ayant plus la puissance de l'arc voltaïque, mais se trouvant réduit aux proportions de l'éclairage domestique." Louis Figuier, *L'Art de l'éclairage* (Paris: Jouvet et Cie., 1882), 244.

127. "On obtient ainsi une illumination de peu de puissance, mais qui est précisément ce que l'on recherchait pour l'éclairage de l'intérieur des maisons." Figuier, *L'Art de l'éclairage* 245.

128. Preece, "Electric Lighting at the Paris Exhbition," quoted in du Moncel and Preece, *Incandescent Electric Lights,* 141. See also Georges Fournier (Chimiste-Électricien), *La Lumière Électrique dans les appartements* (Paris: Bernard Tignol, 1885).

129. Graeme Gooday, *Domesticating Electricity: Technology, Uncertainty and Gender, 1880–1914* (London: Pickering & Chatto, 2008), 159 and following.

130. Gooday, *Domesticating Electricity,* 178.

131. Gooday, *Domesticating Electricity,* 179.

132. Wolfgang Schivelbusch, *Disenchanted Night: The Industrialization of Light in the Nineteenth Century*, trans. Angela Davies (Berkeley: University of California Press, 1988), 180–81.

133. "Mieux une maison est éclairée, plus elle est insalubre pour les personnes qui l'habitent." Fournier (Chimiste-Électricien), *La Lumière électrique dans les appartements,* 8.

134. "Cette lumière existe-t-elle? Oui, elle existe. Elle existe dans l'emploi des lampes électriques à incandescence." Fournier (Chimiste-Électricien), *La Lumière électrique dans les appartements,* 13.

135. "Le XIXe siècle a développé le transport rapide des personnes et des idées par la vapeur, le télégraphe et le téléphone: dans la lumière électrique, il aura inauguré la lumière de l'avenir." Fournier (Chimiste-Électricien), *La Lumière électrique dans les appartements,* 34.

136. As early as the later 1880s, Vallotton showed an interest in capturing intense and eccentric light effects, as exemplified by *Still Life with Pitcher and Plate* (1887; Baltimore Museum of Art). His bravado painting of fragmented passages of bright yellow light reflected against the shiny shoulders of a brass vessel is a tour de force. A painted illusion of similarly refractory effects in a still life will not be found elsewhere.

137. Claude Monet's *Interior, After Dinner* (1868–69; National Gallery of Art) is a useful counterpoint. It is an exemplary naturalist nighttime interior in its quiet assessment of the subtleties of diffuse lamplight (from a ceiling oil lamp) plus firelight in the context of a quiet social gathering. Ashley Miller Dunn (Northwestern University seminar paper, December 2010) learned that in his *Livre de raison*, Vallotton emphasized his interest in lamplight by referring to the painting as "Intérieur. vestibule. effet de lampe." Marina Ducrey, *Félix Vallotton, 1865–1921: L'Œuvre peint* (Lausanne: Fondation Félix Vallotton, 2005), vol. 2, 304–5.

138. "Non, le problème de l'éclairage des peintures par la lumière électrique n'est pas encore résolu." Ernest Chesneau, "L'Art et l'électricité," *L'Art: Revue hebdomadaire illustrée* 7, no. 4 (1881): 173.

139. "Elle décompose sensiblement les rayons jaunes et dénature d'une façon absolue l'aspect des tableaux, en particulier des paysages où domine la couleur verte." Chesneau, "L'Art et l'électricité," 173.

140. "L'éclairage par incandescence est plus particulièrement représenté par les systèmes Edison et Swan. Ils offrent l'un et l'autre de très grands avantages." Chesneau, "L'Art et l'électricité," 174.

Chapter Four

Epigraph: Michel Melot, *The Impressionist Print*, trans. Caroline Beamish (New Haven, CT: Yale University Press, 1996), 8.

1. Camille Pissarro's graphic work deserves the same consideration, but because he does not address *éclairage*, I do not discuss his prints. Michel Melot, "Camille Pissarro et l'invention de l'estampe

originale," in *Camille Pissarro: Impressions gravées* (Pontoise: Musée Tavet-Delacour and Paris: Somogy Éditions d'art, 2017), 9–31.

2. Jean-Paul Bouillon, "L'Estampe impressioniste: Chronique bibliographique," *Revue de l'art* 86 (1989): 66.
3. Rick Brettell, "Peintres-graveurs: Degas, Pissarro ou la genèse d'une tradition moderniste," in Nicole Minder, *Degas et Pissarro: Alchimie d'une rencontre* (Vevey, Switzerland: Cabinet Cantonal des Estampes, Musée Jenisch, 1998), 8–9.
4. Nicole Minder, *Degas et Pissarro*, 25.
5. Michel Melot, "Mary Cassatt, An Artist between Two Worlds," in *Mary Cassatt Impressions* (Giverny: Musée d'Art Américain and the Terra Foundation for American Art, 2005), 87. The Société des Aquafortistes, founded by Alfred Cadart in 1862, also attempted to move etching in that direction.
6. Peter Parshall, "Why Study Prints Now? Or, the World According to Bartsch," inaugural lecture, Association of Print Scholars, Graduate Center, City University of New York, September 25, 2015; published in *Art in Print* 6, no. 3 (September–October 2016), http://artinprint.org/article/why-study-prints-now-or-the-world-according-to-bartsch/
7. Parshall, "Why Study Prints Now?" n.p.
8. See Amanda T. Zehnder, "Forty Years of Artistic Exchange," and Marc Rosen and Susan Pinsky, "The Medium as Muse: Innovations and Intersections in Printmaking," both in Kimberly A. Jones, *Degas Cassatt* (Washington, DC: National Gallery of Art, 2014), 2–19, 142–44; 100–111, 145–46.
9. The most comprehensive discussion of the project remains Douglas Druick and Peter Zegers, "Degas and the Printed Image, 1856–1914: Part II, The Peintre-Graveur as Peintre-Entrepreneur, 1875–80," in Sue Welsh Reed and Barbara Stern Shapiro, eds., *Edgar Degas: The Painter as Printmaker* (Boston: Museum of Fine Arts, 1984), xxviii–lv, especially xxxix–li.
10. Michel Melot, *The Impressionist Print* (New Haven, CT: Yale University Press, 1996), 151; Melot, "Mary Cassatt, an Artist between Two Worlds," 88.
11. Melot, *The Impressionist Print*, 151 and 159.
12. Peter Parshall, "A Darker Side of Light: Prints, Privacy, and Possession," in *The Darker Side of Light: Arts of Privacy, 1850–1900* (Washington, DC: National Gallery of Art in association with Lund Humphries, 2009), 5.
13. Melot, *The Impressionist Print*, 158.
14. "[L]e nom qu'ils avaient choisi pour [la publication d'estampes d'avant-garde], *Le Jour et la Nuit*, résume admirablement cette concentration sur les valeurs 'atmosphériques,' mais de pure forme, du noir et blanc." Jean-Paul Bouillon, *Félix Bracquemond, 1833–1914* (Paris: Somogy Éditions d'Art, 2003), 19.
15. Barter's ideas are quoted in Zehnder, "Forty Years of Artistic Exchange," 8–9.
16. Druick and Zegers, "Degas and the Printed Image," xlvi.
17. Druick and Zegers, "Degas and the Printed Image," xlix. See the chart in Line Clausen Pedersen, ed., *Degas' Method* (Copenhagen: Ny Carlsberg Glyptotek, 2013), 264, which tracks his subject matter, and shows a big uptick in "Cabaret motifs" between about 1878 and 1880.
18. Clement Greenberg, "On the Role of Nature in Modernist Painting" (1949), in *Art and Culture: Critical Essays* (Boston: Beacon Press, 1961), 171.
19. Barbara Stern Shapiro, "Mary Cassatt's Color Prints and Contemporary French Printmaking," in Nancy Mowll Mathews and Barbara Stern Shapiro, eds., *Mary Cassatt: The Color Prints* (New York: Harry N. Abrams, in association with Williams College Museum of Art, 1989), 58. Brettell argues that the extreme class and ethnic diversity of the Impressionist group, combined with the ongoing industrialization of the image in this era, explains their antitraditional attitude toward the printed reproduction. Brettell, "Peintres-graveurs," 8–9.
20. George Shackelford, "Pas de deux: Mary Cassatt and Edgar Degas," in ed. Judith Barter, *Mary Cassatt: Modern Woman* (Chicago: Art Institute of Chicago; Harry N. Abrams, 1998), 109–43.
21. Quoted in Parshall, "A Darker Side of Light: Prints, Privacy, and Possession," 12.

22. Peter Parshall discusses another critical component of the practice, printing as separation from direct engagement, in his "Degas and the Closeted Image," in Pedersen, ed., *Degas' Method,* 153–70.
23. Emily Dickinson, P593.
24. For a related argument, see R. Stanley Johnson, *Mary Cassatt Retrospective Exhibition* (Chicago: R. S. Johnson Fine Art, 1997).
25. The most important interrogation of this relationship remains the edited volume, *Modernism and Modernity: The Vancouver Conference Papers*, ed. Benjamin H. D. Buchloh, Serge Guilbaut, and David Solkin (Halifax: Press of NSCAD, 1983, 2004), based upon the 1981 conference of the same name at the University of British Columbia.
26. "Les années 1879–80 constituent . . . une date charnière pour le développement des arts graphiques." Minder, *Degas et Pissarro*, 15.
27. Andreas Blühm and Louise Lippincott, *Light! The Industrial Age, 1750–1900: Art and Science, Technology and Society* (Pittsburgh: Carnegie Museum of Art, 2000), 172.
28. My argument explicitly echoes and was shaped by Lyn Nead's claims about gas streetlight in Victorian London and Tim Clark's opinions about steam in modernist painting. Lynda Nead, "Gas and Light," in *Victorian Babylon: People, Streets and Images in Nineteenth-Century London* (New Haven, CT: Yale University Press, 2000); T. J. Clark, "Modernism, Postmodernism and Steam," *October* 100 (Spring 2002): 154–74.
29. See Sasha Archibald, "Blinded by the Light," *Cabinet* (Spring 2006), issue 21, "Electricity."
30. Henri de Parville, *L'Électricité et ses applications: Exposition de Paris* (Paris: G. Masson, 1882), 354: "Nous sommes habitués à nous représenter la lumière électrique sous forme de foyers éblouissants, scintillants, durs à l'oeil, bruyants, changeant sans cesse d'intensité, aux tons variables et blafards. Ici, au contraire, on a devant soi une lumière qui a été en quelque sorte civilisée, accommodée à nos habitudes, mise à notre portée; chaque bec éclair comme du gaz, main comme un gaz qu'il eût fallu inventer, un gaz donnant une lumière d'une fixité parfaite, gaie et brillante sans gêner la rétine."
31. This passionate immersion should also be seen in the context of the "etching revival," under way by the 1850s and 1860s, and lasting until the Great War. See Elizabeth Helsinger et al., *The "Writing" of Modern Life: The Etching Revival in France, Britain and the U.S., 1850–1940* (Chicago: Smart Museum of Art, University of Chicago, 2008).
32. "Puis, quelle étude des effets de la lumière!" J.-K. Huysmans, "L'Exposition des Indépendants en 1880," in *L'Art moderne* (Paris: G. Charpentier, 1883), 116.
33. "Le temps d'apprentissage est terminé et les voilà qui paraissent, en public, sur les pointes, dans des coups de gaz, dans des jets de lumière électrique." Huysmans, "L'Exposition des Indépendants en 1880," 115.
34. Richard Kendall, *Degas by Himself: Drawings, Prints, Paintings, Writings* (Edison, NJ: Chartwell Books, 1994), 112.
35. Walter Benjamin, *The Arcades Project*, trans. Rolf Tiedemann (Cambridge, MA: Belknap Press of Harvard University Press, 1999), 570. The story was first published in *Gil Blas*, October 1882.
36. Kendall, *Degas by Himself*, 116.
37. Minder, *Degas et Pissarro*, 17. "Les interventions directes sur la matrice peuvent se faire avec n'importe quel outil plus dur que le cuivre. Ainsi, l'anecdote veut que Degas, retenu chez son ami Rouart par le mauvais temps et ne disposant pas de ses outils de graveur, ait pris ce qui lui semblait le plus adéquat, en l'occurrence la tige de charbon d'une lampe à arc électrique, un conglomérat de charbon souvent appelé 'crayon électrique.'" Minder, *Degas et Pissarro*, 28n37: "Racontée par Alexis Rouart à Marcel Guérin, rapportée pour la première fois par Delteil (no 39), puis par Guérin 1997, 61."
38. Rosen and Pinsky, "The Medium as Muse," 108.
39. Melot, *The Impressionist Print*, 158.
40. Warm thanks to Jodi Hauptmann, senior curator, Department of Drawings and Prints, Museum of

Modern Art, New York, for inviting me to write about these two prints for her exhibition catalogue. Hollis Clayson, "Darkness and the Light of Lamps,"in Jodi Hauptmann, ed. *Degas: A Strange New Beauty* (New York: Museum of Modern Art, 2016), 84–87.

41. Eugenia Parry Janis, *Degas Monotypes* (Cambridge, MA: Fogg Art Museum and Garland Publishing, 1968), pl. 8; and Richard R. Brettell and Suzanne Folds McCullagh, *Degas in the Art Institute of Chicago* (New York: Harry N. Abrams, 1984), 86, noted the changed setting years ago.
42. Frank Géraldy, "L'électricité au théâtre," *La Lumière électrique* 14 (July 1880): 284. The first theater to convert entirely to electric light was in San Francisco in 1879; the Savoy in London became the first in Europe in 1881. Blühm and Lippincott, *Light!* 166. Paris theaters were obliged to electrify in 1887.
43. Blühm and Lippincott, *Light!* 166.
44. Barbara Stern Shapiro, "Mary Cassatt's Color Prints and Contemporary French Printmaking," in Nancy Mowll Mathews and Barbara Stern Shapiro, eds., *Mary Cassatt: The Color Prints* (New York: Harry N. Abrams, in association with Williams College Museum of Art, 1989), 58.
45. Melot, *The Impressionist Print*, 225. Cassatt's fluency was richly affirmed by Kimberly Jones's 2014 exhibition at the National Gallery of Art (Washington, DC), *Degas Cassatt*.
46. Melot, *The Impressionist Print*, 225.
47. When this print hung in a 2012 exhibit in the Block Museum at Northwestern University, I was in the gallery trying unsuccessfully to figure it out when Martha Tedeschi, former deputy director for art and research, Art Institute of Chicago, current director of the Harvard University Art Museums, and senior scholar of nineteenth-century prints, came through the door. We conferred, and even she could not understand the technologies that Cassatt used in this bewildering print.
48. Melot, "Mary Cassatt, An Artist between Two Worlds," 84.
49. Melot, "Mary Cassatt, An Artist between Two Worlds," 85.
50. Melot, "Mary Cassatt, An Artist between Two Worlds," 86.
51. Cassatt's exclusion from public amusements as a bourgeois lady is well explained by Griselda Pollock's classic essay, "Modernity and the Spaces of Femininity," in *Vision and Difference: Femininity, Feminism and the Histories of Art* (London: Routledge, 1988), 50–90.
52. Pastel and gouache with metallic paint on tan paper; no. 55 in the 1879 Impressionist exhibition. See Judith Barter, ed., *Mary Cassatt: Modern Woman*, no. 19, p. 246; and Ruth Berson, *The New Painting: Impressionism 1874–1886, Documentation*, vol. 2, *Exhibited Works* (Seattle: University of Washington Press, 1996), no. 55, pp. 109, 127.
53. Etching, soft-ground etching, and aquatint, state 4 (first and last states [fourth] were shown at the 1880 [fifth] Impressionist exhibition). Berson, *The New Painting*, vol. 2, *Exhibited Works*, nos. 25, 147, 161.
54. Soft-ground and aquatint, state 2. Marc Rosen and Susan Pinsky, *Mary Cassatt: Prints and Drawings from the Collection of Ambroise Vollard* (New York: Adelson Galleries, 2008), nos. 11, 40.
55. "Mlle Cassatt qui est Américaine, je crois, nous peint des Françaises; mais, dans ses habitations si parisiennes, elle met le bienveillant sourire du *at home*; elle dégage, à Paris, ce qu'aucun de nos peintres ne saurait exprimer, la joyeuse quiétude, la bonhomie tranquille d'un intérieur." J.-K. Huysmans, "L'Exposition des Indépendants en 1881," *L'Art modern* (Paris: G. Charpentier, 1883), 233–34.
56. "The Painter of Children," *Literary Digest*, July 10, 1926, 27, quoted in Pamela Ivinski, "'So Firm and Powerful a Hand': Mary Cassatt's Techniques and Questions of Gender." in Ingrid Pfeiffer and Max Hollein, eds., *Women Impressionists* (Frankfurt: Schirn Kunsthalle, 2008), 187.
57. Among the related prints are these: *Under the Lamp* (c. 1882), soft-ground etching and aquatint in black on cream wove paper, 19.2 × 21.8 cm (image/plate); 23.7 × 32.1 cm (sheet), Albert H. Wolf Memorial Collection, 1938.33, Art Institute of Chicago; Adelyn Breeskin, *Mary Cassatt: A Catalogue Raisonné of the Graphic Work,* 2nd rev. ed. (Washington, DC: Smithsonian Institution, 1979), 71 I/II; Breeskin, *The Graphic Work of Mary Cassatt: A Catalogue Raisonné* (New York: H. Bittner, 1948), 71 I/II. *Reading the Newspaper, No. 2* (c. 1883), soft-ground and aquatint on cream laid paper, 13.5 × 16.5 cm (image/

plate); 23.8 × 30.5 cm (sheet), Art Institute of Chicago, Kate S. Buckingham Fund, 1965.870; Breeskin, *Mary Cassatt*, 73 II/III; Breeskin, *Graphic Work of Mary Cassatt*, 73 II/III. Aspects of this analysis were published as Hollis Clayson, "Mary Cassatt's Lamp," in Hollis Clayson and André Dombrowski, *Is Paris Still the Capital of the Nineteenth Century? Essays on Art and Modernity, 1850–1900* (New York: Routledge, 2016), 257–83.

58. Credit goes to André Dombrowski for this idea.
59. Archibald, "Blinded by the Light," 97.
60. Several of my interlocutors suggest that the trapezoidal framed object at upper right may be a skylight (affording a glimpse of a sliver of the moon, if not a reflection in the glass?), suggesting the transfer of a library scenario to a *chambre de bonne* at the top of the house. Servants borrowed the lamp? It came to them as a hand-me-down?

61. James Allen Smith, *In the Public Eye: A History of Reading in Modern France, 1800–1940* (Princeton, NJ: Princeton University Press, 1991), 144. See also Kathryn Brown, *Women Readers in French Painting 1870–1890: A Space for the Imagination* (Surrey, UK: Ashgate, 2012).
62. And indeed, like a lighthouse, it is not a networked light. Gaslight, by contrast, was one of the signature networks of the city and of the era. Like the Jablochkoff candles that I claim Cassatt's metaphorical globes referenced, arc lights were also not networked in the full sense of the word. A small group of them—between nine and twelve—ran from a generator. Only with incandescent light technology was electric light finally "subdivided," to use the pseudoscientific terminology of the day. So, technically speaking, Edison did not invent electric light; he subdivided it. On the fragility of the networked city, see Peter Soppelsa, "Finding Fragility in Paris: The Politics of Infrastructure after Haussmann," *Proceedings of the WSFH* 37 (2009): 233–47.
63. François Loyer, *Paris Nineteenth Century: Architecture and Urbanism*, trans. Charles Lynn Clark (New York: Abbeville Press, 1988), 189; Frederick A. Sweet, *Miss Mary Cassatt, Impressionist from Pennsylvania* (Norman: University of Oklahoma Press, 1966), 206–7.
64. Graeme Gooday, *Domesticating Electricity: Technology, Uncertainty and Gender, 1880–1914* (London: Pickering & Chatto, 2008), 170.
65. Without the assistance of my "lamp friends" on both sides of the Atlantic, I would never have figured this out. My most fervent gratitude goes to Keith Letsche of West Chicago, Illinois, and Ara Kebapçioğlu of Lumière de l'Œil, Paris, both pillars of the international Rushlight Club. Monsieur Ara (as he is called) believes that Cassatt's lamp may well have been bought from Oppenheimer Frères, 19–21 rue de Cléry, the major specialist in Paris for such "oriental" items at the time. Personal communication, July 16, 2010.
66. I respectfully disagree with Barbara J. MacAdam, who writes: "The tall table lamp, which was situated in the artist's library, provides an important formal element in this and many images from the period. Cassatt was enthralled by the visual effects of artificial illumination, especially gaslight, which in the mid-nineteenth century transformed the appearance and utility of both public and private settings at nighttime. In the aquatint, the lamp's illumination creates flat, abstract patterns of light and dark that give the image particular graphic strength. The dramatic contrast between *the luminous gaslight globe* and the dark surroundings vividly recalls prints that Degas created during the same time that explore lighting effects in the cafés, streets, and theaters of Paris" Barbara J. MacAdam et al., *Marks of Distinction: Two Hundred Years of American Drawings and Watercolors from the Hood Museum of Art* (New York: Hudson Hills Press, 2005), 104 (emphasis added).
67. Personal communication, Keith Letsche, July 6, 2010.
68. Ross Murray, *Warne's Model Housekeeper: A Manual of Domestic Economy in All Its Brands* (London: Frederick Warne & Co., 1883), 39–40. See also Philippe Deitz, *Histoire des luminaires: Histoire des hommes* (Liège, Belgium: Éditions du Perron, 2009), 220–33; *Encyclopedia Britannica* (New York: Werner Company, 1893), vol. 14, 246. According to Keith Letsche (personal communication, October 14, 2011), they were not trimmed when lit. Along with filling them up and all other lamp management tasks, trimming was usually done during daylight and in the kitchen or other servant areas.

69. Murray, *Warne's Model Housekeeper*, 39–40.
70. Quoted in Émilie Ruffin-Maisonneuve, ed., *Sous la Lampe: Peintures de 1830 à 1930* (Saint-Maur, France: Musée de Saint-Maur, Villa Médicis, 2010), 14–15.
71. Simone Delattre, *Les Douze heures noires: La nuit à Paris au XIXe siècle* (Paris: Albin Michel, 2003), 117.
72. Ruffin-Maisonneuve, ed., *Sous La Lampe*, cat. 6, p. 21. See also, for example, William Henry Lippincott's painting *Infantry in Arms* (1887; Philadelphia, Pennsylvania Academy of the Fine Arts, 1922.10), whose elegant interior includes two moderator lamps on the mantelpiece.
73. On the silhouetted figures in the wood engraving, a conspicuous feature that contrasts with Cassatt's handling of the figures in her etchings, see Nancy Forgione, "'The Shadow Only': Shadow and Silhouette in Late Nineteenth-Century Paris," *Art Bulletin* 81, no. 3 (September 1999): 490–512. The stark circularity and brightness of Cassatt's globes similarly anticipate James Tissot's renderings of the arc lights in the Hippodrome in both painted and etched form: *Les Femmes de Paris: Ces dames des chars*, oil on canvas (c. 1883–85; Providence, RISD Museum), and etching and drypoint (1885; private collection).
74. On Cassatt's social identity and cultural nationality, see Hollis Clayson, "Voluntary Exile and Cosmopolitanism in the Transatlantic Arts Community, 1870–1914," in Christian Führmeister, Hubertus Kohle, and Veerle Thielemans, eds., *American Artists in Munich: Artistic Migration and Cultural Exchange Processes* (Berlin: Deutscher Kunstverlag, 2009), 15–26, especially 20–24.
75. I owe this definition of "engagement" to Catherine Roach, "The Turner Inheritance," *Art History* 34, no. 3 (June 2011): 595.

Chapter Five

1. Henry James, "John S. Sargent," in *Picture and Text* (New York: Harper & Brothers, 1893), quoted in Marc Simpson, *The Rockefeller Collection of American Art* (San Francisco: Fine Arts Museums of San Francisco, 1994), 219.
2. This chapter draws in part upon two of my published essays: Hollis Clayson, "Outsiders: American Painters and Cosmopolitanism in the City of Light, 1871–1914," in Frédéric Monneyron and Martine Xiberras, eds., *La France dans le regard des États-Unis/France as Seen by the United States* (Perpignan: Presses Universitaires de Perpignan, 2006), 57–71; and Hollis Clayson, "Voluntary Exile and Cosmopolitanism in the Transatlantic Arts Community, 1870–1914," in Christian Fuhrmeister, Hubertus Kohle, and Veerle Thielemans, eds., *American Artists in Munich: Artistic Migration and Cultural Exchange Processes* (Berlin: Deutscher Kunstverlag, 2009), 15–26.
3. And not just Curran, Hassam, and Prendergast. There are other relevant pictures, not discussed here. Two are in the collection of the Terra Foundation of American Art, Daniel J. Terra Collection: Theodore Earl Butler, *Place de Rome at Night* (1905), oil on canvas, image: 59.7 × 73.0 cm (23½ × 28¾ in.), frame: 78.4 × 91.4 cm (30⅞ × 36 in.), 1994.16; and Edward Willis Redfield, *France* (1898–99), oil on canvas, image: 79.4 × 102.1 cm (31¼ × 40³⁄₁₆ in.), frame: 98.1 × 121.3 cm (38⅝ × 47¾ in.), 1992.126.
4. William Sharpe, "The Nocturne in *Fin-de-siècle* Paris," in Barbara T. Cooper and Mary Donaldson-Evans, eds., *Modernity and Revolution in Late Nineteenth-Century France* (Newark: University of Delaware Press, 1992), 123. A statement made by the director of the Wadsworth Athenaeum in 1940 points in the same direction by identifying a conundrum in the iconography of Impressionist pictures: "The Impressionists, those acknowledged portraitists of light, found the night scene on the whole not as sympathetic as one would have supposed." A[rthur] Everett Austin Jr., *Night Scenes: An Exhibition of the Wadsworth Athenaeum* (Hartford, CT: Wadsworth Athenaeum, 1940), 4.
5. The watery materiality of "oceanic interculture" emphasizes the physicality of all that ocean travel and makes the term preferable in many ways to "the Atlantic World." Joseph Roach, *Cities of the Dead: Circum-Atlantic Performance* (New York: Columbia University Press, 1996), xi.
6. The excellent anthology, *Foreign Artists and Communities in Modern Paris, 1870–1914*, ed. Karen L. Carter and Susan Waller (Burlington, VT: Ashgate, 2015), covers allied material but has a contrasting

remit. Their book, which covers the migration of artists to Paris for professional opportunities, and treats arrivals in the French capital from Western and Eastern Europe, Asia, and South America as well as North America, focuses upon immigrants and expatriates.

7. To paraphrase Peter Wollen, "The Cosmopolitan Ideal in the Arts," in George Robertson et al., eds., *Travellers' Tales: Narratives of Home and Displacement* (London: Routledge, 1994), 189.
8. Edward Saïd, "No Reconciliation Allowed," in Andre Aciman, ed., *Letters of Transit: Reflections on Exile, Identity, Language, and Loss* (New York: New Press, 1999), 106.
9. Georg Simmel, "The Stranger" (1908), in Donald N. Levine, ed., *Georg Simmel on Individuality and Social Forms* (Chicago: University of Chicago Press, 1971), 145.
10. Theodor Adorno, *Minima Moralia: Reflections from a Damaged Life*, trans. Ren. Jephcott (London: Verso, 1974), 33.
11. Edward Saïd, "Reflections on Exile," *Granta* 13 (Autumn 1984): 166.
12. Saïd, "Reflections on Exile," 167.
13. Saïd, "Reflections on Exile," 171–72.
14. Letter to Grace Norton, quoted in Malcolm Bradbury, "Second Countries: The Expatriate Tradition in American Writing," *Yearbook of English Studies* 8 (1978): 31.
15. My take on Hassam's perspective on Paris differs from those of others previously published, including, for example, H. Barbara Weinberg's: "Hassam's Paris paintings disclose a bourgeois gentility that echoes his lifestyle and typifies the cheerful euphemism that helps define American Impressionism." Weinberg, "Hassam in Paris, 1886–1889," in *Childe Hassam, American Impressionist* (New Haven, CT: Yale University Press, 2004), 60.
16. Bradbury, "Second Countries," 19. See also Malcolm Bradbury, *Dangerous Pilgrimages: Trans-Atlantic Mythologies and the Novel* (London: Penguin, 1995).
17. Bradbury, "Second Countries," 20.
18. Bradbury, "Second Countries," 23.
19. Bradbury, "Second Countries," 29.
20. Bradbury, "Second Countries," 38.
21. Edith Wharton, *French Ways and Their Meaning* (Lee, MA: Berkshire House, 1997), xiii.
22. Wharton, *French Ways*,18.
23. Wharton, *French Ways*, 59.
24. Wharton, *French Ways*, 63.
25. Wharton, *French Ways*, 69.
26. Wharton, *French Ways*, 30.
27. Kirsten Swinth, *Painting Professionals: Women Artists and the Development of Modern American Art, 1870–1930* (Chapel Hill: University of North Carolina Press, 2001), 37.
28. Wharton, *French Ways*, 65.
29. "Les dieux s'en vont. Nous devenons décidément des Américains." Jules Claretie, *La Vie à Paris 1882*, 2nd ed. (Paris: Victor Havard Éditeur, n.d.), 54.
30. "Il est bien évident que l'*américanisme* fait chez nous des progrès nouveaux, presque chaque jour. . La poussée de curiosité, d'anxiété, de fièvre, qui s'est produite vers Maurice Vignaux de l'école de Toulouse, et George Slosson, maître des *bandes américaines* est un symptôme de cette modification apportée peu à peu à notre tempérament national. C'est par infiltrations que l'humeur yankee pénètre en nous. Mais le *Yankeesme* nous gagne et, de toutes les façons, nous transforme totalement." Claretie, *La Vie à Paris 1882*, 54–55.
31. "Avez-vous jamais, en traversant l'avenue de l'Opéra qui, elle-même, prend des aspects de coin de Londres ou de New-York avec ses grands magasins et ses offices, levé les yeux vers le ciel et les toits? Tour un réseau de fils télégraphiques, un tas de portées musicales dont quelque cheminée figure la clef, se croisent et s'entrecroisent au-dessus des têtes. Le télégraphe sauté d'un pâté de maisons à un autre, de tuyau en tuyau et découpe, là-haut, ses lignes grêles, droites comme les rayures d'un cahier d'écolier et qui sont comme les toiles d'araignée de la civilisation. Ce télégraphe qui, pareil à un fil à

beurre coupe de la sorte l'horizon et le ciel, et finira par donner à Paris l'aspect de San-Francisco, c'est l'avant-coureur de toute une suite ininterrompue de progrès matériels destinés à modifier singulièrement le paysage parisien. Nous avons déjà le tramway à vapeur; nous aurons le railway aérien, les locomotives-omnibus, des tunnels sous les Variétés, des ponts suspendus devant le Gymnase: tout un Paris chauffée au gaz, éclairée a l'électricité, élevé au téléphone et promené à la vapeur. . . . Oui, regardez bien encore ce Paris livré à tout l'*utilitarisme* américain." Claretie, *La Vie à Paris 1882*, 55–56.

32. Kenneth Silver, "*At the Sculpture Exhibition* by Charles Curran," *Yale University Art Gallery Bulletin*, Summer 1974, 21.
33. Undated Richard York Gallery flyer describing an oil on board for sale, Curran's *Portrait of Wm. J. Whittemore* (1887).
34. Jane Ward Faquin, "Charles Courtney Curran, A Biography," in *Charles Courtney Curran: Seeking the Ideal* (Memphis: Dixon Gallery and Gardens, 2014), 19.
35. Faquin, "Charles Courtney Curran, A Biography," 22.
36. Michael David Zellman, *American Art Analog*, vol. 1, *1688–1842* (Langhorne, PA: Chelsea House Publishers, in association with American Art Analog, 1986), 583. The many prizes he won after his return to the United States are listed in Glenn B. Opitz, ed., *Mantle Fielding's Dictionary of American Painters, Sculptors and Engravers*, 2nd ed. (Poughkeepsie, NY: Apollo, 1986), 194–95. According to the New Britain Museum of American Art website (http://ink.nbmaa.org): "Active as an organizer as well as an exhibitor, Curran served as the assistant director of the American art exhibitions at both the Paris Exposition and the Pan-American Exposition. He was elected to full membership in the National Academy in 1904 and served as recording secretary of the National Academy from 1910 to 1920, when he was elected corresponding secretary, a position he held until 1941. Curran taught at Pratt Institute, the Art Students League, and Cooper Union and was a member of the New York Water Color Club, American Water Color Society, Society of American Artists, Lotos Club, and Salmagundi Club." Grace and Charles were married for fifty-four years. The painter frequently used members of his wife's family and their own family as models. (Jean Ward Faquin, *Charles Courtney Curran: Seeking the Ideal*, 23).
37. Jane Ward Faquin, *Charles Courtney Curran: Seeking the Ideal*, 42. His interest in nighttime was not reserved exclusively for the French capital. In addition to his two Paris nocturnes, he painted a Chicago nocturne: *Administration Building of the 1893 Columbian Exposition in Chicago at Night* (1893; 8¾ × 13 in.). Faquin, *Charles Courtney Curran: Seeking the Ideal*, 28, 70.
38. Debra Force, *Images de la France: American Artists in France, 1880–1925* (New York: Debra Force Fine Art, 2002), 14.
39. Robin Jaffee Frank, *A Private View: American Paintings from the Manoogian Collection* (New Haven, CT: Yale University Art Gallery and the Detroit Institute of Arts, 1991), 27.
40. In 1868 Morris bought the exclusive advertising concession from the City of Paris and founded La Société Fermière des Colonnes Morris.
41. François Loyer, *Paris Nineteenth Century: Architecture and Urbanism*, trans. Charles Lynn Clark (New York: Abbeville Press, 1988), 306.
42. Force, *Images de la France*, 14.
43. Faquin, *Charles Courtney Curran: Seeking the Ideal*, 11, 44.
44. William H. Gerdts, "Three Themes," in *Childe Hassam: Impressionist* (New York: Abbeville Press, 1999), 121.
45. Gerdts, "Three Themes," 134.
46. For exhaustive accounts of Hassam's time in Paris, see Ulrich W. Hiesinger, "Paris, 1886–89," in *Childe Hassam: American Impressionist* (Munich: Prestel-Verlag, 1994), 31–60 and 177–79; and H. Barbara Weinberg, "Hassam in Paris, 1886–1889," in *Childe Hassam, American Impressionist* (New Haven, CT: Yale University Press, 2004), 53–85.
47. Hiesinger, *Childe Hassam: American Impressionist*, 31. Hassam and his wife had a very comfortable bourgeois household setup compared to most young American students who were considerably less well housed.

48. Hiesinger, *Childe Hassam*, 179.
49. Hiesinger, *Childe Hassam*, 179.
50. "Nul ne peut entrer dans l'atelier s'il ne parle français." Bennard B. Perlman, *Robert Henri: His Life and Art* (Mineola, NY: Dover Publications, 1991), 12.
51. Weir's many complaints about the French during his Paris student days are quoted and discussed in my essay: Hollis Clayson, "Enthralled and Dismayed by Paris: Julian Alden Weir in the Transatlantic World," in Marian Wardle, ed., *The Weir Family, 1820–1920: Expanding the Traditions of American Art* (Hanover, NH: University Press of New England, 2011), 54–74.
52. "The limited chances for success caused tension between French students and foreigners." H. Wayne Morgan, ed., *An American Art Student in Paris: The Letters of Kenyon Cox, 1877–1882* (Kent, OH: Kent State University Press, 1986), 102.
53. According to John Davis, shared with me in conversation, May 23, 2014.
54. Deborah Chotner, *American Paintings from the Manoogian Collection* (Washington, DC: National Gallery of Art, 1989), 150.
55. Julia Alexander has argued that Hassam's *A Paris Nocturne* belongs to a "socially ambiguous genre," by virtue of playing on the social ambiguity of a seemingly upper-class woman being shown "in a somewhat compromising setting." I see none of that in the painting. She also argues: "Hassam's decision to exploit the nocturnal, and therefore man-made, light heightens the enigmatic atmosphere of an everyday street-scene." Julia Alexander, *A Private View: American Paintings from the Manoogian Collection* (New Haven, CT: Yale University Art Gallery, 1993), 21–22
56. *Les Grands Boulevards* (Paris: Musée Carnavalet, 1985), 208, no. 431.
57. Chotner, *American Paintings from the Manoogian Collection*, 150.
58. *Paris at Twilight* (1887; oil on canvas, 64.7 × 54 cm [25½ × 21¼ in.], private collection) is an even quieter city scene that again relies on a carriage working a street in a generic city setting at the exceptionally delicate tonal moment of twilight. Social incident is kept to an absolute minimum.
59. Hassam made lithographs of the same scene.
60. See also Prendergast, *Early Evening, Paris* (c. 1892; monotype, sheet 34.3 24.8 cm [13½ × 9¾ in.], Terra Foundation for American Art, Daniel J. Terra Collection, 1992.81); and *Evening Shower, Paris* (1892–94, oil on panel, 34.2 × 20.9 cm [13½ × 8¼ in], Michael Altman Fine Art, New York, NY.).
61. Stéphane Guégan, *A Passion for France: The Collection of Marlene and Spencer Hays* (Paris: Musée d'Orsay, 2013), 104.
62. Gerdts, "Three Themes," 129. Gerdts helpfully enumerates the liberties that Hassam took with the topography of the square including placing a sunset glow to the north. Gerdts, "Three Themes, 224. The title "Fairyland" deserves to be taken seriously.
63. William Sharpe argues that the nocturne became a privileged means for picturing New York that could eschew its physical and social filth. William Sharpe, *New York Nocturne: The City after Dark in Literature, Painting, and Photography, 1850–1950* (Princeton, NJ: Princeton University Press, 2008), chap. 2.
64. Lenard R. Berlanstein, *Big Business and Industrial Conflict in Nineteenth-Century France: A Social History of the Parisian Gas Company* (Berkeley: University of California Press, 1991), 36.
65. The watercolor *Horse Drawn Cabs at Evening, New York* (c. 1890; 35.6 × 45.1 cm [14 × 17¾ in.], Terra Foundation of American Art) bears a very strong family resemblance to the Paris pictures already inventoried.
66. *Nocturne, RR Crossing, Chicago* (1893; opaque watercolor on paper, 40.6 × 29.8 cm [16 × 11¾ in.], Boston, Museum of Fine Arts) also endows the big-city night with a threatening and spectral quality. Headlights become ghostly eyes barely piercing an all-but-impenetrable blue night. *Rainy Night* (c. 1895; transparent and opaque watercolor over graphite on wove paper, 28.5 × 21 cm [11.2 × 8.3 in.], Fine Arts Museums of San Francisco) depicts a cacophony of urban encounters at night: a clash between the black-clad members of a clanging energetic crowd thrown into relief by the corona of piercing white light that surrounds them.

67. Hassam, *Rainy Night under the Electric Light, Madison Square, Looking South*, and *Rainy Night on Broadway* (n.d.), from Childe Hassam, *Three Cities* (New York: R. H. Russell, 1899). See also figure 5.16.
68. *The Chestnut Vender, Place Pigalle*, and *Flower Girls* (n.d.), from Hassam, *Three Cities*. See also figure 5.17.
69. The Notre Dame scene in the book (here, fig. 5.17) is very closely based upon his 1888 oil painting, *Notre Dame Cathedral, Paris, 1888* (unframed 44.5 × 54.6 cm [17½ × 21½ in.], Detroit Institute of the Arts).
70. Ben Highmore, "Cultural History's Crumpled Handkerchief," *Art History* 25, no. 5 (November 2002): 702.

Chapter Six

Epigraph: Edvard Munch, Manuscript T2770, Oslo, Munch Museum Archives, quoted in Reinhold Heller, *Munch: His Life and Work* (Chicago: University of Chicago Press, 1984), 64.

1. Jay Clarke has convincingly discussed Munch's embeddedness in manifold European art networks, his internationalism and absorption of "influences." Jay Clarke, *Becoming Edvard Munch: Influence, Anxiety and Myth* (New Haven, CT: Yale University Press with the Art Institute of Chicago, 2009). On his stays in France (Paris, Saint-Cloud, Le Havre, and Nice), see Rodolphe Rapetti et al., *Munch et la France* (Paris: Réunion des Musées Nationaux, 1991). In the 1991 catalogue, Arne Eggum pinpoints 1880 as the decisive year for the Scandinavian turn to Paris. He calls it "l'année de la rupture, "l'année de de la grande mutation," or "l'année où les artistes norvégiens se détournèrent de l'Allemagne pour regarder vers Paris." Arne Eggum, "Le naturalisme français, l'impressionisme et le jeune Munch," in Rapetti et al., *Munch et la France*, 33. Louise Lippincott concurs: "Beginning in the late 1870s, more and more Scandinavian Painters headed toward Paris rather than Munich, Berlin, or Rome. . . . Although they enrolled in established art schools, and preferred to exhibit in the official Salons, they also saw the work of Monet, Pissarro, Caillebotte, Manet, Bastien-Lepage, and Raffaëlli." Lippincott, *Edward Munch: Starry Night* (Malibu, CA: Getty Museum Studies on Art, 1988), 13. For a recent study of another Paris-based Scandinavian artist, see my essay: Hollis Clayson, "Anders Zorn's Etched Portraits of American Men, or the Trouble with French Masculinity," in Temma Balducci, Heather Jensen, and Pamela Warner, eds., *Interior Portraiture and Masculine Identity in France, 1870–1914* (Aldershot, UK: Ashgate, 2011), 177–95. James Ganz recently wrote that the widely courted Zorn felt the sting of French xenophobia. James A. Ganz, "Etching in the City of Light: Anders Zorn among the Post-Impressionists," in Johan Cederlund et al., *Anders Zorn, Sweden's Master Painter* (New York: Skira Rizzoli with the Fine Arts Museums of San Francisco, 2013), 82.
2. Stéphane Guégan gives Bonnat's teaching (Caillebotte was also his student) a thorough going-over in his "Ecce Homo," in *Gustave Caillebotte: The Painter's Eye* (Washington, DC: National Gallery of Art, 2015), 99–107, 262–63.
3. Rodolphe Rapetti ("Munch et Paris: 1889–1891," in Rapetti et al., *Munch et la France*) illustrates several of Munch's very good-quality figure drawings (*académies* of both male and female models) done in Bonnat's studio, 65–73. According to Rapetti, Munch's closest compatriots were Kalle Löchen and Valentin Kielland. Kielland and Jörgen Sörensen also enrolled in Bonnat's studio. Once Munch was planted out in Saint-Cloud, Rapetti names Frits Thaulow, Thorolf Holmboe, Jörgen Sörgensen. and Jonas Lie as his companions from time to time; all were "artists or intellectuals of the Scandinavian colony of Paris." Zorn observed that the Scandinavians and the Americans were the ones most subject to French antiforeigner sentiment. Ganz, "Etching in the City of Light," 82.
4. An early picture, *Morning* was on show on the Champs de Mars. About the display of this 1884 painting in Paris in 1889, Eggum observed, "The Norwegian adjudicating committee considered it inadvisable to allow him to exhibit a more radical picture." Arne Eggum, *Edvard Munch: Paintings, Sketches, and Studies* (New York: C. N. Potter, distributed by Crown Publishers, 1984), 61.
5. The most detailed account of Munch's life and work in Saint-Cloud is Reinhold Heller, "Edvard Munch's 'Night,' tThe Aesthetics of Decadence, and the Content of Biography," *Arts Magazine* 53, no.

2 (October 1978): 80–105. Heller notes (89) that Munch had already moved from Paris out to Neuilly, but by the end of November, finding Neuilly's climate raw and cold, he moved back into the city. The subsequent move west to Saint-Cloud was then part of a pattern of restlessness, but it was only in Saint-Cloud that he really began to paint in an innovative way. According to Arne Eggum, Munch located there in order to escape from the French capital during an outbreak of cholera. Arne Eggum, *Edvard Munch: Paintings, Sketches, and Studies*, 62. Rapetti cites Eggum when he also speaks of cholera. Rapetti, "Munch et Paris: 1889–1891," in Rapetti et al., *Munch et la France*, 65. However, according to Frédérique Krupa, *Paris: Urban Sanitation before the 20th Century*, chapter entitled "Parisian Sanitation from 1789–1900," the last cholera outbreak in Paris was in 1892. http://www.translucency.com/frede/parisproject/sanit1789_1900.html. And Heller says it was an 1889 outbreak of influenza. Heller, "Edvard Munch's 'Night,'" 91. Rapetti notes that the impecuniousness of the Munch family back in Kristiania after the patriarch's death may also have prompted young Munch's move to cheaper lodgings. Rapetti, "Munch et Paris: 1889–1891," in Rapetti et al., *Munch et la France*, 65.

6. William J. Rolfe, revised and enlarged by William D. Crockett, *A Satchel Guide to Europe*, 44th ed. (Boston: Houghton Mifflin, 1924), 161. I tried and comically failed to retrace Munch's steps and point of view from his rented room on a ferociously rainy May 24, 2014. The hyperpaved, twenty-first-century Saint-Cloud with its thrusting autoroute (the A21) alongside the river took me completely by surprise.
7. Rapetti et al., *Munch et la France*, 64. But there is no evidence that other fellow-artist countrymen lived in Saint-Cloud beyond Munch's roommate Emanuel Goldstein.
8. See Hollis Clayson, "Threshold Space: Parisian Modernism Betwixt and Between (1869 to 1891)," in Janet McLean, ed., *Impressionist Interiors* (Dublin: National Gallery of Ireland, 2008), 14–29.
9. "Threshold people" is Victor Turner's term, quoted by Mark Ledbury, "Greuze in Limbo," in Philip Conisbee, ed., *French Genre Painting in the Eighteenth Century* (New Haven, CT: Yale University Press, 2007), 184. In Munch's etching of the same scenario, the contrasts are, of course, tonal rather than chromatic.
10. Ledbury, "Greuze in Limbo," 179.
11. This is a proposition against the grain of much scholarship on the function and valence of light in these pictures. For example: "In *Night in Saint-Cloud*, Munch transformed [Caspar David] Friedrich's distant light from a Romantic symbol of aspiration into a fin-de-siècle image of separation and loneliness." Lippincott, *Edward Munch: Starry Night*, 33.
12. Christoph Asendorf, *Batteries of Life: On the History of Things and Their Perception in Modernity*, trans. Don Reneau (Berkeley: University of California Press, 1993), 155.
13. On this defining trait, we could again quote Martin Bressani and Marc Grignon: "L'alternance de la lumière et de l'ombre est à la base du spectacle visuel au XIXe siècle." Martin Bressani and Marc Grignon, "Les Fantasmagories du gaz d'éclairage à Paris au XIXe siècle," in Libero Andreotti, ed., *Spielraum: Walter Benjamin et l'architecture* (Paris: Éditions de la Villette, 2011), 9.
14. It is not sure where this work is set, despite the familiarity it reflects with Paris pictures. Leading candidates are clubs in Munch's hometown, Kristiania: the Flora Salon at the Tivoli, or Bazarhallen, popularly known as Basserallen, a beer-hall with cabaret opened in 1885 on the initiative of Munch's friend Frits Thaulow. Gerd Woll, *Edvard Munch: Complete Paintings, Catalogue Raisonné*, vol. 1, *1880–1897* (London: Thames & Hudson, 2009), no. 128, p. 144.
15. Edvard Munch, "Impressions from a Ballroom, New Year's Eve in Saint-Cloud, 1889," quoted in Charles Harrison and Paul Wood, eds., *Art in Theory, 1815–1900, An Anthology of Changing Ideas* (Oxford: Wiley-Blackwell, 1998), 1041.
16. Michael Fried, *Manet's Modernism, or the Face of Painting in the 1860s* (Chicago: University of Chicago Press, 1996), 259–61.
17. Arne Eggum quotes Munch's asseveration in French: "mon adieu définitif à l'impressionnisme ou au réalisme." Eggum, "Le Naturalisme français, l'impressionnisme et le jeune Munch," in Rapetti et al., *Munch et la France*, 51.

18. See, for example, Rodolphe Rapetti, "Munch et Paris: 1889–1891," in Rapetti et al., *Munch et la France*, 90. And Lippincott writes: "*Night in Saint-Cloud* captures Munch's constant feeling of isolation and alienation, the result of being a penniless Norwegian in Paris, of being an artist rejected by his native audience, of being a lover passed over by his beloved, and, finally, of being the rebellious but grieving son of a father who had just died. Even though Emanuel Goldstein [Munch's roommate and poet] may have posed for the shadowy figure at the window, the painting is the first of Munch's many psychological self-portraits." Lippincott, *Edvard Munch: Starry Night*, 31–32. Heller makes the most elaborate case for the painting's autobiographical reverberations. "The painting is quite literally autobiographical, representing an actual scene in Munch's life, enveloped in the blue haze, the 'comforting fog' of death, silence, and loneliness." He also argues that the blue tones are not mimetic but rather "the Decadents' *frisson bleu de l'angoisse*." Heller, "Edvard Munch's 'Night,'" 90 and 87.
19. I am not discussing, for example, his several figure pictures from the same months, set in a café and very much in thrall to Jean-François Raffaëlli.
20. Heller ("Edvard Munch's 'Night,'" 89) stresses the artist's chagrined reaction to his father's death and its link to the painting: "Alone in Paris, unable to return home, unable to be with his family, unable to see his father again, Edvard Munch desperately wrote a series of autobiographical notations recording his own despairing reaction to the news of his father's sudden death, and *Night* forms the visual counterpart to these written impressions." Heller reports (104n97) that Munch lived at 12, quai de Saint-Cloud. Rapetti ("Munch et Paris: 1889–1891," in Rapetti et al., *Munch et la France*, 65–66) concurs on the address but provides a bit more information about the building. The rented room was in Hôtel-Restaurant du Belvédère (12, quai de Saint-Cloud), and Munch made the move with his friend the Danish poet, Emanuel Goldstein. Concerning their ability to communicate, Danish and Norwegian are effectively and functionally the same language. Since a thoroughfare by that name does not now exist, it is reasonable to assume that it is the contemporary quai-side street, rue de Saint-Cloud, wedged between the Seine and the Parc de Saint-Cloud, where Munch frequently strolled. The postcard photo reproduced in Rapetti (66) confirms the riverside location of Munch's dwelling. The building does not exist in any recognizable form today.
21. The *catalogue raisonné* includes eight, nos. 194–201. Woll, *Edvard Munch Complete Paintings, Catalogue Raisonné*, vol. 1, *1880–1897*. Rapetti, "Munch et Paris: 1889–1891," in Rapetti et al., *Munch et la France*, 73–84, is good on this topic.
22. Cary Hollinshead-Strick of the American University of Paris, "Catching Bodies to Recompose Society: *Les Filets de St Cloud*," paper presented at the Nineteenth-Century French Studies conference, Richmond, VA, October 2013. The author and I communicated regarding Munch's Saint-Cloud pictures, and she suggested that the young Norwegian's bridge may have been the next one south, le Pont de Sèvres, but it is too far downriver and hidden by its curve to have been Munch's motif.
23. The cluster includes the following works: *La Nuit* (*Night in Saint-Cloud*) (1890; oil on canvas, Oslo National Gallery); *Night in Saint-Cloud* (1892; oil on paper, private collection); *La Nuit* (*Night in Saint-Cloud*) (1893; oil on canvas, Oslo, private collection); *La Nuit* (*Night in Saint-Cloud*) (1893; pastel, Oslo, private collection); *Clair de lune* (*Moonlight, Night in Saint-Cloud*) (1895; drypoint, Art Institute of Chicago; fig. 6.11).
24. Rodolphe Rapetti considers it to be a highly original and arresting picture, and the first mature work Munch completed in France, inasmuch as it instances his departure from Scandinavian naturalism and his integration of diverse influences. Rapetti et al., *Munch et la France*, 90. Reinhold Heller agrees: "It is a fixed pivot point from which Munch's mature art proceeds … [it] brought to Norwegian art the concerns of a developing European anti-naturalism, an art of personal ideas. Moods, and the emotive stirring of the psyche." Heller, "Edvard Munch's 'Night,'" 80, 83.
25. Louise Lippincott, for example, has stressed the dejection of the figure and the consequent melancholy of the painting's mood by arguing that the framed landscape reinforces the sense of the sitter's separation from the "distant warmth and light of gathered humanity across the river." Lippincott, *Edvard Munch: Starry Night*, 31. Heller is even more emphatic about the separation: "From the life of

the city and the movement of the river he depicted in other paintings during the early spring months of 1890, he is totally removed and isolated by the intervening surface plane of the window. . . . Life within is separated from life without." Heller, "Edvard Munch's 'Night,'" 100.

26. I thank Dr. Lasse Jacobsen of the Munch Museum, Oslo, for her very kind provision of scans of this letter, and Joseph Hammond for making contact with her on my behalf.
27. Heller, "Edvard Munch's 'Night,'" 93.
28. Arne Eggum, *Edvard Munch: Paintings, Sketches, and Studies,* 64. Eggum (63) suggests that it may have been a draft of a reply to a letter from a woman, Aase Nørregaard.
29. Rapetti, *Munch et la France*, 91.
30. Jay Clarke refers to blue-violet as an "indigenous color," and mentions the frequent recourse to blue mood pictures among artists of Munch's generation. Clarke, *Becoming Edvard Munch*, 18 and 36. Lippincott discusses the frequent trope of the Blue Hour picture in the North. Lippincott, *Edvard Munch: Starry Night*, 27–28. Heller writes: "What Munch visualizes is the Decadents' *frisson bleu de l'angoisse*, not distress itself but the chilling quiveringly blue presentiment of it." Heller, "Edvard Munch's 'Night,'" 87.

31. Personal communication from Keith Letsche, Rushlight Club member and lighting expert, August 2, 2011.
32. In November 2012, when I lectured on this painting in Oslo, Arne Eggum, dean of Munch scholars, suggested that the lights out the window were unmistakably those of the Eiffel Tower. I have struggled to determine whether the 1889 structure would have been visible from Munch's room on the *première étage* and whether it was illumined in 1890. The tower is easily seen during the day from the heights of the Parc Saint-Cloud, but my investigatory walks in Saint-Cloud suggest that while Munch was of course aware of it, it was not visible out his own window. Clearly there is no reference to the Tour Eiffel in either figure 6.8 or 6.10. I am now pursuing research on the lighting of the tower between 1889 and 1900, and its cultural representation. Insofar as the literature on the tower's illumination only discusses it as the light of spectacle par excellence during the Expositions Universelles, it is time to investigate the everyday (every night) lighting of the tower.
33. Charles Rice, *The Emergence of the Interior: Architecture, Modernity, Domesticity* (London: Routledge, 2007), 35.
34. Victoria Rosner, *Modernism and the Architecture of Private Life* (New York: Columbia University Press, 2005), 61–62.
35. Victor Turner, *Ritual Process: Structure and Anti-Structure* (Chicago: University of Chicago Press, 1969), 95.
36. Exemplary studies of doors as thresholds are Georg Simmel, "Bridge and Door," *Theory Culture Society* 11, no. 5 (1994): 5–10; Claude Gandelman, "Penetrating Doors," in *Reading Pictures, Viewing Texts* (Bloomington: Indiana University Press, 1991); and Mark Ledbury, "Greuze in Limbo: Being 'Betwixt and Between,'" *Studies in the History of Art* 72 (2007). On visual representations organized around the window, see Sabine Rewald, *Rooms with a View: The Open Window in the 19th Century* (New York: Metropolitan Museum of Art, 2011). That Munch's windows are closed makes all the difference.
37. Simmel, "Bridge and Door," 5.
38. Benjamin, *Arcades Project*, 423.
39. Imogen Racz, *Art and the Home: Comfort, Alienation and the Everyday* (London: I. B. Tauris, 2015), 34. And see Baudelaire's 1863 poem, "Les Fenêtres."
40. Claude Gandelman, "Penetrating Doors," 41.
41. See Jonathan Crary on absentmindedness in his *Suspensions of Perception: Attention, Spectacle, and Modern Culture* (Cambridge, MA: MIT Press, 1999), passim.
42. Isobel Armstrong, *Victorian Glassworlds: Glass Culture and the Imagination, 1830–1880* (Oxford: Oxford University Press, 2008), 7.

43. Simone Delattre, *Les douze heures noires. La nuit à Paris au XIXe siècle* (Paris: Albin Michel, 2000), 146.
44. Full credit to Elizabeth Benjamin for this clever idea.
45. Wolfgang Schivelbusch, *The Railway Journey: Trains and Travel in the Nineteenth Century*, trans. Anselm Hollo (New York: Urizen Books, 1979), 25. See also Anne M. Lyden, *Railroad Vision: Photography, Travel, and Perception* (Los Angeles: J. Paul Getty Museum, 2004), which contains no photos that resemble Munch's painted scenario.
46. Anne Distel et al., *Gustave Caillebotte: Urban Impressionist* (Chicago: Art Insitute of Chicago, 1994), no. 59, 149. I discussed this work by Caillebotte in very similar terms in Clayson, "Threshold Space," 2008.
47. Distel et al., *Gustave Caillebotte: Urban Impressionist*, 148.
48. Distel et al., *Gustave Caillebotte: Urban Impressionist*, 143.
49. In this spirit, Rodolphe Rapetti writes, "The man's posture (he is the painter's younger brother René), hands in his pockets, apparently staring at a female silhouette in the street, the armchair facing the window, the deserted city, the motionless carriage—all speak of idleness, of time wasted." Rapetti, "Paris Seen from a Window," in Distel et al., *Gustave Caillebotte: Urban Impressionist*, 142.
50. Priscilla Parkhurst Ferguson, *Paris as Revolution: Writing the Nineteenth-Century City* (Berkeley: University of California Press, 1994), 80–114.
51. For the record, Caillebotte also painted *Man in a Top Hat Seated near a Window* in 1880 (Algiers, Musée des Beaux-Arts). It was not exhibited during his lifetime. Although it is ostensibly closer to Munch's painting, I do not discuss it here because the man in Caillebotte's picture looks across the interior *not* at or through the windowed doors to his left.

Conclusion

1. Van Morrison, "Caravan," *Moondance*, Warner Bros., album released January 27, 1970.
2. See my discussion of the problematic use of "case study" in the humanities. Hollis Clayson, *Paris in Despair: Art and Everyday Life under Siege (1870–71)* (Chicago: University of Chicago Press, 2002), 11–12, 361.
3. Electric light did not illuminate homes until the very end of the 1800s, by which time gas illumination was also in wider use, but electricity did not take over until well into the 1900s. Brian Bowers opens his discussion of improved gas lighting for the home pointedly: "If gaslight had continued to mean the light from a gas flame then electricity would quickly have replaced gas as the preferred means of getting light." Brian Bowers, *Lengthening the Day: A History of Lighting Technology* (Oxford: Oxford University Press, 1998), 127. I discuss the complexity of indoor lighting preferences in connection with Mary Cassatt's intaglio prints in chapter 4. Simone Delattre's wisdom is ever fresh: "The bourgeois nest prefers the flame." Simone Delattre, *Les Douze heures noires. La nuit à Paris au XIXe siècle* (Paris: Albin Michel, 2000), 117. A comprehensive selection of images of the flame-lit interior appears in Bernadette Boustany et al., *Sous la lLampe, Peintures de 1830 à 1930* (La Varenne: Musée de Saint-Maur, 2010).
4. Lenard R. Berlanstein, *Big Business and Industrial Conflict in Nineteenth-Century France: A Social History of the Parisian Gas Company* (Berkeley: University of California Press, 1991), 21.
5. Berlanstein, *Big Business and Industrial Conflict*, 19.
6. Matthew Beaumont's observation that Joachim Schlör's influential *Nights in the Big City* traced the "shifting relations of 'night and violence, night and sexuality, night and pleasure, night and solitude' in the context of the social history of metropolitan streets" underscores the specificity of the representational focus on pleasure in nocturnal settings examined here. Matthew Beaumont, "Foreword," in Joachim Schlör, *Nights in the Big City: Paris, Berlin, London, 1840–1930*, trans. Pierre Gottfried Imhof and Dafydd Rees Robers (London: Reaktion, 2016), 10. Schlör's book appeared in German in 1991

and was first published in English (by Reaktion Books) in 1998. For a consideration of the broader compass of themes explored by American nocturnists (adumbrated in my chapter 5), see Hélène Valance, *Nuits américaines: L'art du nocturne aux États-Unis, 1890–1917* (Paris: PUPS, 2015).

7. Gustave Kahn is referring to Whistler in his introduction to *Les Dessins de Georges Seurat* (Paris: Bernheim-Jeune, 1928), vol. 1, n.p.
8. The ambiguities referenced here are not those that solicit the imaginative contribution of the onlooker, as studied by Dario Gamboni, *Potential Images: Ambiguity and Indeterminacy in Modern Art* (London: Reaktion, 2004).
9. Andreas Blühm and Louise Lippincott, *Light! The Industrial Age, 1750–1900: Art and Science, Technology and Society* (Pittsburgh: Carnegie Museum, 2000), 188.
10. For its grand opening in 1889, the Moulin Rouge hung Lautrec's *At the Cirque Fernando* (1888; Art Institute of Chicago), near the entrance. Reinhold Heller, *Toulouse-Lautrec: The Soul of Montmartre* (Munich: Prestel-Verlag, 1997), 57.
11. Heller, *Toulouse-Lautrec*, 58.
12. Katherine M. Kuenzli reports that Lautrec's first poster was the winning entry in a competition to which Pierre Bonnard also submitted an entry. *The Nabis and Intimate Modernism: Painting and the Decorative at the Fin de Siècle* (Burlington, VT: Ashgate, 2010), 67.
13. My gratitude goes to Keith Letsche and Ara Kebapçioğlu for securing the lighting identification.
14. A final point about illumination in Lautrec's graphic work: the light projected upon the flowing costume of the gyrating midwestern American dancer, Loïe Fuller, in Paris—often wrongly dubbed the Electric Salomé and the subject of Lautrec's exceptional 1893 lithographs colored with metallic ink—was also fueled by gas. Full credit goes to the sleuthing skills of Jamie Holeman, a former graduate student in History at Northwestern University. Jamie Holeman, "*La fée gazeuse?* Light Blindness and the Historiography of Loie Fuller," seminar paper, Art History 450, December 6, 2010. The lithographs are reproduced in *Pariser Nächte—Henri de Toulouse-Lautrec* (Bremen: Edition Braus, 1994), 200–201.
15. Nor did I linger upon the drawings of Georges Seurat (1859–91), penetrating investigations of the subtleties of light and shadow in the artificially lit interior and outdoor spaces of the French capital. To put it baldly, it was the nonpareil subtlety of his drawings that kept them out of my purview. The decisiveness of Jonathan Crary's reading of Seurat's artwork also discouraged me from having my own say. Jonathan Crary, "1888: Illuminations of Disenchantment," in *Suspensions of Perception: Attention, Spectacle, and Modern Culture* (Cambridge, MA: MIT Press, 1999), 149–280.
16. This book title of the future is meant to evoke Walter Benjamin's *Charles Baudelaire: A Lyric Poet in the Era of High Capitalism*, trans. Harry Zohn (London: New Left Books), 1973.
17. Tim Edensor, *From Light to Dark: Daylight, Illumination and Gloom* (Minneapolis: University of Minnesota Press, 2017), 165. Edensor quotes Craig Koslofsky, *Evening's Empire: A History of Night in Early Modern Europe* (Cambridge: Cambridge University Press, 2011).
18. J.-K. Huysmans, review of the 1882 Salon, "Appendice," in *L'Art moderne* (Paris: G. Charpentier, 1883), 271–72. "Le sujet est bien moderne et l'idée de M. Manet de mettre ainsi sa figure de femme, dans son milieu, est ingénieuse; mais que signifie cet éclairage? Ça, de la lumière de gaz et de la lumière électrique? allons donc, c'est un vague plein air, un bain de jour pâle!—Dès lors, tout s'écroule—les Folies-Bergère ne peuvent exister et n'existent que le soir; ainsi comprises et sophistiquées, elles sont absurdes. C'est vraiment déplorable de voir un homme de la valeur de M. Manet sacrifier à de tels subterfuges et faire, en somme, des tableaux aussi conventionnels que ceux des autres!"

Index

Page numbers in italics indicate illustrations.